The NEUROBIOLOGY of COCAINE

Cellular and Molecular Mechanisms

Edited by
Ronald P. Hammer, Jr.

CRC Press
Boca Raton New York London Tokyo

Library of Congress Cataloging-in-Publication Data

The neurobiology of cocaine : cellular and molecular mechanisms /
Ronald P. Hammer, Jr., editor.
p. cm.
Includes bibliographical references and index.
ISBN 0-8493-8311-0 (alk. paper)
1. Cocaine--Physiological effect. 2. Neuropharmacology. I. Hammer, Ronald P.
[DNLM: 1. Cocaine--pharmacology. 2. Neurotransmitters--metabolism. 3. Central Nervous System--drug effects. QV 113 N494 1995]
QP801.C68N48 1995
615′.785--dc20
DNLM/DLC
for Library of Congress 95-7021
CIP

Direct all inquiries to CRC Press, Inc., 2000 Corporate Blvd., N.W., Boca Raton, Florida 33431.

International Standard Book Number 0-8493-8311-0
Library of Congress Card Number 95-7021
Printed in the United States of America 1 2 3 4 5 6 7 8 9 0
Printed on acid-free paper

Athina Markou

DEDICATION

To those neuroscientists and researchers whose determined efforts have brought us this far in the Decade of the Brain.

PREFACE

Cocaine is one of several alkaloids present in the leaves of the *Erythroxylon coca* plant, whose stimulant properties have been known for centuries. Pharmacological studies began more than a hundred years ago when Karl Koller first described cocaine's local anesthetic properties and Sigmund Freud its psychoactive and euphorogenic characteristics. Several epidemic waves resulted from various attempts to popularize the drug, with use peaking again in the 1980s. Fortunately, programmatic initiatives by the National Institute on Drug Abuse produced a surge of research effort which has borne fruit in recent years, clarifying the reinforcing and addictive potential of cocaine. The curbing of cocaine abuse among "recreational" cocaine users in the 1990s is one useful consequence, but equally important is the elucidation of the mechanisms of action of cocaine. These mechanisms, and the valuable new insights they provide into normal brain functioning, are the topics of this book. The reader should note that the reports contained herein are limited to effects on the central nervous system. A variety of direct peripheral effects, some of which now are known to have lethal consequences in drug abusers, are not discussed. Many of the most exciting recent developments are described, however, including the brain circuitry underlying reward and behavioral sensitization induced by cocaine, the neurochemical and functional aspects of cocaine withdrawal, the impact of cocaine on the dopamine transporter, and the cascade of signal transduction underlying cocaine-induced psychomotor activation.

The book is divided into three sections: Mechanisms of Cocaine Reinforcement, Cellular Effects of Cocaine, and Molecular Approaches to Studying Cocaine Activation. These sections are not exclusive; different aspects of subjects such as sensitization and withdrawal are discussed throughout the text. It is not necessary to read the chapters sequentially, and readers are encouraged to select chapters based upon their own interests. The authors represent many of the foremost research groups investigating topics related to cocaine mechanisms, but some issues necessarily are omitted. As each chapter contains a complete discussion of the related issues, some redundancy in the work as a whole is inevitable.

In the first section on mechanisms of cocaine reinforcement, we begin with a discussion by Graham and Porrino on the mechanisms underlying cocaine self-administration using an animal paradigm which mimics human self-administration. These studies characterize the regional substrates underlying initiation and maintenance of cocaine reinforcement in the brain. This regional theme continues in a chapter by Hammer and colleagues, whose metabolic studies illustrate the brain circuits underlying sensitization, tolerance, and withdrawal following cocaine administration. Emmett-Oglesby then examines the behavioral models used to assess psychostimulant action, and describes analyses of the motivational and subjective effects of cocaine. The hormonal effects of cocaine and the role of neuroendocrine responses in the behavioral actions of the drug are examined in a chapter by Kuhn and Little. Cocaine action in the human brain is reviewed by Volkow and Fowler in a discussion of imaging studies in patients and their comparison to data obtained from animal experiments. The results suggest that animal experiments provide adequate models for cocaine effects in humans, with some distinctions resulting from the complexity of human brain and behavior.

The next section on cellular effects begins with a discussion by Kalivas on the neural basis of behavioral sensitization to cocaine, an important phenomenon of augmented responses following repeated drug exposure in animals which may be related to stereotyped, compulsive behaviors and relapse exhibited by many chronic cocaine abusers. The physiological responses of dopamine and serotonin neurons are reviewed, respectively, in chapters by White and colleagues and Cunningham. The difference between cellular responses to cocaine in naive animals and in those treated repeatedly with cocaine illustrates an important component of sensitization. Unterwald next addresses the effects of acute and repeated cocaine administration on endogenous opioid systems located in reward centers and their efferent targets. The neurochemical responses in these regions during withdrawal following chronic cocaine treatment then are examined in a chapter by Weiss and colleagues. Finally, regulation of the dopamine transporter by cocaine is reviewed by Zahniser and colleagues.

The last section explores molecular studies of cocaine action, focusing on the recently cloned dopamine transporter molecule and on alteration of signal transduction pathways. Kuhar and colleagues discuss the pharmacology, biochemistry, and molecular structure of the dopamine transporter as affected by cocaine. The effects of cocaine on immediate-early gene response in striatal neurons are reviewed by Graybiel and colleagues. Finally, Fitzgerald and Nestler examine the regulation of signal transduction pathways by acute and chronic cocaine administration. It is clear from these molecular studies that chronic cocaine exposure produces effects which are different from and more persistent than those produced by acute cocaine exposure.

By elucidating the cellular and molecular alterations that must be targeted for optimal pharmacotherapy, these studies ultimately will yield better approaches to treatment of cocaine abuse, including relapse prevention during withdrawal. Such new approaches are essential to alleviate the devastating impact of cocaine on the lives of abusers. In addition, these works elucidate the normal biology of brain reward systems, and help to define the biological substrates underlying motivation and motor activity—processes which are essential for survival. In spite of the successes reported in this volume, we still have a long way to go…

Ronald P. Hammer, Jr.
Boston, Massachusetts

ABOUT THE EDITOR

Ronald P. Hammer, Jr., Ph.D., is a Professor of Psychiatry, Anatomy and Cellular Biology and Pharmacology and Experimental Therapeutics at the Tufts University School of Medicine, and is Director of the Laboratory of Research in Psychiatry at the New England Medical Center in Boston. He received his Doctor of Philosophy degree in Anatomy and Neuroscience from the University of California at Los Angeles in 1980.

Dr. Hammer was appointed as Staff Fellow at the National Institute of Mental Health in Bethesda, Maryland, in 1981, and as Senior Staff Fellow in 1983. The Intramural Research Program at the NIMH provided inspiration toward the exploration of neural mechanisms, including metabolic alteration, signal transduction and regulation of gene expression, topics which are discussed by various authors in this volume. Departure to a gentler climate came in 1984 with an appointment as Associate Professor of Anatomy, Pharmacology and Cell, Molecular and Neural Sciences at the University of Hawaii School of Medicine. In 1993, Dr. Hammer returned to his Massachusetts roots as a Professor at Tufts University.

Dr. Hammer is the recipient of a Research Center Development Award from the National Institutes of Health, and research awards from the National Institute on Drug Abuse, the National Institute of Child Health and Human Development, the National Science Foundation and several private foundations. He has also received the University of Hawaii Regents' Medal for Excellence in Research. He is a member of the American Association for the Advancement of Science, the Society for Neuroscience, the International Society of Developmental Neuroscience, the Endocrine Society, the Society for Neuroendocrinology, the American Association of Anatomists, the American Neuropsychiatric Association, and the Society for Biological Psychiatry, and has published extensively in the neuroscience literature.

CONTRIBUTORS

Sabina Berretta
Department of Brain and Cognitive Sciences
Massachusetts Institute of Technology
Cambridge, Massachusetts

John W. Boja
Neuroscience Branch
NIDA Addiction Research Center
Baltimore, Maryland

Wayne A. Cass
Department of Anatomy and Neurobiology
University of Kentucky Medical Center
Lexington, Kentucky

Catherine Cerruti
Neuroscience Branch
Addiction Research Center
Balrimore Maryland

Kathryn A. Cunningham
Department of Pharmacology and Toxicology
University of Texas Medical Branch
Galveston, Texas

Bulent Elibol
Noroloji Departmani
Hacettepe Universitesı Tip
Sikhiye, Ankara, Turkey

Michael W. Emmett-Oglesby
Department of Pharmacology
University of North Texas Health Science Center at Fort Worth
Fort Worth, Texas

Lawrence W. Fitzgerald
Departments of Psychiatry and Pharmacology
Yale University School of Medicine
New Haven, Connecticut

Joanna S. Fowler
Chemistry Department
Brookhaven National Laboratory
Upton, New York

Greg A. Gerhardt
Psychiatry
University of Colorado Health Science Center
Denver, Colorado

John H. Graham
Department of Neurobiology and Anatomy
The Bowman Gray School of Medicine
Winston-Salem, North Carolina

Ann M. Graybiel
Department of Brain and Cognitive Sciences
Massachusetts Institute of Technology
Cambridge, Massachusetts

Ronald P. Hammer, Jr.
Department of Psychiatry
Tufts University School of Medicine
Boston, Massachusetts

Douglas J. Henry
Department of Pharmacology
University of Washington
Seattle, Washington

Xiu-Ti Hu
Department of Neuroscience
Finch University of Health Sciences
Chicago Medical School
North Chicago, Illinois

Peter W. Kalivas
Alcohol and Drug Abuse Program
Washington State University
Pullman, Washington

Michael J. Kuhar
National Institutes of Health (NIH)
Intramural Research Program (IRP)
Neuroscience Branch
Baltimore, Maryland

Cynthia M. Kuhn
Department of Pharmacology
Duke University Medical Center
Durham, North Carolina

John Lever
The Johns Hopkins Medical Institute
Division of Nuclear Medicine
Baltimore, Maryland

Patrick Little
Department of Pharmacology
Duke University Medical Center
Durham, North Carolina

Fu-Chin Liu
Department of Brain and Cognitive Sciences
Massachusetts Institute of Technology
Cambridge, Massachusetts

Athina Markou
Scripps Research Institute
La Jolla, California

Rosario Moratalla
Department of Brain and Cognitive Sciences
Massachusetts Institute of Technology
Cambridge, Massachusetts

Eric J. Nestler
Department of Psychiatry and Pharmacology
Yale University School of Medicine
New Haven, Connecticut

Loren H. Parsons
The Scripps Research Institute
La Jolla, California

Amrat Patel
Neuroscience Branch
Addiction Research Center
Baltimore, Maryland

Nancy Pilotte
Neuroscience Branch
Addiction Research Center
Baltimore, Maryland

Linda J. Porrino
Department of Physiology and Pharmacology
The Bowman Gray Schoool of Medicine
Winston-Salem, North Carolina

William L. Thomas, Jr.
Department of Anatomy and Reproductive Biology
University of Hawaii School of Medicine
Honolulu, Hawaii

Ellen M. Unterwald
Department of Psychiatry
New York University Medical Center
New York, New York

Nora D. Volkow
Medical Department
Brookhaven National Laboratory
Upton, New York

Friedbert Weiss
Department of Neuropharmacology
The Scripps Research Institute
La Jolla, California

Francis J. White
Department of Neuroscience
Finch University of Health Sciences
The Chicago Medical School
North Chicago, Illinois

Blanche B. Young
Department of Anatomy and Cell Biology
Tufts University School of Medicine
Boston, Massachusetts

Nancy R. Zahniser
Department of Pharmacology
University of Colorado Health Science Center
Denver, Colorado

Xu-Feng Zhang
Department of Pharmacology
University of Washington
Seattle, Washington

TABLE OF CONTENTS

Section I

MECHANISMS OF COCAINE REINFORCEMENT

Chapter 1

Neuroanatomical Basis of Cocaine Self-Administration

John H. Graham and Linda J. Porrino

CONTENTS

I. INTRODUCTION

A wide variety of chemical substances have been classified as abused drugs on the basis of their excessive or inappropriate use. These substances have in common the ability to produce pleasurable feelings or sensations in those who use them. It is commonly believed that it is this property, the capacity to elicit euphoria, that is responsible for their abuse. Among the most widely abused of these substances is the psychostimulant, cocaine. Cocaine is known to produce behavioral activation,[1] serve as a discriminative stimulus, and support self-administration.[2] Cocaine has a high affinity for dopamine, serotonin, and norepinephrine transporters.[3] The ability of cocaine to elevate synaptic levels of monoamines by blocking their reuptake apparently causes the behavioral and subjective effects of the drug in animals and humans.[4-7]

Drugs that are self-administered by humans generally have been shown to be self-administered by animals, as well.[8] Intravenous self-administration of drugs by animals, therefore, provides an excellent model for assessing the reinforcing properties of a substance, and it also can be used as a model to study the neuroanatomical substrates that mediate self-administration. In this chapter, we will review findings from studies that have helped to identify the brain regions and circuits underlying cocaine self-administration. Although many different methods have been applied to this problem, this report will focus mainly on the application of the quantitative [^{14}C]2-deoxyglucose (2DG) autoradiographic method to the characterization of the anatomical substrates of cocaine self-administration.

II. LESIONING STUDIES

Many experiments have utilized discrete lesions of particular brain regions in an effort to disrupt cocaine self-administration by theoretically diminishing the reinforcing effects. Some of the earliest studies made use of electrolytic lesions; however, electrolytic lesions destroy all neuronal elements (e.g., cell bodies and fibers of passage) within a region and,

1-8493-8311-0/95/$0.00+$.50

therefore, make it difficult to attribute changes in self-administration behavior to particular neurochemical or neuroanatomical substrates. Lesions produced by neurotoxins allow for more precise destruction of certain neuronal populations within a region. The use of 6-hydroxydopamine (6-OHDA), for example, permits the destruction of dopaminergic terminals (presynaptic), and 5,7-dihydroxytryptamine eliminates serotonergic terminals. Excitotoxins such as kainic acid or ibotenic acid destroy neuronal cell bodies. Such lesions have provided much insight into the neurochemistry and neuroanatomy involved,[9] although a complete review of such studies is beyond the scope of this chapter. Many of these studies have focused on the role of the nucleus accumbens and its dopaminergic inputs originating in the ventral tegmental area. A series of studies by Roberts and colleagues have shown that dopaminergic depletion in either region attenuates the reinforcing effects of cocaine.[10-12] Numerous studies have confirmed and extended these findings, clearly establishing that dopaminergic transmission within the nucleus accumbens is a critical substrate for cocaine self-administration.

The nature of the neural circuits involved, especially those which include the nucleus accumbens, however, has not been studied extensively. For example, what are the efferent pathways from the nucleus accumbens through which the actions of cocaine are translated into self-administration behavior? One of the major targets of the nucleus accumbens is the ventral pallidum. This region has been shown in electrophysiological and behavioral studies to be an important connection for the mediation of the locomotor activation related to pharmacological and electrical stimulation of the accumbens.[13,14] Lesions of the ventral pallidum block the activating effects of cocaine and other psychostimulants, whereas lesions of other targets of the nucleus accumbens are without consequence.[15] Furthermore, Hubner and Koob[11] have demonstrated that the ventral pallidum plays an important role in both cocaine and heroin self-administration. In this study, animals trained to self-administer cocaine received ibotenic acid lesions of the ventral pallidum. The effect of destroying these cell bodies was to decrease both the drug intake on a fixed ratio schedule, and to decrease the break point on a progressive ratio schedule. These results indicate a decrease in the reinforcing efficacy of cocaine in the lesioned animals. Further studies indicate that subregions of the ventral pallidum may be distinguished by their apparent effects on cocaine self-administration. Ibotenic acid lesions of the sublenticular extended amygdala and the subcommissural ventral pallidum decreased cocaine self-administration on a fixed ratio schedule in rats.[16] However, lesions in these two areas did not produce the same effects on progressive ratio responding. Lesions of the sublenticular extended amygdala lowered the breakpoint, while lesions of the subcommissural ventral pallidum did not affect responding on this type of schedule. The nucleus accumbens projects to both these regions, and these results suggest that the sublenticular region of the extended amygdala may play a more significant role in cocaine self-administration.

Two distinct regions of the nucleus accumbens, the core and the shell, project to these subregions of the ventral pallidum. The core most prominently innervates the subcommissural ventral pallidum, while the shell projects mainly to the sublenticular region.[17,18] Furthermore, a recent report of intracranial self-stimulation in the ventral pallidum indicates that cocaine can potentiate reinforcement in this area; cocaine (i.p.) increased rate of responding over a range of pulse frequencies.[19] The role of the ventral pallidum and its subdivisions as the efferent system through which the effects of cocaine are mediated appears clear from these studies. The potential role of other efferent and afferent projections to the nucleus accumbens, however, requires further exploration.

III. INTRACRANIAL SELF-ADMINISTRATION STUDIES

Another approach to identifying the substrates of cocaine reinforcement has been the use of intracranial self-administration to discrete brain regions.[20,21] In such studies, animals have a cannula directly implanted into a specific brain region and then are trained to self-administer drugs into that particular brain area via the cannula. As would be expected, animals do not maintain responding when cannulae are placed in most regions, indicating that such areas are not critical to the reinforcing effects of the drug. In contrast, maintanence of responding is a strong indication that the cannula is placed in an area involved in the reinforcing effects of cocaine. Rats will self-administer cocaine into the medial prefrontal cortex, but not the nucleus accumbens.[20,21] Such self-administration is dose-dependent with an inverted U-shaped dose-intake relationship. 6-OHDA lesions to remove the dopaminergic innervation of the self-administration site attenuated intake to the level and pattern observed with vehicle treatment,[20] and these animals reinitiated responding if cocaine was replaced with dopamine. These data are consistent with the hypothesis that cocaine initiates reinforcing neuronal activity in the medial prefrontal cortex by a direct action on dopamine nerve endings, since removal results in a loss of reinforcing efficacy. These experiments are particularly noteworthy because they establish the importance of parts of the dopaminergic mesocorticolimbic system beyond the nucleus accumbens and ventral tegmental area as substrates of cocaine self-administration.

IV. METABOLIC MAPPING STUDIES OF THE EFFECTS OF COCAINE

One limitation of these techniques is that they do not describe circuits, but only specific parts of a circuit. Given the broad spectrum of neurochemical, physiological, and behavioral actions of cocaine, it is likely that the overall response is a product of multiple processes at several anatomical loci. In order to determine the neural substrates of these actions, it is necessary to identify neural events in circuits and pathways throughout the brain. Herein lies the advantage of metabolic mapping. The 2DG method allows measurement of local rates of cerebral glucose utilization in all regions of the central nervous system simultaneously. The rate at which glucose is consumed in the brain is a measure of the amount of energy used, because under normal physiological conditions, glucose is virtually the exclusive substrate for energy metabolism in the brain. In turn, the amount of energy used is a direct reflection of the amount of work done, which, in the brain, is in the form of electrical or synaptic activity.[22] By measuring changes in glucose utilization within a particular area using the 2DG method, changes in functional activity can be inferred. However, the most significant advantage of the 2DG method is that it affords a means to quantitatively measure a dynamic biochemical process in conscious, freely moving animals. Because this procedure utilizes autoradiography, high spatial resolution is permitted so that small anatomical regions and subregions can be identified.

A. ACUTE STUDIES

With the 2DG method, it is possible to study the effects of cocaine in a variety of behavioral paradigms (e.g., locomotor behavior, self-administration, and intracranial self-stimulation). Early studies mapping the effects of cocaine focused on acute administration—identifying those brain areas affected by an initial exposure to cocaine in drug-naive animals.[23,24] Studies in our laboratory utilized the intravenous route of administration because its kinetics best approximate the kinetics of the routes of administration most frequently chosen by human cocaine abusers: intranasal, intrapulmonary, and intravenous. These studies have shown that low doses of cocaine (0.5 and 1.0 mg/kg, i.v.) produce a very selective pattern of changes in cerebral glucose metabolism. Only the medial prefrontal cortex, nucleus accumbens (Figure 1), and substantia nigra pars reticulata exhibit increased rates of cerebral metabolism, while decreases are restricted to the lateral habenula.[24] Higher doses (5.0 mg/kg) produce pronounced elevations of glucose utilization throughout the brain; changes in the nigrostriatal motor system as well as in the mesocorticolimbic system are most prominent (Figure 1).

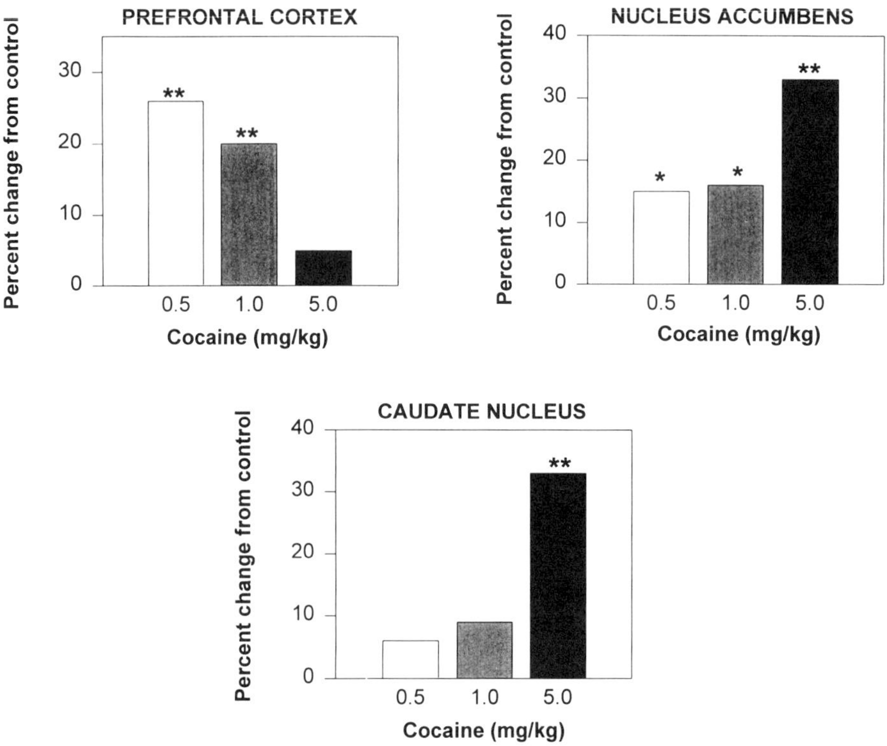

Figure 1. Dose-dependent effects of acute intravenous cocaine administration on rates of local cerebral glucose utilization in the rat prefrontal cortex (upper left), nucleus accumbens (upper right) and caudate-putamen (lower). Data are plotted as percent change from vehicle injected controls. Data derived from Porrino et al., 1988.[24]

These findings are consistent with studies that have implicated the nucleus accumbens in the reinforcing effects of cocaine and other psychostimulants.[10,25-27] Although these data certainly confirm the importance of the nucleus accumbens in mediating the effects of cocaine, they suggest that the medial prefrontal cortex is also important. As discussed earlier, this region is unique in its ability to support direct intracranial self-administration of cocaine.[20,21] Although lesions of the prefrontal cortex region do not abolish responding maintained by intravenous infusion of cocaine, the prefrontal cortex may be involved in the initiation or acquisition of cocaine self-administration rather than in its maintenance.[20]

A further demonstration of the importance of the medial prefrontal cortex, nucleus accumbens, and olfactory tubercle is provided by metabolic mapping studies comparing the effects of cocaine administered intravenously and intraperitoneally. In these studies, functional activity in these regions was altered only when cocaine was administered intravenously, regardless of the dose. The pharmacokinetics or the rate of delivery of cocaine often is considered to be an important determinant of the powerful euphorogenic effects of the drug in humans.[28] The fact that the intravenous route of administration (by which cocaine rapidly reaches peak concentrations in plasma and brain) singularly affects functional activity in the nucleus accumbens and medial prefrontal cortex supports the notion that these regions may be uniquely involved in the rewarding effects of cocaine. Once again, metabolic mapping of the acute effects of cocaine has shown that portions of limbic cortex along with the ventral striatum underlie the actions of this psychostimulant.

B. COCAINE SELF-ADMINISTRATION

Our laboratory also has investigated the effects of cocaine self-administration on cerebral metabolism. These studies are important for two reasons. First, cocaine is acting as a reinforcer, because it is presented contingent upon the animal's response. Therefore, we are examining the reinforcing effects of cocaine. Second, animals are receiving cocaine chronically over repeated sessions. As a result, we are examining the long-term consequence of chronic exposure to cocaine, in addition to the acute pharmacological effects occurring within a self-administration session. In these studies, rates of local cerebral glucose utilization were measured in animals trained to self-administer cocaine on a fixed ratio 2 schedule of reinforcement during daily sessions. Tracer was injected after either 2 or 12 cocaine self-infusions.

These studies revealed a signficantly different pattern of changes in glucose utilization in the dopaminergic mesocorticolimbic system as compared to the pattern associated with the acute, noncontingent administration of cocaine described above. In contrast to the acute administration, chronic cocaine self-administration resulted in decreased rates of metabolism in a number of limbic structures, including the medial prefrontal cortex, anterior cingulate cortex, and nucleus accumbens in self-administering animals when compared to yoked saline animals.[29] It is interesting to note that cocaine withdrawal following studies in self-administering in rats also show decreased rates of glucose metabolism in the anterior cingulate cortex,[30] while abstinence after chronic, noncontingent cocaine produces decreases in the medial prefrontal cortex.[31] Both of these areas receive dopaminergic innervation from the ventral tegmental area, although this is not the predominant neurochemical innervation in these areas.[32] The similarity of effects between withdrawal studies in which cocaine is not present at the time of the experiment, and self-administration studies in which cocaine is on board, suggests that these changes may represent some form of adaptation to chronic cocaine exposure, rather than a direct pharmacological effect of cocaine.

Acute cocaine produces significant increases in local cerebral glucose utilization in the nucleus accumbens at low and high doses.[24] Although the nucleus accumbens did not display significant changes in our studies, the rostral nucleus accumbens does display a pattern of changes that resembles what we have observed in both the medial prefrontal cortex and the anterior cingulate cortex.[29] Since the nucleus accumbens receives excitatory amino acid projections from prefrontal and anterior cingulate cortex,[33] and parallel metabolic changes are observed in these regions, it is possible that these cortical regions "drive" the response of the rostral nucleus accumbens. Such a conclusion is supported by the fact that intracranial self-administration of cocaine into the medial prefrontal cortex produces an ipsilateral increase of dopamine turnover in the nucleus accumbens.[34] It has been suggested that because the accumbens receives input from a number of mesocorticolimbic regions, it may be uniquely positioned to translate motivational information into motor activation through its connections with the nigrostriatal motor system.[35] Indeed, the nucleus accumbens itself supports intracranial self-stimulation.[36] It does not, however, support intracranial self-administration of cocaine, indicating that the nucleus accumbens may play a role in maintenance of self-administration, as opposed to its initiation.[20] The olfactory tubercle is another limbic region that shows enhanced glucose utilization with acute intravenous cocaine administration.[37] However, in self-administering animals, we observed only a marginally significant increase in the olfactory tubercle.[29] Withdrawal studies also indicate a role for the olfactory tubercle in cocaine self-administration,[30] but the nature of this involvement has not yet been characterized.

The ventral tegmental area, which sends dopaminergic projections to the nucleus accumbens, olfactory tubercle, and medial prefrontal cortex has not been shown to be altered by acute cocaine.[24] Nor did we see changes in regional glucose utilization in our

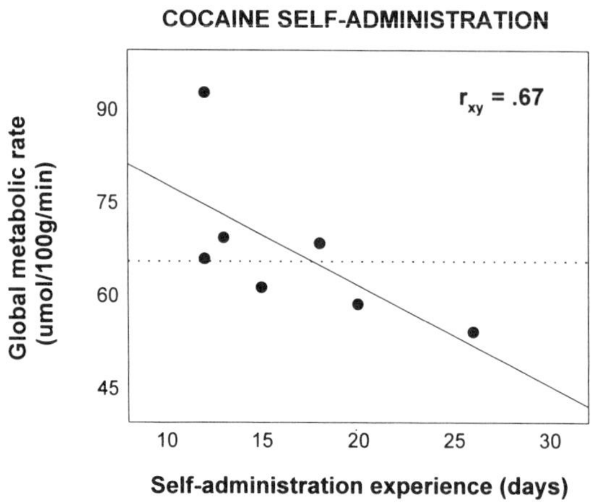

Figure 2. Correlation between number of days of cocaine self-administration and global cerebral metabolic rates (μmol/100 g/min) in rats self-administering cocaine after 2 infusions (n = 7). Value of Pearson product moment correlation coefficient is shown in the upper right. The dashed line indicates global rates of metabolism in normal yoked saline treated rats (n = 5).

studies with cocaine self-administering rats. Nevertheless, the ventral tegmental area has been shown to play an important role in cocaine self-administration. Lesions of this area with 6-OHDA produced a long-lasting diminution of cocaine intake in animals trained to self-administer cocaine.[12] However, self-administration of cocaine directly into the ventral tegmental area does not support drug-maintained responding.[20] This supports the idea that it is the regions containing terminals of ventral tegmental area dopaminergic projections which are important for maintaining responding in cocaine self-administration.

Recent studies have begun to focus on the role of serotonin in the reinforcing effects of cocaine. This is particularly important because cocaine displays high affinity for the serotonin transporter,[3] as well as for 5-HT_3 receptors.[38] For example, fluoxetine, a selective serotonergic reuptake blocker, can attenuate cocaine self-administration[39,40] and lesions of the medial forebrain bundle or the amygdala with 5,7-dihydroxytryptamine can increase the reinforcing efficacy of cocaine as measured by a progressive ratio schedule.[9] Furthermore, cocaine enhances serotonergic autoregulation through 5-HT_{1A} receptors on serotonergic cell bodies.[41] In addition, chronic, noncontingent cocaine enhances the inhibitory responses of serotonergic neurons to cocaine, fluoxetine or the 5-HT_{1A} agonist 8-OH-DPAT.[42] Chronic cocaine self-administration decreased rates of glucose metabolism in the median raphe (unpublished observation), although the acute administration of cocaine did not.[24] This suggests that serotonin may either be involved in self-administration itself or that the influence of serotonergic systems may increase with greater exposure to cocaine.

In these chronic, self-administration studies, the effects of cocaine are different from those observed following acute administration in a number of regions. Acute cocaine increases glucose utilization in most limbic regions, whereas chronic cocaine self-administration produces decreases in many of these same areas, indicating changes that may extend beyond the pharmacological effects of the cocaine that is present during the 2DG procedure. This is supported by the fact that greater decreases in global metabolic rates were observed after increased experience with cocaine self-administration (Figure 2). These decreases probably reflect the adaptational response that cocaine produces along these various pathways, an adaptation to the acute neurochemical and behavioral effects of cocaine in association with learning that may occur with self-administration behaviors.

C. OTHER EFFECTS OF COCAINE

Cocaine not only produces reinforcing effects, but also produces a great deal of locomotor behavior in self-administering rats—much more than required to press the lever for infusions. As the session progresses, and cocaine intake increases, this behavioral activation shifts toward focused stereotypies, which are interrupted only when the animal returns to the lever for another infusion. Although it is thought that the nigrostriatal system is not involved directly in the reinforcing effects of cocaine, since lesions of the caudate-putamen do not disrupt cocaine self-administration,[43] the substantia nigra and extrapyramidal areas receiving striatal projections may play a key role in the motor effects of self-administered cocaine. In fact, the extrapyramidal response to self-administered cocaine, as measured by the 2DG method, is quite similar to the acute effects of cocaine upon these pathways, unlike the results observed in the mesocorticolimbic system.

The caudate-putamen, the dominant target of dopaminergic neurons in the substantia nigra pars compacta, shows sensitivity to the acute administration of both high and low cocaine doses (1.0 and 5.0 mg/kg, i.v.), which increases glucose utilization in its dorsomedial, dorsolateral, and ventral regions.[37] Chronic, self-administered cocaine does not significantly alter rates of glucose metabolism in these three areas.[29] Acute cocaine increases local cerebral glucose metabolism in both the pars reticulata and the pars compacta of the substantia nigra. The pars compacta, which projects to the caudate-putamen, increases only in response to high doses of acute cocaine,[24] while no significant changes are seen following chronic self-administration.

Although the caudate-putamen does not display any significant alterations in rates of glucose metabolism, a number of areas that receive striatal efferent projections display such changes. The substantia nigra pars reticulata, the globus pallidus, and the entopeduncular nucleus all have increased rates of local cerebral glucose utilization in response to chronic cocaine self-administration.[29] In particular, the substantia nigra pars reticulata, which receives projections from both the dorsal and ventral striatum, exhibits such increases. Cocaine has been shown to increase dopamine levels in the caudate-putamen,[44,45] and such changes might intensify the GABAergic transmission from the striatum to the pars reticulata. In acute experiments, the pars reticulata shows increases in glucose utilization in response to either low or high doses of i.v. cocaine.[24] The entopeduncular nucleus, which receives efferents from the caudate-putamen as well as the from the rostral pole and core of the accumbens,[18] shows increases in rate of glucose utilization in response to high doses of acute cocaine.[24,37] Chronic, self-administered cocaine also substantially increases these rates in the entopeduncular nucleus. In the globus pallidus, acute cocaine administration increases rates of glucose utilization in response to either high or low doses of i.v. cocaine.[24,37] Likewise, our studies with chronically self-administered cocaine show increases in glucose utilization in the globus pallidus.[29] The globus pallidus in turn projects to the subthalamic nucleus.[46] In acute cocaine 2DG studies, local cerebral glucose utilization in the subthalamic nucleus only increased in response to high doses of cocaine.[24,37] In our chronic self-administration studies, cocaine produces large increases in rates of glucose metabolism in this area. Changes in glucose utilization in these regions (i.e., substantia nigra pars reticulata, globus pallidus, entopeduncular nucleus, and subthalamic nucleus) probably are a reflection of the motor activating effects of cocaine, particularly, the focused stereotypies that occur after a number of infusions.

D. ISOLATING THE REINFORCING EFFECTS OF COCAINE

In order to establish the identity of the circuits directly involved in mediating the reinforcing effects of cocaine, the 2DG method was applied in rats self-administering the drug during an experimental session.[47] Rates of glucose utilization in these rats were compared to littermate control rats receiving yoked infusions of either cocaine or saline. Response-dependent administration of cocaine (self-administration) produced large, widespread elevations of glucose utilization throughout the mesocorticolimbic system including the prefrontal cortex, nucleus accumbens, and olfactory tubercle (Figure 3) compared to yoked saline controls. Increases also were evident in the nigrostriatal motor system.

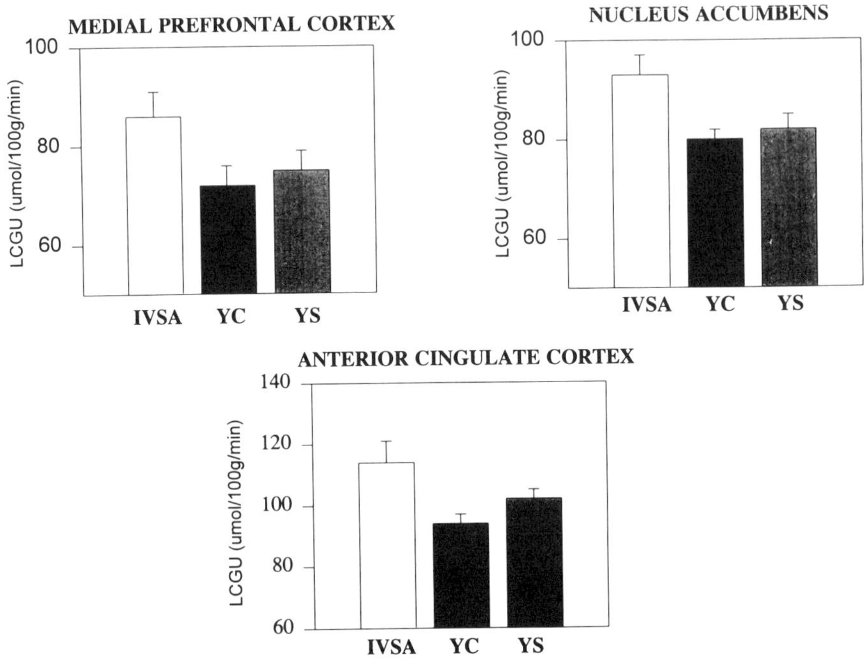

Figure 3. Rates of local cerebral glucose utilization (LCGU in µmol/100 g/min) following intravenous cocaine self-administration (IVSA), passive infusions of cocaine according to the same schedule as self-administering rats (yoked control, YC), and similar saline infusions (yoked saline, YS) in the medial prefrontal cortex (upper left), nucleus accumbens (upper right), and the anterior cingulate cortex (lower).

This increase of metabolic rates is in contrast with the findings of our other self-administration studies. In these self-adminstration studies, however, rats had less self-administration experience and their global metabolic rates were generally well above saline controls.

In contrast, passive cocaine infusion or response-independent cocaine presentation increased functional activity compared to yoked saline animals in portions of the nigrostriatal motor system, identical to those seen in self-administering animals. The similarity of the changes in the nigrostriatal motor system reflect the similarity of the stereotypies and other motor activating effects elicited by cocaine, regardless of the context of administration. However, functional activity was not altered in any forebrain limbic structures. The differences in the two patterns of changes in functional activity clearly demonstrate the differences between and the importance of the form of presentation of the drug. Cocaine may be a reinforcer when self-administered, but not when infused passively. The changes seen in self-administering animals, but not in the passively infused animals, are related directly to the reinforcement associated with cocaine, rather than to its pharmacological effects alone.

V. METABOLIC MAPPING STUDIES IN NONHUMAN PRIMATES

The circuitry involved in the acute and chronic effects of cocaine also has been examined in nonhuman primates.[48] Although only a limited number of monkeys have been studied, preliminary results indicate that the acute intravenous administration of cocaine

(1 mg/kg) produces significant changes in both cortical and subcortical regions. More specifically, cocaine-depressed rates of metabolism in the core and shell of the nucleus accumbens as well as in the olfactory tubercle. In addition, other dopaminergically innervated areas, namely the anterior putamen and dorsal caudate, both show depressed metabolism in response to cocaine. The anatomical distribution of these changes, although not the direction, is similar to that seen following the acute intravenous administration of a moderate dose of cocaine to naive rats.[24,37] The cerebral metabolic response in the primate, however, is far more complex than that seen in the rat. Functional activity is altered in a number of brain regions, especially cortical regions, which were unaffected in rat studies regardless of dose. Although there was a general tendency for rates of cortical metabolism to be lower in cocaine-treated monkeys, significant changes were observed in the medial orbital cortex and in the anteroventral portion of the insula. In addition, metabolism was decreased in the subiculum and the entorhinal cortex.

This pattern of metabolic alterations is noteworthy because its topographical distribution closely parallels the distribution of cortical and hippocampal projections to the nucleus accumbens. There is a striking overlap between the anatomical distribution of territories projecting to the nucleus accumbens and the location of the reduction in metabolism seen in the cortex and the hippocampal complex. The decreased rates of metabolism in these afferents might be responsible in part for the significant decreases in glucose utilization observed in the core and shell of the nucleus accumbens. These data emphasize the significance of corticostriatal circuitry in mediating the effects of cocaine in primates.

A preliminary 2DG experiment in a monkey having a prolonged history of cocaine self-administration indicates a pattern of cerebral metabolic changes quite different from that seen with acute exposure (unpublished observations; Huston-Lyons, Friedman, Nader, and Porrino). As with acute exposure, rates of metabolism were depressed in the orbital cortex, but effects also were prominent in the gyrus rectus and subcallosal cingulate gyrus, areas unaffected by acute cocaine exposure. Large decreases also were seen in the temporal pole. Metabolic changes in this area are correlated positively with a measure of euphoria in humans,[49] and this site has been shown to have a high density of cocaine binding sites in human and monkey brains.[50,51] As seen in the acute experiments, the regions of cortex affected by cocaine coincide with those that project to the nucleus accumbens. Areas not projecting to the ventral striatum did not show the same pattern of metabolic changes. Further differences between the metabolic response in this self-administering monkey and those treated acutely with cocaine are evident in the ventral striatum. Alterations of glucose utilization were found only in the most rostral parts of the nucleus accumbens, its shell and the olfactory tubercle, but not in the core. It appears, therefore, that there is a shift in the topography of the functional response to cocaine with chronic self-administration. The shift in the ventral striatum parallels that in cortex (e.g., gyrus rectus and cingulate cortex), where significant changes in glucose utilization become evident in areas projecting mainly to shell. These effects resemble those seen in the rat. Although the sample is too small in this primate study, the data suggest that behavioral context and drug history are important determinants of the functional response to cocaine.

VI. CONCLUSIONS

One of the essential advantages of imaging with the 2DG method is the ability to visualize changes in brain functional activity throughout the entire central nervous system simultaneously. Using this method, we have been able to identify the broad range of neuroanatomical regions and circuits through which the effects of cocaine are mediated. What becomes evident from these studies is that which component brain regions that make up these circuits depends on the dose, route of administration, behavioral history, and context of presentation of cocaine.

The data in the imaging studies described in this review clearly support the importance of the nucleus accumbens as central to cocaine self-administration; this has been shown in numerous other studies using a multitude of other methodologies. These imaging studies, however, also have stressed the significance of limbic cortical regions in the mediation of this behavior. The nucleus accumbens may be the focal point in which the integration of information from a number of regions critical for cocaine self-administration, including the prefrontal cortex, cingulate cortex, and hippocampal formation, occurs.

ACKNOWLEDGMENTS

This work was supported by USPHS Research Awards R01 DA07522 and P50 DA06634.

REFERENCES

1. Scheel-Kruger, J., Braestrup, C., Nielson, M., Golembioska, K., and Mogilnicka, E., Cocaine: discussion on the role of dopamine in the biochemical mechanism of action, in *Cocaine and Other Stimulants,* Ellinwood, E. H. and Kilbey, M. M., Eds., Plenum Press, New York, 1977, 373.
2. Woolverton, W. L. and Balster, R.L., Behavioral pharmacology of local anesthetics: reinforcing and discriminative stimulus effects, *Pharmacol. Biochem. Behav.,* 16, 491, 1982.
3. Ritz, M. C., Lamb, R. J., Goldberg, S. R., and Kuhar, M. J., Cocaine receptors on dopamine transporters are related to self-administration of cocaine, *Science,* 237, 1219, 1987.
4. Broderick, P. A., Cocaine's colocalized effects on synaptic serotonin and dopamine in ventral tegmentum in a reinforcement paradigm, *Pharmacol. Biochem. Behav.,* 42, 889, 1992.
5. Dackis, C. A. and Gold, M. S., Treatment strategies for cocaine detoxification, in *Cocaine: Pharmacology, Physiology and Clinical Strategies,* Lakoski, J. M., Galloway, M. P., and White, F. J., Eds., CRC Press, Boca Raton, 1992, 441.
6. Hurd, Y. L., Weiss, F., Koob, G., and Ungerstedt, U., The influence of cocaine self-administration on *in vivo* dopamine and acetylcholine neurotransmission in rat caudate-putamen, *Neurosci. Lett.,* 109, 227, 1990.
7. Hernandez, E. F. and Hoebel, B. G., Food reward and cocaine increase extracellular dopamine in the nucleus accumbens as measured by microdialysis, *Life Sci.,* 42, 1705, 1988.
8. Weeks, J. R., Experimental morphine addiction: method for automatic intravenous injections in unrestrained rats, *Science,* 138, 143, 1962.
9. Roberts, D. C. S., Neural substrates mediating cocaine reinforcement: the role of monoamine systems, in *Cocaine: Pharmacology, Physiology and Clinical Strategies,* Lakoski, J. M., Galloway, M. P., and White, F. J., Eds., CRC Press, Boca Raton, 1992, 73.
10. Roberts, D. C. S., Koob, G. F., Klonoff, P., and Fibiger, H. C., Extinction and recovery of cocaine self-administration following 6-hydroxydopamine lesions of the nucleus accumbens, *Pharmacol. Biochem. Behav.,* 12, 781, 1980.
11. Hubner, C. B. and Koob, G. F., The ventral pallidum plays a role in mediating cocaine and heroin self-administration in the rat, *Brain Res.,* 508, 20, 1990.
12. Roberts, D. C. S. and Koob, G. F., Disruption of cocaine self-administration following 6-hydroxydopamine lesions of the ventral tegmental area in rats, *Pharmacol. Biochem. Behav.,* 17, 901, 1982.
13. Mogenson, G. J. and Wu, M., Subpallidal projections to the mesencephalic locomotor region investigated with a combination of behavioral and electrophysiological recording techniques, *Brain Res. Bull.,* 16, 383, 1986.
14. Yang, C. R. and Mogenson, G. J., Ventral pallidal neuronal responses to dopamine receptor stimulation in the nucleus accumbens, *Brain Res.,* 489, 237, 1989.
15. Swerdlow, N. R. and Koob, G. F., Lesions of the dorsomedial nucleus of the thalamus, medial prefrontal cortex and pedunculopontine nucleus: effects on locomotor activity mediated by nucleus accumbens-ventral pallidal circuitry, *Brain Res.,* 412, 233, 1987.

16. Robledo, P. and Koob, G. F., Two discrete nucleus accumbens projection areas differentially mediate cocaine self-administration in the rat, *Behav. Brain Res.*, 55, 159, 1993.
17. Heimer, L., Zahm, D. S., Churchill, L., Kalivas, P. W., and Wohltmann, C., Specificity in the projection patterns of accumbal core and shell in the rat, *Neuroscience*, 41, 89, 1991.
18. Zahm, D. S. and Heimer, L., Specificity in the efferent projections of the nucleus accumbens in the rat: comparisons of the rostral pole projection patterns with those of the core and shell, *J. Comp. Neurol.*, 327, 220, 1993.
19. Panagis, G. and Spyraki, C., The effects of cocaine and morphine on the rewarding efficacy of ventral pallidal intracranial self stimulation (ICSS), *Neuropsychopharmacology*, 9, 166, 1993.
20. Goeders, N. E. and Smith, J. E., Cortical dopaminergic involvement in cocaine reinforcement, *Science*, 221, 773, 1983.
21. Goeders, N. E., Dworkin, S. I., and Smith, J. E., Neuropharmacological assessment of cocaine self-administration into the medial prefrontal cortex, *Pharmacol. Biochem. Behav.*, 24, 1429, 1986.
22. Sokoloff, L., The radioactive deoxyglucose method: Theory, procedure, and applications for the measurement of local glucose utilization in the central nervous system, in, *Advances in Neurochemistry*, Agranoff, B. W. and Aprison, M. H., Eds., Plenum Press, New York, 1982, 7.
23. London, E. D., Wilkerson, G., Goldberg, S. R., and Risner, M. E., Effects of L-cocaine on local cerebral glucose utilization in the rat, *Neurosci. Lett.*, 68, 73, 1986.
24. Porrino, L. J., Domer, F. R., Crane, A. M., and Sokoloff, L., Selective alterations in cerebral metabolism within the mesocorticolimbic dopaminergic system produced by acute cocaine administration in rats, *Neuropsychopharmacology*, 1, 109, 1988.
25. DeWit, H. and Wise, R. A., Blockade of cocaine reinforcement in rats with the dopamine receptor blocker pimozide, but not the noradrenergic blockers phentolamine and phenoxybenzamine, *Can. J. Psychol.*, 31, 195, 1977.
26. Roberts, D. C. S., Corcoran, M. E., and Fibiger, H. C., On the role of ascending catecholamine systems in intravenous self-administration of cocaine, *Pharmacol. Biochem. Behav.*, 6, 615, 1977.
27. Lyness, W. H., Friedle, N. M., and Moore, K. E., Destruction of dopaminergic nerve terminals in nucleus accumbens: effects on D-amphetamine self-administration, *Pharmacol. Biochem. Behav.*, 11, 553, 1979.
28. Resnick, R. B., Kestenbaum, R. S., and Schwartz, L. K., Acute systemic effects of cocaine in man: a controller study by intranasal and intravenous routes, *Science*, 195, 696, 1977.
29. Graham, J. H., LaRosee, J. R., Dworkin, S. I., Samson, H. H., and Porrino, L. J., Functional effects of chronic cocaine self-administration in rats, in *Soc. Neurosci. Abstr.*, 19, 1863, 1993.
30. Hammer, R. P., Pires, W. S., Markou, A., and Koob, G. F., Withdrawal following cocaine self-administration decreases regional cerebral metabolic rate in critical brain reward regions, *Synapse*, 14, 73, 1993.
31. Clow, D. W. and Hammer, R. P., Cocaine abstinence following chronic treatment alters cerebral metabolism in dopaminergic reward regions: bromocriptine enhances recovery, *Neuropsychopharmacology*, 4, 71, 1991.
32. Roth, R. H., Wolf, M. E., and Deutch, A.Y., Neurochemistry of midbrain dopamine systems, in *Psychopharmacology: The Third Generation of Progress*, Meltzer, H.Y., Ed., Raven Press, New York, 1987, 81.
33. Brog, J. S., Salyapongse, A., Deutch, A. Y., and Zahm, D. S., The patterns of afferent innervation of the core and shell in the "accumbens" part of the rat ventral striatum: immunohistochemical detection of retrogradely transported fluoro-gold, *J. Comp. Neurol.*, 338, 255, 1993.
34. Goeders, N. E. and Smith, J. E., Intracranial cocaine self-administration into the medial prefrontal cortex increases dopamine turnover in the nucleus accumbens, *J. Pharmacol. Exp. Ther.*, 265, 592, 1993.
35. Koob, G. F., Drugs of abuse: anatomy, pharmacology and function of reward pathways, *Trends Pharmacol. Sci.*, 13, 177, 1992.
36. Rolls, E. T., Contrasting effects of hypothalamic and nucleus accumbens septi self-stimulation on brain stem single unit activity and cortical arousal, *Brain Res.*, 31, 275, 1971.
37. Porrino, L. J., Functional consequences of acute cocaine treatment depend on route of administration, *Psychopharmacology*, 112, 343, 1993.
38. Kilpatrick, G. J., Jones, B. J., and Tyers, M. B., Identification and distribution of 5-HT3 receptors in rat brain using radioligand binding, *Nature*, 330, 746, 1987.
39. Carroll, M. E., Lac, S. T., Asencio, A., and Kragh, R., Fluoxetine reduces intravenous cocaine self-administration in rats, *Pharmacol. Biochem. Behav.*, 35, 237, 1990.

40. Richardson, N. R. and Roberts, D. C. S., Fluoxetine pretreatment reduces breaking points on a progressive ratio schedule reinforced by intravenous cocaine self-administration in the rat, *Life Sci.*, 49, 833, 1991.
41. Cunningham, K. A. and Lakoski, J. M., The interaction of cocaine with serotonin dorsal raphe neurons: single unit extracellular recording studies, *Neuropsychopharmacology*, 3, 41, 1990.
42. Cunningham, K. A., Paris, J. M., and Goeders, N. E., Chronic cocaine enhances serotonin autoregulation and serotonin uptake binding, *Synapse*, 11, 112, 1992.
43. Koob G. F., Vaccarino, F. J., Amalric, M., and Bloom, F. E., Positive reinforcement properties of drugs: search for neural substrates, in *Brain Reward Systems*, Engle, J. and Oreland, L., Eds., Raven Press, New York, 1987, 35.
44. Church, W. H., Justice, J. B., and Byrd, L. D., Extracellular dopamine in rat striatum following uptake inhibition by cocaine, nomifensine and benztropine, *Eur. J. Pharmacol.*, 139, 345, 1987.
45. Hurd, Y. L. and Ungerstedt, U., Cocaine: an *in vivo* microdialysis evaluation of its acute action on dopamine transmission in rat striatum, *Synapse*, 3, 48, 1989.
46. Gerfen, C. R., The neostriatal mosaic: multiple levels of compartmental organization, *Trends Neurosci.*, 15, 133, 1992.
47. Dworkin, S. I., Porrino, L. J., and Smith, J. E., Pharmacology of basal forebrain involvement in reinforcement, in *The Basal Forebrain*, Napier, T. C., Ed., Plenum Press, New York, 1991, 327.
48. Huston-Lyons, D., Friedman, D., Nader, M., Grant, K., and Porrino, L. J., Acute intravenous administration of cocaine decreases cerebral glucose metabolism in the monkey, in *Comm. Prob. Drug Dep., Abstr.*, 1994, 598.
49. London, E. D., Cascella, N. G., Wong, D. F., Phillips, R. L., Dannals, R. F., Links, J. M., Herring, R., Grayson, R., Jaffe, J. H., and Wagner, H. N., Cocaine-induced reduction of glucose utilization in human brain, *Arch. Gen. Psychiatry*, 47, 567, 1990.
50. Biegon, A., Dillon, K., Volkow, N. D., Hitzemann, R. J., Fowler, J. S., and Wolf, A. P., Quantitative autoradiography of cocaine binding sites in human brain postmortem, *Synapse*, 10, 126, 1992.
51. Kaufman, M. J. and Madras, B. K., Distribution of cocaine recognition sites in monkey brain: II. Ex vivo autoradiography with [^{3}H]CFT and [^{125}I]RTI-55, *Synapse*, 12, 99, 1992.

Chapter 2

Regional Metabolic Manifestations of Cocaine Exposure: Sensitization, Tolerance, and Withdrawal

Ronald P. Hammer, Jr., Blanche B. Young, and William L. Thomas, Jr.

CONTENTS

I. INTRODUCTION

Cocaine has been used widely as a psychostimulant in Western societies for more than 100 years. Chronic use and abuse of cocaine have peaked in recent years, stimulating an extensive search for the mechanisms underlying reinforcing potency of this drug. Furthermore, recent clinical studies have described a behavioral phenomenon of withdrawal,[1,2] which suggests the existence of a dependence state following chronic cocaine use. However, dependence and withdrawal from cocaine clearly differ from the classical descriptions derived from studies of opiate withdrawal, which manifests as a physical state affecting autonomic functioning. Instead, cocaine withdrawal seems to alter the motivational state affecting emotive and behavioral functioning. However, the outcome and impact of cocaine dependence are no less severe than those of other abused substances (e.g., alcohol and narcotics).

The pharmacology, physiology, and chemistry of cocaine have been studied extensively. The route, dose, and paradigm of drug treatment often have varied in these studies, producing inconsistent results. Initial studies focused almost exclusively on the effects of acute cocaine treatment, which blocks dopamine reuptake,[3] increases extracellular dopamine levels in ventral striatal regions,[4,5] and stimulates locomotor activity[6] in experimental animals. Serotonin (5-HT) and norepinephrine reuptake also are reduced,[7,8] but the binding to and blockade of the dopamine transporter seems to underlie the reinforcing efficacy of cocaine.[9]

Psychomotor activation induced by acute cocaine probably is mediated by the activity of direct and/or indirect striatal projections to the entopeduncular nucleus (EP)/substantia nigra pars reticulata (SNr) or the pallidum,[10] respectively. The induction of immediate early gene expression, which heralds synaptic activation of a neuron,[11] in the striatum by cocaine is dependent on dopamine D_1 receptor stimulation[12,13] (see Chapter 13) in the presence of 5-HT.[14] These striatal projection neurons which contain predominantly D_1 receptors[15] innervate both the EP and SNr.[16] In fact, neurons of both the dorsal and ventral striatum are known to project to the EP,[17] the rodent homologue of the internal segment

1-8493-8311-0/95/$0.00+$.50

of the globus pallidus (GP), which communicates with motor cortical areas via a thalamic relay. Furthermore, both regions can affect the mesencephalic locomotor region directly.[10,18] Thus, the efferent circuitry from striatal regions which exhibit maximal cocaine effects can impact motor control functions directly.

The reinforcing effects of cocaine appear to involve the mesocorticolimbic dopamine circuit, in which dopaminergic projections from the ventral tegmental area (VTA) innervate the medial prefrontal cortex (MPf; prelimbic and infralimbic regions) and anterior cingulate cortex, as well as the nucleus accumbens (NAc; medial "shell" and lateral "core" regions), olfactory tubercle, and amygdaloid nuclei. A complex circuitry interconnects these regions: neurons of the prelimbic MPf and anterior cingulate cortex innervate the NAc core, while the infralimbic MPf supplies the NAc shell. NAc core neurons innervate the dorsolateral portion of the ventral pallidum (VP) and the rostral portion of the lateral hypothalamus, while the NAc shell projects to the ventromedial VP and projects diffusely throughout the lateral hypothalamus. Neurons of both VP regions supply the mediodorsal thalamus, while the dorsolateral VP also innervates the subthalamus and substantia nigra; and mediodorsal thalamic neurons innervate the prelimbic MPf.[19-22] Thus, a recurrent loop is present which can extend limbic cortical activity through the NAc shell to the VP and the thalamus, back to the prelimbic MPf and the NAc core, and out through extrapyramidal motor control centers. The limbic nature of these regions suggests that the ensuing output more closely involves motivation than voluntary motor control.

The earliest data describing differential effects of chronic cocaine came from behavioral studies, which revealed that repeated daily treatment produces sensitization of locomotor responses.[23] Behavioral sensitization occurs only if the treatment and test environments are similar, suggesting that the conditioned response is context-dependent.[24,25] Such responses probably are an important component of normal learning, which depends on reinforcement following the presentation of behaviorally relevant stimuli. Cocaine could, then, "short circuit" this normal mechanism by inducing reinforcement in the absence of additional relevant stimuli, resulting in a conditioned response or habit formation. Furthermore, it has been suggested that sensitization might underlie the progressive increase of behavioral pathology observed in psychostimulant abusers, which can recur with subsequent drug challenge many years later.[26]

The mechanism underlying this behavioral sensitization probably involves mesolimbic dopamine neurons, whose spontaneous activity increases following repeated cocaine treatment[27] (see Chapter 7). Repeated daily treatment sensitizes dopamine response in the NAc[28,29] (see Chapter 6), an effect which is correlated positively with motor activity.[30,31] The relative dopaminergic response varies with length of treatment, since the basal level of dopamine in the NAc initially increases,[32] then declines below normal with continuing treatment[33] and during withdrawal.[34,35] This occurs despite the fact that regional brain cocaine concentration is elevated after chronic treatment,[36] due to increased absorption.[37]

That chronic exposure ultimately attenuates dopaminergic response in the NAc implies the development of tolerance. However, the particular treatment paradigm (i.e., dose, frequency, and route of administration) can affect the outcome. Continuous cocaine infusion by osmotic minipump produces tolerance of [^{3}H]dopamine uptake in the dorsal and ventral striatum[38] and locomotor activity,[39-41] and attenuates cocaine-induced dopamine efflux from striatal slices.[42] Furthermore, repeated injection of cocaine numerous times daily produces tolerance to cocaine self-administration,[43] and attenuates extracellular dopamine levels, especially in the ventral striatum.[44]

In contrast, cocaine challenge following repeated daily treatment generally produces locomotor sensitization and increases extracellular dopamine levels in the ventral striatum, but only after a sufficient period of drug abstinence.[45,46] This delay could be due to an alteration of dopamine transporter function following repeated cocaine treatment (see Chapter 11). This initial abstinence period is similar to the "crash" phase of early withdrawal, while the subsequent period of sensitized response to drug challenge might be related to the "craving" phase which follows early withdrawal in human cocaine abusers.[1,2]

We have attempted to elucidate the neural circuitry underlying these phases of cocaine-induced behavior (i.e., sensitization, tolerance, and withdrawal) using a metabolic marker which permits the quantitative assessment of regional synaptic activity. This autoradiographic procedure requires the administration of a pulse of [^{14}C]2-deoxyglucose (2DG), followed by the collection of serial plasma samples across a 45-min exposure period for subsequent determination of plasma glucose and deoxyglucose levels. After rapid removal, freezing, and sectioning of brains, exposure of sections to X-ray film with calibrated radiostandards, and regional quantitative densitometric analyses, we utilize the operational equation of Sokoloff[47] to determine the regional cerebral metabolic rate for glucose ($rCMR_{glc}$) in a given region of interest defined on an image of the Nissl-stained section which produced the autoradiograph, if necessary. The 2DG procedure provides a unique, global view of metabolic activity due to the proportional transport of 2DG with glucose, and its subsequent intracellular trapping in the form of 2DG-6-phosphate.[47] This view is generated, in great part, at the level of synaptic terminals,[48] due to the relatively greater surface-to-volume ratio present in synapses compared to cell bodies.[49] Thus, this metabolic sample is the product of activity requiring glucose in numerous afferent projections to the region of interest, without regard to the neurotransmitter content or its postsynaptic effect.

The utility of this procedure in pharmacological studies is established widely.[50] Its major advantages are that regional metabolic activity can be determined simultaneously across various brain regions during a period of efficacy in an unrestrained, behaving animal. In the case of rewarding drugs, such behaviors may be indicative of drug reinforcement (e.g., lever pressing for, frequency of, and threshold dose which supports self-administration). Coincident alterations following drug exposure in regions which are connected anatomically suggest the involvement of functional brain circuits in these drug effects. Thus, the efferent pathways and regional targets involved in psychostimulant reinforcement (see Chapter 1) or other drug-induced behaviors can be determined and correlated. Furthermore, the pharmacology underlying these effects can be examined using selective agonist or antagonist (pre)treatments. Finally, $rCMR_{glc}$ and other markers, such as selective pharmacologic receptor or transporter binding density, can be compared in adjacent sections from the same brain,[51] and significant correlations between these measures might imply a causative relationship.

II. SENSITIZATION OF METABOLIC ACTIVITY BY REPEATED COCAINE TREATMENT

Sensitization causes an augmented response to a conditioned stimulus. Repeated cocaine treatment induces locomotor sensitization, which consists of enhanced motor exploratory output and stereotyped behaviors. Sensitization can occur rapidly, as a single dose of cocaine (40 mg/kg, i.p.) can augment locomotor behavior upon subsequent treatment (10 mg/kg, i.p.), as long as both treatments are given in the same environment.[25]

We have used a modification of this rapid sensitization paradigm to determine whether $rCMR_{glc}$ is sensitized in response to repeated cocaine challenge.[52] Rats were injected with 30 mg/kg cocaine or saline vehicle i.p. and placed in a plastic "shoebox" cage for 1 hr. The following morning, the animals were implanted with femoral arterial and venous cannulae under fluothane anesthesia and allowed to recover for at least 4 hr, after which they were injected again with 30 mg/kg cocaine or saline vehicle and placed in the same environment. Thus, three treatment groups were generated according to their treatment on the first and second days, as follows: saline-saline treatment, saline-cocaine treatment, and cocaine-cocaine treatment.

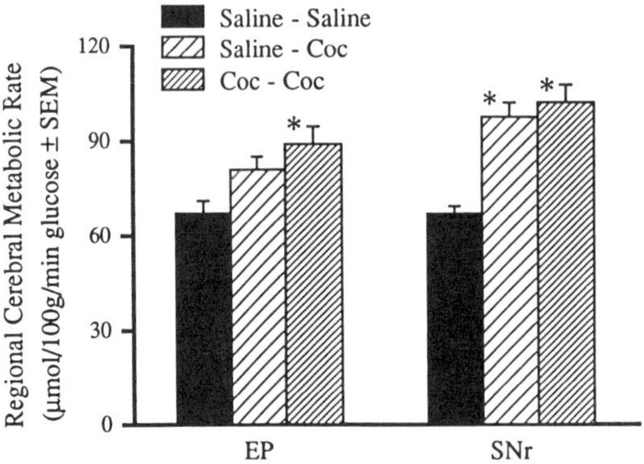

Figure 1. Effect of cocaine (30 mg/kg, i.p.) treatment on regional cerebral metabolic rate for glucose ($rCMR_{glc}$) in the entopeduncular nucleus (EP) and substantia nigra pars reticulata (SNr). Rats were injected daily for 2 days with either saline vehicle (Saline-Saline) or cocaine (Cocaine-Cocaine), or were treated with saline on day 1 and cocaine on day 2 (Saline-Cocaine) in the same environment on both days. Metabolic response was sensitized in both regions. * – significant ($p \leq 0.05$) difference from SS group by ANOVA followed by Newman-Keuls *post hoc* test. Data from Thomas and Hammer.[52]

The results reveal that repeated cocaine treatment significantly increased $rCMR_{glc}$ only in the EP, GP and SNr, wherein $rCMR_{glc}$ was significantly higher after two cocaine treatments (Figure 1). In contrast, the lateral habenula exhibited reduced $rCMR_{glc}$ following cocaine treatment. Thus, $rCMR_{glc}$ response to cocaine appears to be sensitized, but only in extrapyramidal regions innervated by direct striatal projection neurons.

We also have shown that pretreatment with SCH 23390, a dopamine D_1 receptor antagonist, dose-dependently reduced $rCMR_{glc}$ in the EP and SNr to levels equivalent to those observed in the absence of cocaine (Figure 2).[53] The cocaine-induced striatal extracellular dopamine response, which is sensitized by cocaine in the rapid sensitization paradigm,[31] probably stimulates dopamine D_1 receptors to activate EP/SNr projection neurons causing these neurons to fire, ultimately producing an increase of synaptic activity in their terminals located in the EP and SNr. These striatal neurons are thought to contain GABA, dynorphin, and Substance P,[54] which presumably act postsynaptically to inhibit firing of EP and SNr neurons projecting to the thalamus. Interestingly, pretreatment with κ-opioid receptor agonists blocks both cocaine-induced locomotor sensitization[55] and immediate early gene expression.[56] This suggests that cocaine's induction of dynorphin,[57,58] which binds to κ-opioid receptors, regulates immediate early gene expression[59] as well as the activity of direct striatal projection neurons. Thus, enhanced synaptic transmission which underlies the dopamine D_1-dependent sensitization of $rCMR_{glc}$ in EP

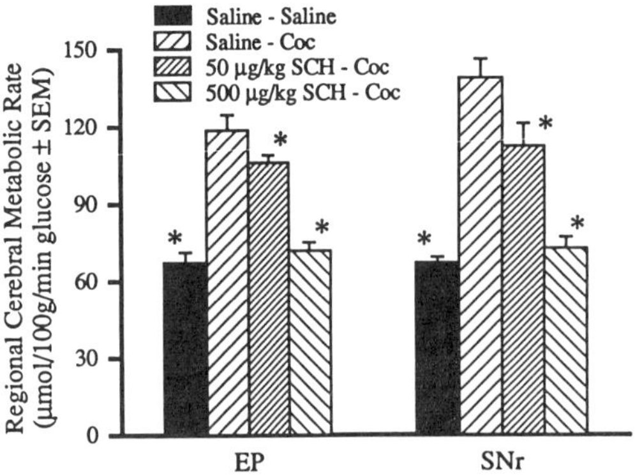

Figure 2. Effect of SCH 23390 (50 or 500 µg/kg, i.p.) or saline vehicle pretreatment on regional cerebral metabolic rate for glucose ($rCMR_{glc}$) in the entopeduncular nucleus (EP) and substantia nigra pars reticulata (SNr) following cocaine (30 mg/kg, i.p.) treatment. Cocaine (Saline-Coc) increased $rCMR_{glc}$ compared to saline vehicle alone (Saline-Saline), while SCH 23390 pretreatment (SCH-Coc) produced a dose-dependent reduction of $rCMR_{glc}$ in both regions. * – significant ($p \leq 0.05$) difference from SC group by ANOVA followed by Newman-Keuls *post hoc* test. Data from Thomas et al.[53]

Figure 3. Effect of repeated daily cocaine (10 mg/kg, i.p.) or saline vehicle treatment on regional cerebral metabolic rate for glucose ($rCMR_{glc}$) with or without cocaine (30 mg/kg, i.p.) challenge on the final treatment day. $rCMR_{glc}$ was reduced following chronic treatment in mesocorticolimbic regions, including the nucleus accumbens (NAc) and medial prefrontal cortex (MPf), and subsequent challenge had no effect. In contrast, $rCMR_{glc}$ was unaffected by chronic treatment, while cocaine challenge significantly increased $rCMR_{glc}$ in the substantia nigra pars reticulata (SNr). * – significant ($p \leq 0.05$) difference from SS group by Duncan's Multiple Range test. Data from Hammer and Cooke.[65]

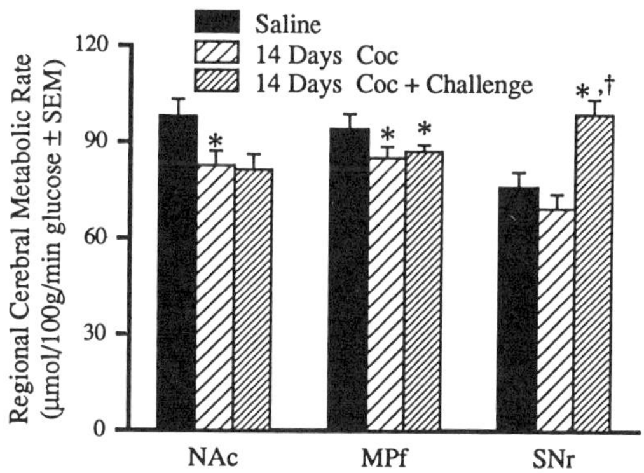

and SNr provides a likely substrate for the behavioral sensitization produced by cocaine, especially since $rCMR_{glc}$ in both EP and SNr is correlated positively with locomotor activity following acute cocaine treatment.[60]

The phenomenon of sensitization is separable into acquisition and subsequent expression phases, and distinct neural substrates are involved in each of these phases (see Chapter 6). Whereas acquisition of behavioral sensitization is dependent upon psychostimulant action in the VTA, its expression involves the NAc.[61] Although the process of sensitization of metabolic activity occurs rapidly, the alteration of $rCMR_{glc}$ in extrapyramidal motor regions suggests that this process is related to expression of locomotor behavior rather than to the acquisition of sensitization.

III. CHRONIC COCAINE TREATMENT PRODUCES METABOLIC "TOLERANCE"

Chronic cocaine use now is known to cause persistent changes of neural function due to physiological, pharmacological, and even structural alterations. Although acute cocaine administration increases $rCMR_{glc}$ in mesolimbic[60] and extrapyramidal[62,63] regions of drug-naive rat brain, such treatment appears to *reduce* $rCMR_{glc}$ in cortical and other regions in patients with a history of cocaine abuse.[64] We wondered whether this animal model was appropriate, and whether drug experience altered the metabolic response to subsequent cocaine exposure. Therefore, we treated rats for 14 days i.p. with either saline vehicle or 10 mg/kg cocaine on the final 1, 3, 7, 10, or 14 treatment days, then examined $rCMR_{glc}$.[65] The results reveal that this low dose of cocaine increased $rCMR_{glc}$ slightly following acute treatment, but was without effect following 3 to 7 days of treatment, and significantly reduced $rCMR_{glc}$ after 14 days of treatment in mesocorticolimbic and a few other regions including primary sensory cortex, but not in striatal target regions, such as the EP and SNr (Figure 3). Furthermore, a higher dose (30 mg/kg, i.p.) on the final day following chronic 10 mg/kg treatment significantly increased $rCMR_{glc}$ in extrapyramidal terminal sites, but not in mesolimbic regions (Figure 3). Thus, regional metabolic activity ultimately was reduced in mesolimbic regions and some cortical regions after repeated cocaine treatment compared to drug-naive animals. Metabolic activation of extrapyramidal targets was possible following chronic treatment, but only by administering a higher dose of cocaine.

The time course of these effects is remarkably similar to the attenuation and ultimate reduction of extracellular dopamine levels in the NAc following repeated administration of the same cocaine dose.[33] In addition, immediate early gene expression, which is induced in the striatum by acute cocaine exposure, is attenuated following chronic treatment,[59,66] consistent with a relative reduction of activity in striatal circuits. These metabolic data are consistent with those obtained in rats self-administering cocaine, in which $rCMR_{glc}$ was reduced early in the self-administration session in mesocorticolimbic regions, including the MPf and anterior cingulate cortex, and increased with enhanced motor activity later in the session in extrapyramidal regions[67] (see Chapter 1). Such effects could be related to the involvement of these cortical regions in the initiation of cocaine reinforcement[68] by mediating dopamine response in the NAc.[69] Furthermore, greater experience with cocaine self-administration produced greater decline of global $rCMR_{glc}$,[67] which could represent a conditioned response to chronic cocaine exposure and/or an adaptive response to learning in the self-administration paradigm. In any case, the regional metabolic effects produced by chronic cocaine appear to be unrelated to the type of paradigm utilized, i.e., noncontingent (experimenter-administration) or contingent (self-administration), and are consistent between experimental animal treatment and human drug abuse.

The observed reduction of $rCMR_{glc}$ following chronic cocaine in mesocorticolimbic regions suggests that tolerance of metabolic activity might develop with drug experience. This could provoke subsequent administration of higher doses in order to regain original levels of psychomotor (and metabolic) activation. If these effects represent an adaptive response, however, then the metabolic response should be compared to the basal $rCMR_{glc}$ level in the absence of cocaine, which might differ from the original level in saline-treated control animals. Since both the noncontingent and contingent paradigms described above examined $rCMR_{glc}$ in the presence of cocaine, we cannot determine whether basal levels were altered by chronic treatment. In any case, the persistent metabolic alterations which occur with chronic cocaine treatment could represent a substrate for dependence, which appears to affect mesocorticolimbic regions selectively.

IV. COCAINE WITHDRAWAL

Vitally important, but poorly understood, are the changes which occur during withdrawal following chronic cocaine administration. Indeed, the persistent effects induced by chronic cocaine frequently lead to relapse during withdrawal in human users. Yet little is known about the mechanisms underlying withdrawal and its behavioral sequelae. The realization that cocaine increases striatal dopamine content led to the hypothesis that dopamine levels are depleted with chronic use.[70] We now know that chronic treatment attenuates the increase of extracellular dopamine in the striatum,[32,71] and that dopamine levels ultimately decrease during withdrawal.[34,35] The response of mesolimbic dopamine circuits also is affected, as the threshold for self-stimulation of lateral hypothalamic dopamine fibers increases during early withdrawal following cocaine self-administration,[72] and this effect is correlated with the amount of cocaine consumed. Thus, the mesolimbic dopamine pathway may be desensitized during early withdrawal, possibly resulting in anhedonia. Finally, dopamine transporter mRNA expression in VTA neurons projecting to the NAc[73] and dopamine transporter binding in the NAc[74] are reduced during late (10 days), but not early, withdrawal after chronic cocaine treatment. Thus, a selective impairment of the mesolimbic dopamine pathway is present during withdrawal after chronic cocaine administration.

In order to determine which brain regions might be involved, we examined $rCMR_{glc}$ during withdrawal following chronic contingent and noncontingent cocaine administration, with a special emphasis on the efferent targets of those regions in which dopamine pathways are altered. We have examined the regional metabolic effects of time course of withdrawal, as well as the effects of various pharmacotherapies on metabolic recovery.

A. METABOLIC CHANGES DURING COCAINE WITHDRAWAL

If mesocorticolimbic dopamine function is altered selectively during withdrawal, then one might expect to observe effects on metabolic activity in neural circuits whose functions are crucial to reinforcement, including those regions targeted by dopamine projections and the areas innervated by these regions. Various studies have shown this to be the case. After 3 days of withdrawal following 14 days of repeated daily cocaine treatment (10 mg/kg, i.p.), $rCMRg_{lc}$ was reduced in the MPf and various portions of the "extended amygdala"[20] (e.g., the NAc, VP, lateral hypothalamus, and central nucleus of the amygdala), as well as in extrapyramidal target regions (e.g., the GP, EP, and SNr).[75,76] Furthermore, the effect within mesolimbic circuits is quite selective; the NAc core and its efferent targets, the dorsolateral VP and rostral portion of the lateral hypothalamus, but not the targets of the NAc shell, exhibited reduced $rCMR_{glc}$ (Figure 4).[76] This suggests that the NAc core is more affected than the NAc shell during early cocaine withdrawal. Such a selective effect on NAc core pathways may be especially important in mediating locomotor activity.

We then examined regional metabolic effects of withdrawal following cocaine self-administration in order to determine whether contingent and noncontingent drug treatment produced similar results. We examined $rCMR_{glc}$ either 6 or 72 hr following a 12-hr self-administration binge, and compared the results to those obtained in rats previously trained to self-administer cocaine or naive rats placed into an empty operant chamber for 12 hr. The results are similar to those obtained following noncontingent cocaine administration, except that the reduction of $rCMR_{glc}$ was more widespread.[77] Metabolic activity was reduced during withdrawal in mesolimbic areas, septal and amygdaloid nuclei, piriform, cingulate, sensory and motor cortices, as well as in several extrapyramidal regions (e.g, rostral caudatoputamen and EP). Certain mesolimbic circuits may be involved selectively, as the NAc core and its projection to the rostral portion of the lateral hypothalamus were affected, while NAc shell projections were not. These effects almost always were more severe at 72 hr than at 6 hr after binge exposure. Furthermore, intermediate $rCMR_{glc}$ values often were observed in cocaine-trained animals in the absence of binge exposure. This suggests that cocaine experience can influence the magnitude of the regional metabolic reduction, as the trained animals last received cocaine approximately 30 hr earlier. In addition, the amount of cocaine consumed can affect the extent of metabolic depression, as metabolic response was correlated negatively with the amount of cocaine consumed during binge exposure in striatal and cortical regions. Thus, drug "history" can determine the severity of regional brain metabolic changes, and might influence the extent of the effect.

This persistent and progressive reduction of $rCMR_{glc}$ in mesolimbic circuits is especially interesting, since this same paradigm is known to increase the threshold for intracranial self-stimulation 6, but not 72, hr after withdrawal following cocaine self-administration,[72] and extracellular dopamine levels are reduced in the NAc at 6, but not at 72, hr after withdrawal.[35] The alteration of self-stimulation threshold suggests the presence of a deficit of the dopaminergic response to a stimulus, which is confirmed by the measurement of dopamine level in the NAc. However, $rCMR_{glc}$ assesses basal metabolic activity which depends on dopamine as well as other neurotransmitters and, therefore, might not recover as early as does dopamine level. The circuits which affect $rCMR_{glc}$ and ultimately influence behavior apparently recover more slowly after a cocaine self-administration binge. The reduction of basal metabolic rate observed during cocaine withdrawal becomes more severe with time during early withdrawal, despite the apparent recovery of certain behavioral-motivational responses.

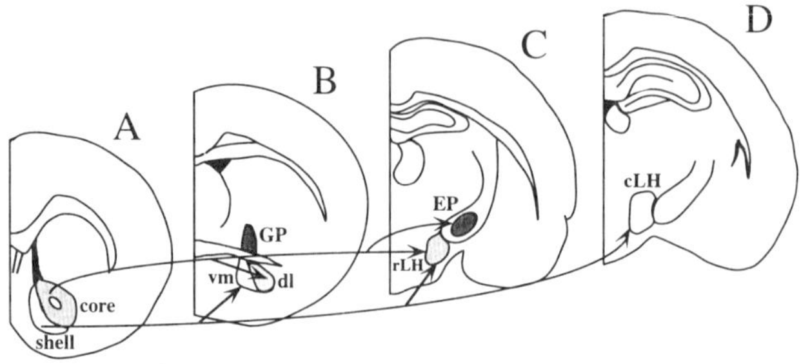

Figure 4. Efferent circuitry of cocaine withdrawal shown on hemisections of the rat brain at levels equivalent to +1.2 (A), –0.26 (B), –2.3 (C) and –3.6 (D) mm from bregma.[97] Neurons of the nucleus accumbens core (A) project to the dorsolateral (dl) portion of the ventral pallidum (B) and rostral portion of the lateral hypothalamus (rLH) and entopeduncular nucleus (EP in C)[20-22], all of which exhibited reduced regional cerebral metabolic rate for glucose ($rCMR_{glc}$) after 3 days withdrawal following chronic cocaine treatment, as did the globus pallidus (GP in B). (Light gray illustrates those regions in which $rCMR_{glc}$ was reduced after 3 days withdrawal.) In contrast, neither the ventromedial (vm) portion of the ventral pallidum (B), nor the caudal portion of the lateral hypothalamus (cLH), which are innervated by the nucleus accumbens shell (A), showed significant reduction of $rCMR_{glc}$. Metabolic activity remained depressed only in the GP and EP after 7 days withdrawal. (Dark gray illustrates those regions in which $rCMR_{glc}$ is reduced after both 3 and 7 days withdrawal.) Data from Young et al.[76]

B. TIME COURSE OF METABOLIC RECOVERY

We have examined $rCMR_{glc}$ over the time course of withdrawal using a paradigm of noncontingent administration of 10 mg/kg/day cocaine for 14 days.[78] It should be noted that this dose is considerably less than that utilized in the contingent withdrawal studies described above,[77] wherein 20 to 40 mg/kg were self-administered by rats daily during training and 80 to 140 mg/kg were self-administered during a 12-hr drug binge. This dose difference could underlie the greater magnitude of effects observed in the self-administration study. In contrast, our studies of withdrawal following noncontingent administration of 10 mg/kg/day cocaine for 14 days probably illustrate metabolic effects on the brain regions most sensitive to change. Such regions could represent the major substrates underlying behavioral sequelae of drug withdrawal: motivational deficits, hypoactivity, anhedonia, and depression.[72]

This investigation of time course replicates and extends our earlier results. Three days after discontinuation of cocaine treatment, $rCMR_{glc}$ decreased significantly in mesolimbic circuits (e.g., NAc, olfactory tubercle, central amygdaloid nucleus, ventral pallidum, and rostral portion of the lateral hypothalamus), extrapyramidal areas (e.g., GP and SNr), and a few other regions. By 7 days, $rCMR_{glc}$ in mesolimbic regions was no longer significantly reduced, while $rCMR_{glc}$ in the GP, EP, and SNr remained significantly lower than in drug-naive rats (Figure 4). All these regions exhibited significant recovery by 21 days withdrawal. Once again, metabolic activity is reduced in both mesolimbic and extrapyramidal circuits during early withdrawal; however, this effect persists in extrapyramidal circuits longer than in mesolimbic circuits. These changes are remarkably persistent considering the brief period of low-dose exposure to cocaine. In patients with a history of recent cocaine abuse, a similar reduction of $rCMR_{glc}$[79] and regional cerebral blood flow[80,81] has been observed, especially in mesocorticolimbic brain regions, lasting from 1 week to 4 months after last drug use. It should be noted that these patients frequently abuse other drugs in addition to cocaine, so the severity of metabolic effects may be greater and the time course of recovery longer than in animal models.

What are the implications of these regional differences? It would seem that $rCMR_{glc}$ returns to "normal" more rapidly in mesolimbic circuits than in extrapyramidal circuits. The sustained extrapyramidal effects could underlie a persistent, relative reduction of locomotor activity, since metabolic and locomotor activity are correlated in the EP and SNr following cocaine treatment.[60] The metabolic alterations in mesolimbic regions are more difficult to interpret, however, since $rCMR_{glc}$ values obtained in the drug-naive state might no longer represent basal levels in animals undergoing withdrawal. In fact, $rCMR_{glc}$ might have been reduced for as long as 7 days in 3-day withdrawal rats (since treatment day 10[65]). Therefore, the relative increase of $rCMR_{glc}$ compared to basal levels in mesolimbic regions could be described as an increase of activity in motivational circuits rather than "recovery" to normal levels. This increase could manifest as a lessening of depressive symptoms, or it could represent increased motivation toward drug seeking (i.e., craving). One way to distinguish these effects might be to examine the metabolic response to cocaine challenge during withdrawal, and to compare the response between regions and across time.

C. PHARMACOTHERAPY DURING COCAINE WITHDRAWAL

Given the implications of metabolic changes during "recovery", one should exercise caution in interpreting the results of pharmacotherapy which might induce a rapid increase of $rCMR_{glc}$ in mesolimbic regions. Nevertheless, pharmacotherapeutic trials can tell us much about the pharmacology underlying $rCMR_{glc}$ changes during cocaine withdrawal. With this in mind, we examined the effects of several compounds which affect dopamine systems either directly or indirectly.

The dopamine depletion hypothesis suggests that direct dopamine receptor stimulation might effectively alleviate symptoms of cocaine withdrawal.[82] Therefore, bromocriptine, a dopamine receptor agonist with relative preference for dopamine D_2 receptors, has been utilized as a treatment during cocaine withdrawal with some success,[83,84] although this compound has some undesirable side effects.[85] In addition, bromocriptine pretreatment reduces cocaine self-administration in rats,[86] and reverses the elevation of intracranial self-stimulation threshold during cocaine withdrawal.[87] It seemed likely, therefore, that daily treatment with bromocriptine during withdrawal could reverse regional metabolic reductions present during withdrawal. In fact, bromocriptine (10 mg/kg) was partially effective, eliminating the reduction of $rCMR_{glc}$ in the MPf, NAc, and VTA.[75] Direct dopamine receptor stimulation reversed the metabolic effect of cocaine withdrawal, which suggests that dopamine receptor function is involved in the metabolic response.

Since bromocriptine acts predominantly at dopamine D_2 receptors, we examined the effect of cocaine withdrawal and bromocriptine therapy on these receptors directly using *in vitro* receptor autoradiography of [^{3}H]spiperone binding in sections adjacent to those in which $rCMR_{glc}$ was measured.[88] The results reveal that D_2 receptor binding increased significantly in the NAc and olfactory tubercle during cocaine withdrawal. Moreover, chronic (3-day) treatment with bromocriptine during withdrawal reversed these increases of dopamine D_2 receptor labeling, and this effect was correlated negatively with the effect on $rCMR_{glc}$ in the NAc ($r = -0.597$, $p \leq 0.05$). Thus, the compensatory downregulation of D_2 receptors which is induced rapidly by bromocriptine during cocaine withdrawal might be related to its effect on $rCMR_{glc}$ in the NAc.

Treatment with a direct dopamine receptor agonist was only partially effective in returning $rCMR_{glc}$ to pretreatment levels. Reasoning that treatment with an indirect dopamine agonist might produce less compensatory receptor alteration and, therefore, might be more effective, we examined regional metabolic effects after treatment of rats during cocaine withdrawal with a σ-receptor antagonist, BMY 14802, which is known to release dopamine *in vivo*[89] and antagonize the locomotor stimulant effect of cocaine.[90]

However, this compound is not selective. It also binds to 5-HT_{1A} receptors, and its enhancement of mesolimbic and nigrostriatal dopaminergic neuronal activity probably is unrelated to its σ-receptor properties, as a selective σ-agonist, DTG, has no effect on either of these populations of dopamine neurons.[91] Therefore, we also examined the effect of gepirone, a selective 5-HT_{1A} partial agonist, on $rCMR_{glc}$. The results revealed that both compounds had similar biphasic, dose-dependent effects, whereby low doses prevented the reduction of $rCMR_{glc}$ in mesolimbic and extrapyramidal regions and high doses exacerbated these effects.[76] These effects probably were due to the action of these compounds on 5-HT_{1A} receptors, as gepirone treatment produced a greater effect in the same direction as did BMY 14802 treatment. The apparent biphasic dose-response suggests the presence of a "window of efficacy". This could be artifactual, however, since high treatment doses substantially increased plasma glucose levels, which might not be completely corrected by adjusting the lumped constant in the operational equation of the deoxyglucose method.[92]

These 5-HT_{1A} agonists could alter $rCMR_{glc}$ in several ways. Mesolimbic and nigrostriatal dopamine neurons could be activated, as is known to occur after acute BMY 14802 administration,[91] causing exocytosis of dopamine-containing vesicles at terminals. In addition, 5-HT_{1A} receptor stimulation could regulate the release of dopamine at synaptic terminals, especially since 5-HT_{1A} function may be supersensitive during withdrawal following chronic cocaine treatment.[93,94] These compounds could increase the firing of a subset of striatal neurons in which activity is modulated by 5-HT.[14,95] Thus, 5-HT_{1A} receptor activation could alter metabolic activity by acting as an indirect dopamine agonist or by acting directly on neurons responsive to 5-HT. In any case, serotonergic modulation of metabolic activity during cocaine withdrawal is apparently more effective than direct dopaminergic stimulation, suggesting that serotonergic pharmacotherapy might be a productive treatment during cocaine withdrawal.

V. CONCLUSIONS

Analyses of the regional metabolic effects of chronic cocaine in animals have revealed much about the neural circuits underlying the mechanism of action of cocaine. Initial exposure to cocaine activates extrapyramidal circuits which are probably involved in motor behavior. Repeated exposure initially sensitizes this response, which may serve to intensify the effects of cocaine in beginning users. The route of administration and pharmacokinetics are important, as intravenous treatment increases $rCMR_{glc}$ in mesolimbic structures, while intraperitoneal treatment does not.[96] Thus, the pharmacokinetic properties of the route of administration preferred by many cocaine abusers could optimize the effects on reward circuits in beginning users, especially if sensitization occurs in mesolimbic regions after intravenous (or intranasal/inhalation) treatment.

Continued treatment leads to the development of a metabolic pattern which might represent tolerance to the initial stimulatory effects of cocaine. The eventual reduction of $rCMR_{glc}$ in mesolimbic regions could develop as a conditioned response in which cellular signalling mechanisms are altered. Dose and frequency of administration seem to be important in the development of tolerance, which occurs most readily following continuous, multiple daily, or low dose treatments. Interestingly, the acute reduction of $rCMR_{glc}$ in humans with a history of cocaine abuse is associated with reinforcement,[64] suggesting that a novel messenger might be induced following chronic cocaine exposure. In fact, although the induction of several immediate early genes is attenuated in striatal neurons by chronic cocaine treatment, continued upregulation of the AP-1 binding site suggests that a novel fos-related antigen might be involved in the persistent reinforcing effects.[66] Thus, the development of tolerance could promote the administration of larger doses while allowing continued reinforcement.

The relative reduction of $rCMR_{glc}$ after chronic treatment becomes even more severe during withdrawal, especially in the efferent targets of projections from the NAc core. Both mesolimbic and extrapyramidal regions are affected during the early "crash" phase, but metabolic reduction persists longer in locomotor regions innervated by the dorsal striatum. Intravenous cocaine self-administration produces similar regional metabolic reductions, albeit of greater magnitude than those produced by noncontingent cocaine administration; the magnitude of these effects is related to the amount of drug taken during unlimited-access self-administration. Thus, higher doses can lead to greater withdrawal effects, which could result in more drug craving during recovery.

The results following noncontingent cocaine administration in experimental animals match those obtained in humans who have taken the drug, both during withdrawal and after chronic administration. This suggests that the experimental paradigm in animals can be used as an appropriate model of cocaine effects in human brain. If so, then pharmacotherapeutic trials also should yield accurate reflections of human response. Such experiments implicate both dopamine and 5-HT in the regional metabolic response to cocaine withdrawal. In fact, serotonergic therapy apparently facilitates metabolic recovery better than does direct dopaminergic stimulation. These data suggest that 5-HT reuptake inhibitors might reverse cerebral metabolic deficits effectively in humans during cocaine withdrawal.

REFERENCES

1. Gawin, F. H. and Ellinwood, E. H., Cocaine and other stimulants, *N. Engl. J. Med.*, 318, 1173, 1988.
2. Gawin, F. H., Cocaine addiction: psychology and neurophysiology, *Science*, 251, 1580, 1991.
3. Patrick, R. L., Snyder, T. E., and Barchas, J. D., Regulation of dopamine synthesis in rat brain striatal synaptosomes, *Mol. Pharmacol.*, 11, 621, 1975.
4. Pettit, H. O. and Justice, J. B., Dopamine in the nucleus accumbens during cocaine self-administration as studied by in vivo microdialysis, *Pharmacol. Biochem. Behav.*, 34, 899, 1989.
5. Hurd, Y. L. and Ungerstedt, U., Cocaine: an in vivo microdialysis evaluation of its acute action on dopamine transmission in rat striatum, *Synapse*, 3, 48, 1989.
6. Pradhan, S., Roy, S. N., and Pradhan, S. N., Correlation of behavioral and neurochemical effects of acute administration of cocaine in rats, *Life Sci.*, 22, 1737, 1978.
7. Taylor, D. and Ho, B. T., Comparison of inhibition of monoamine uptake by cocaine, methylphenidate and amphetamine, *Res. Commun. Chem. Pathol. Pharmacol.*, 21, 67, 1978.
8. Hadfield, M. G., Mott, D. E. W., and Ismay, J. A., Cocaine: effect of in vivo administration on synaptosomal uptake of norepinephrine, *Biochem. Pharmacol.*, 29, 1861, 1980.
9. Ritz, M. C., Lamb, R. J., Goldberg, S. R., and Kuhar, M. J., Cocaine receptors on dopamine transporters are related to self-administration of cocaine, *Science*, 237, 1219, 1987.
10. Gerfen, C. R., The neostriatal mosaic: multiple levels of compartmental organization, *TINS*, 15, 133, 1992.
11. Morgan, J. I. and Curran, T., Stimulus-transcription coupling in neurons: role of cellular immediate-early genes, *TINS*, 12, 459, 1989.
12. Graybiel, A. M., Moratalla, R., and Robertson, H. A., Amphetamine and cocaine induce drug-specific activation of the c-fos gene in striosome matrix compartments and limbic subdivisions of the striatum, *Proc. Natl. Acad. Sci. USA*, 87, 6912, 1990.
13. Young, S. T., Porrino, L. J., and Iadarola, M. J., Cocaine induces striatal c-fos-immunoreactive proteins via dopaminergic D_1 receptors, *Proc. Natl. Acad. Sci. USA*, 88, 1291, 1991.
14. Bhat, R. V. and Baraban, J. M., Activation of transcription factor genes in striatum by cocaine: Role of both serotonin and dopamine systems, *J. Pharmacol. Exp. Ther.*, 267, 496, 1993.
15. Gerfen, C. R., Engber, T. M., Mahan, L. C., Susel, Z., Chase, T. N., Monsma, F. J., and Sibley, D. R., D_1 and D_2 dopamine receptor-regulated gene expression of striatonigral and striatopallidal neurons, *Science*, 250, 1429, 1990.
16. Kawaguchi, Y., Wilson, C. J., and Emson, P. C., Projection subtypes of rat neostriatal matrix cells revealed by intracellular injection of biocytin, *J. Neurosci.*, 10, 3421, 1990.

17. Zahm, D. S. and Heimer, L., Specificity in the efferent projections of the nucleus accumbens in the rat: comparison of the rostral pole projection patterns with those of the core and shell, *J. Comp. Neurol.*, 327, 220, 1993.
18. Mogenson, G. J. and Wu, M., Subpallidal projections to the mesencephalic locomotor region investigated with a combination of behavioral and electrophysiological recording techniques, *Brain Res. Bull.*, 16, 383, 1986.
19. Zahm, D. S. and Heimer, L., Two transpallidal pathways originating in the rat nucleus accumbens, *J. Comp. Neurol.*, 302, 437, 1990.
20. Heimer, L., Zahm, D. S., Churchill, L., Kalivas, P. W., and Wohltman, C., Specificity in the projection patterns of accumbens core and shell in the rat, *Neuroscience*, 41, 89, 1991.
21. Zahm, D. S. and Brog, J. S., On the significance of subterritories in the "accumbens" part of the rat ventral striatum, *Neuroscience*, 50, 751, 1992.
22. Brog, J. S., Salyapongse, A., Deutch, A. Y., and Zahm, D. S., The patterns of afferent innervation of the core and shell in the "accumbens" part of the rat ventral striatum: immunohistochemical detection of retrogradely transported fluoro-gold, *J. Comp. Neurol.*, 338, 255, 1993.
23. Post, R. M. and Rose, H., Increasing effects of repetitive cocaine administration in the rat, *Nature*, 260, 731, 1976.
24. Post, R. M., Lockfeld, A., Squillace, K. M., and Contel, N. R., Drug-environment interaction: context dependency of cocaine-induced behavioral sensitization, *Life Sci.*, 28, 755, 1981.
25. Weiss, S. R. B., Post, R. M., Pert, A., Woodward, R., and Murman, D., Context-dependent cocaine sensitization: differential effect of haloperidol on development versus expression, *Pharmacol. Biochem. Behav.*, 34, 655, 1989.
26. Kalivas, P. W. and Stewart, J., Dopamine transmission in the initiation and expression of drug- and stress-induced sensitization of motor activity, *Brain Res. Rev.*, 16, 223, 1991.
27. Henry, D. J., Greene, M. A., and White, F. J., Electrophysiological effects of cocaine in the mesoaccumbens dopamine system—repeated administration, *J. Pharmacol. Exp. Ther.*, 251, 833, 1989.
28. Pettit, H. O., Pan, H.-T., Parsons, L. H., and Justice, J. B., Jr., Extracellular concentrations of cocaine and dopamine are enhanced during chronic cocaine administration, *J. Neurochem.*, 55, 798, 1990.
29. Parsons, L. H. and Justice, J. B., Serotonin and dopamine sensitization in the nucleus accumbens, ventral tegmental area, and dorsal raphe nucleus following repeated cocaine adminsitration, *J. Neurochem.*, 61, 1611, 1993.
30. Kalivas, P. W. and Duffy, P., Effect of acute and daily cocaine treatments on extracellular dopamine in the nucleus accumbens, *Synapse*, 5, 48, 1990.
31. Fontana, D., Post, R. M., and Pert, A., Conditioned increases in mesolimbic dopamine overflow by stimuli associated with cocaine, *Brain Res.*, 629, 31, 1993.
32. Weiss, F., Paulus, M. P., Lorang, M. T., and Koob, G. F., Increases in extracellular dopamine in the nucleus accumbens by cocaine are inversely related to basal levels: effects of acute and repeated administration, *J. Neurosci.*, 12, 4372, 1992.
33. Imperato, A., Mele, A., Scrocco, M. G., and Puglisi, A. S., Chronic cocaine alters limbic extracellular dopamine. Neurochemical basis for addiction, *Eur. J. Pharmacol.*, 212, 299, 1992.
34. Parsons, L. H., Smith, A. D., and Justice, J. B., Jr., Basal extracellular dopamine is decreased in the rat nucleus accumbens during abstinence from chronic cocaine, *Synapse*, 9, 60, 1991.
35. Weiss, F., Markou, A., Lorang, M. T., and Koob, G. F., Basal extracellular dopamine levels in the nucleus accumbens are decreased during cocaine withdrawal after unlimited-access self-administration, *Brain Res.*, 593, 314, 1992.
36. Cass, W. A. and Zahniser, N. R., Cocaine levels in striatum and nucleus accumbens: augmentation following challenge injection in rats withdrawn from repeated cocaine administration, *Neurosci. Lett.*, 152, 177, 1993.
37. Pan, H. T., Menacherry, S., and Justice, J. B., Differences in the pharmacokinetics of cocaine in naive and cocaine-experienced rats, *J. Neurochem.*, 56, 1299, 1991.
38. Izenwasser, S. and Cox, B. M., Inhibition of dopamine uptake by cocaine and nicotine: tolerance to chronic treatments, *Brain Res.*, 573, 119, 1992.
39. Reith, M. E., Benuck, M., and Lajtha, A., Cocaine disposition in the brain after continuous or intermittent treatment and locomotor stimulation in mice, *J. Pharmacol. Exp. Ther.*, 243, 281, 1987.

40. Inada, T., Polk, K., Purser, C., Hume, A., Hoskins, B., Ho, I. K., and Rockhold, R. W., Behavioral and neurochemical effects of continuous infusion of cocaine in rats, *Neuropharmacology*, 31, 701, 1992.
41. King, G. R., Joyner, C., Lee, T. H., Kuhn, C., and Ellinwood, E. H., Intermittent and continuous cocaine administration: residual behavior states during withdrawal, *Pharmacol. Biochem. Behav.*, 43, 243, 1992.
42. King, G. R., Kuhn, C., and Ellinwood, E. H., Dopamine efflux during withdrawal from continuous or intermittent cocaine, *Psychopharmacology*, 111, 179, 1993.
43. Emmett-Oglesby, M. W., Peltier, R. L., Depoortere, R. Y., Pickering, C. L., Hooper, M. L., Gong, Y. H., and Lane, J. D., Tolerance to self-administration of cocaine in rats: time course and dose-response determination using a multi-dose method, *Drug Alcohol Depend.*, 32, 247, 1993.
44. Maisonneuve, I. M. and Kreek, M. J., Acute tolerance to the dopamine response induced by a binge pattern of cocaine administration in male rats: an in vivo microdialysis study, *J. Pharmacol. Exp. Ther.*, 268, 916, 1994.
45. Segal, D. S. and Kuczenski, R., Repeated cocaine administration induces behavioral sensitization and corresponding decreased extracellular dopamine responses in caudate and accumbens, *Brain Res.*, 577, 351, 1992.
46. Kalivas, P. W. and Duffy, P., Time course of extracellular dopamine and behavioral sensitization to cocaine. I. Dopamine axon terminals, *J. Neurosci.*, 13, 266, 1993.
47. Sokoloff, L., Reivich, M., Kennedy, C., Des Rosiers, M. H., Patlak, C. S., Pettigrew, K. D., Sakurada, O., and Shinohara, M., The [^{14}C] deoxyglucose method for the measurement of local cerebral glucose utilization: theory, procedure, and normal values in the conscious and anesthetized albino rat, *J. Neurochem.*, 28, 897, 1977.
48. Kadekaro, M., Vance, W. H., Terrell, M. L., Gary, J., H., Eisenberg, H. M., and Sokoloff, L., Effects on antidromic stimulation of the ventral root on glucose utilization in the ventral horn of the spinal cord in the rat, *Proc. Natl. Acad. Sci. USA*, 84, 5492, 1987.
49. Schwartz, W. J., Smith, C. B., Davidsen, L., Savaki, H., Sokoloff, L., Mata, M., Fink, D. J., and Gainer, H., Metabolic mapping of functional activity in the hypothalamoneurohypophyseal system of the rat, *Science*, 205, 723, 1979.
50. Sokoloff, L. and Porrino, L., Some fundamental considerations in the application of the deoxyglucose method to pharmacological studies, *Pharmacology of Cerebral Ischemia*, Krieglstein, J., Ed., Elsevier, Amsterdam, 1986, 65.
51. Hammer, R. P., Herkenham, M., Pert, C. B., and Quirion, R., Correlation of regional brain metabolism with receptor localization during ketamine anesthesia: combined autoradiographic 2-[^{3}H]deoxy-D-glucose receptor binding technique, *Proc. Natl. Acad. Sci. USA*, 79, 3067, 1982.
52. Thomas, W. L. and Hammer, R. P., Cocaine sensitization activates extrapyramidal but not mesolimbic circuits, *Biol. Psychiatry*, 33, 134, 1993.
53. Thomas, W. L., Cooke, E. S., and Hammer, R. P., Cocaine stimulates entopeduncular metabolic rate by activating D1 receptors in dorsal and ventral striatum, *Soc. Neurosci. Abstr.*, 18, 1841, 1993.
54. Gerfen, C. R. and Young, W.S., Distribution of striatonigral and striatopallidal peptidergic neurons in both patch and matrix compartments: an in situ hybridization histochemistry and fluorescent retrograde tracing study, *Brain Res.*, 460, 161, 1988.
55. Heidbreder, C. A., Goldberg, S. R., and Shippenberg, T. S., The kappa-opioid receptor agonist U-69593 attenuates cocaine-induced behavioral sensitization in the rat, *Brain Res.*, 616, 335, 1993.
56. Steiner, H. and Gerfen, C. R., Dynorphin regulates *c-FOS* and *ZIF268* gene induction by cocaine in striatal neurons, *Soc. Neurosci. Abstr.*, 19, 133, 1993.
57. Daunais, J. B., Roberts, D. C. S., and McGinty, J. F., Cocaine self-administration increases preprodynorphin, but not c-fos, mRNA in rat striatum, *NeuroReport*, 4, 543, 1993.
58. Spangler, R., Unterwald, E. M., and Kreek, M. J., 'Binge' cocaine administration induces a sustained increase of prodynorphin mRNA in rat caudate-putamen, *Mol. Brain Res.*, 19, 323, 1993.
59. Steiner, H. and Gerfen, C. R., Cocaine-induced *c-fos* messenger RNA is inversely related to dynorphin expression in striatum, *J. Neurosci.*, 13, 5066, 1993.
60. Porrino, L. J., Domer, F. R., Crane, A. M., and Sokoloff, L., Selective alterations in cerebral metabolism within the mesocorticolimbic dopaminergic system produced by acute cocaine administration in rats, *Neuropsychopharmacology*, 1, 109, 1988.

61. Hooks, M. S., Jones, G. H., Liem, B. J., and Justice, J. B., Sensitization and individual differences to intraperitoneal amphetamine, cocaine or caffeine following repeated intracranial amphetamine infusions, *Pharmacol. Biochem. Behav.*, 43, 815, 1992.
62. London, E. D., Wilkerson, G., Ori, C., and Kimes, A. S., Central action of psychomotor stimulants on glucose utilization in extrapyramidal motor areas of the rat brain, *Brain Res.*, 512, 155, 1990.
63. London, E. D., Wilkerson, G., Goldberg, S. R., and Risner, M. E., Effects of L-cocaine on local cerebral glucose utilization in the rat, *Neurosci. Lett.*, 68, 73, 1986.
64. London, E. D., Cascella, N. G., Wong, D. F., Phillips, R. L., Dannals, R. F., Links, J. M., Herning, R., Grayson, R., Jaffe, J. H., and Wagner, H. N., Cocaine-induced reduction of glucose utilization in human brain—a study using positron emission tomography and [Fluorine-18]-fluorodeoxyglucose, *Arch. Gen. Psychiatry*, 47, 567, 1990.
65. Hammer, R. P. and Cooke, E. S., Gradual tolerance of metabolic activity is produced in mesolimbic regions by chronic cocaine treatment, while subsequent cocaine challenge activates extrapyramidal regions of rat brain, *J. Neurosci.*, 14, 4289, 1994.
66. Hope, B., Kosofsky, B., Hyman, S. E., and Nestler, E. J., Regulation of immediate early gene expression and AP-1 binding in the rat nucleus accumbens by chronic cocaine, *Proc. Natl. Acad. Sci. USA*, 89, 5764, 1992.
67. Graham, J. H., LaRosee, J. R., Dworkin, S. I., Samson, H. H., and Porrino, L. J., Functional effects of cocaine self-administration in rats, *Soc. Neurosci. Abstr.*, 19, 1863, 1993.
68. Goeders, N. E. and Smith, J. E., Cortical dopaminergic involvement in cocaine reinforcement, *Science*, 221, 773, 1983.
69. Goeders, N. E. and Smith, J. E., Intracranial cocaine self-administration into the medial prefrontal cortex increases dopamine turnover in the nucleus accumbens, *J. Pharmacol. Exp. Ther.*, 265, 592, 1993.
70. Dackis, D. A. and Gold, M. S., New concepts in cocaine addiction: The dopamine depletion hypothesis, *Neurosci. Biochem. Rev.*, 9, 469, 1985.
71. Hurd, Y. L., Weiss, E., Koob, G. F., Anden, N.-E., and Ungerstedt, U., Cocaine reinforcement and extracellular dopamine overflow in rat nucleus accumbens: an in vivo microdialysis study, *Brain Res.*, 498, 199, 1989.
72. Markou, A. and Koob, G. F., Postcocaine anhedonia. An animal model of cocaine withdrawal, *Neuropsychopharmacology*, 4, 17, 1991.
73. Cerruti, C., Pilotte, N., Uhl, G., and Kuhar, M. J., Reduction in dopamine transporter mRNA after cessation of repeated cocaine administration, *Mol. Brain Res.*, 22, 132, 1994.
74. Sharpe, L. G., Pilotte, N. S., Mitchell, W. M., and De Souza, E. B., Withdrawal of repeated cocaine decreases autoradiographic [^{3}H]mazindol-labelling of dopamine transporter in rat nucleus accumbens, *Eur. J. Pharmacol.*, 203, 141, 1991.
75. Clow, D. W. and Hammer, R. P., Cocaine abstinence following chronic treatment alters cerebral metabolism in dopaminergic reward regions: bromocriptine enhances recovery, *Neuropsychopharmacology*, 4, 71, 1991.
76. Young, B. B., Lee, S. J., and Hammer, R. P., Metabolic enhancement by pharmacotherapy with serotonin$_{1A}$ agonists during withdrawal following chronic cocaine treatment, *Neuropsychopharmacology*, in press, 1995.
77. Hammer, R. P., Pires, W. S., Markou, A., and Koob, G. F., Withdrawal following cocaine self-administration decreases regional cerebral metabolic rate in critical brain reward regions, *Synapse*, 14, 73, 1993.
78. Young, B. B., Cooke, E. S., and Hammer, R. P., The time course of regional brain metabolic recovery during cocaine withdrawal, *Soc. Neurosci. Abstr.*, 18, 1864, 1993.
79. Volkow, N. D., Hitzemann, R., Wang, G.-J., Fowler, J. S., Wolf, A. P., Dewey, S., and Handlesman, L., Long-term frontal brain metabolic changes in cocaine abusers, *Synapse*, 11, 184, 1992.
80. Volkow, N. D., Mullani, N., Gould, K. L., Adler, S., and Krajewski, K., Cerebral blood flow in chronic cocaine users: a study with positron emission tomography, *Br. J. Psychiatry*, 152, 641, 1988.
81. Holman, B. L., Carvalho, P. A., Mendelson, J., Teoh, S. K., Nardin, R., Hallgring, E., Hebben, N., and Johnson, K. A., Brain perfusion is abnormal in cocaine-dependent polydrug users: a study using technetium-99m-HMPAO and ASPET, *J. Nucl. Med.*, 32, 1206, 1991.
82. Dackis, C. A., Gold, M. S., Davies, R. K., and Sweeney, D. R., Bromocriptine treatment for cocaine abuse: the dopamine depletion hypothesis, *Int'l. J. Psych. Med.*, 15, 125, 1985.

83. Giannini, A. J., Baumgartel, P., and DiMarzio, L. R., Bromocriptine therapy in cocaine withdrawal, *J. Clin. Pharmacol.*, 27, 267, 1987.
84. Jaffe, J. H., Cascella, N. G., Kumor, K. M., and Sherer, M. A., Cocaine-induced cocaine craving, *Psychopharmacology,* 97, 59, 1989.
85. Tennant, F. S. and Sagherian, A. A., Double-blind comparison of amantadine and bromocriptine for ambulatory withdrawal from cocaine dependence, *Arch. Intern. Med.,* 147, 109, 1987.
86. Hubner, C. B. and Koob, G. F., Bromocriptine produces decreases in cocaine self-administration in the rat, *Neuropsychopharmacology,* 3, 101, 1990.
87. Markou, A. and Koob, G. F., Bromocriptine reverses the elevation in intracranial self-stimulation thresholds observed in a rat model of cocaine withdrawal, *Neuropsychopharmacology,* 7, 213, 1992.
88. Clow, D. W. and Hammer, R. P., Enhanced D_2 receptor binding following chronic cocaine treatment is reversed by bromocriptine, *Soc. Neurosci. Abstr.,* 16, 745, 1990.
89. Rao, T. S., Cler, J. A., Oei, E. J., Iyengar, S., and Wood, P. L., Increased release of dopamine in vivo by BMY-14802: contrasting pattern to clozapine, *Neuropharmacology,* 29, 503, 1990.
90. Menkel, M., Terry, P., Pontecorvo, M., Katz, J. L., and Witkin, J. M., Selective sigma ligands block stimulant effects of cocaine, *Eur. J. Pharmacol.,* 201, 251, 1991.
91. Zhang, J., Chiodo, L. A., Wettstein, J. G., Junien, J.-L., and Freeman, A. S., Acute effects of *sigma* ligands on the electrophysiological activity of rat nigrostriatal and mesoaccumbal dopaminergic neurons, *Synapse,* 11, 267, 1992.
92. Schuier, F., Orzi, F., Suda, S., Lucignani, G., Kennedy, C., and Sokoloff, L., Influence of plasma glucose concentration on lumped constant of the deoxyglucose method: effects of hyperglycemia in the rat, *J. Cerebral Bl. Flow Metab.,* 10, 765, 1990.
93. Cunningham, K. A., Paris, J. M., and Goeders, N. E., Chronic cocaine enhances serotonin autoregulation and serotonin uptake binding, *Synapse,* 11, 112, 1992.
94. King, G. R., Joyner, C. M., and Ellinwood, E. H., Withdrawal from continuous or intermittent cocaine: behavioral responsivity to 5-HT1 receptor agonists, *Pharmacol., Biochem. Behav.*, 45, 577, 1993.
95. White, F. J., Hu, X. T., and Henry, D. J., Electrophysiological effects of cocaine in the rat nucleus accumbens: microiontophoretic studies, *J. Pharmacol. Exp. Ther.*, 266, 1075, 1993.
96. Porrino, L. J., Functional consequences of acute cocaine treatment depend on route of administration, *Psychopharmacology,* 112, 343, 1993.
97. Paxinos, G. and Watson, C., *The Rat Brain in Stereotaxic Coordinates,* Academic Press, New York, 1986.

Chapter 3

Sensitization and Tolerance to the Motivational and Subjective Effects of Psychostimulants

Michael W. Emmett-Oglesby

CONTENTS

I. INTRODUCTION

Tolerance and sensitization are defined in terms of altered responsiveness to drug effects. They share common roots in that drug effects are modified as a consequence of previous exposure to drugs. In the case of tolerance, drug effects are decreased, which is represented by a shift to the right and perhaps downward of the dose-response curve. Conversely, sensitization produces a leftward, and perhaps an upward, shift of the dose-response curve.

Historically, tolerance to drug effects has been studied more frequently than sensitization; however, in recent years sensitization has received increasing emphasis. In a review of these phenomena, Stewart and Badiani noted that tolerance and sensitization occur to *effects* of drugs, not to drugs per se.[1] This admonition is particularly important: a drug may produce many effects simultaneously, such as alterations in autonomic functions (e.g., decreased body temperature), disruption of behavior, and modification of locomotor activity. Tolerance/sensitization may occur to different extents, and at different rates, to many of these drug effects. What then is a "relevant" variable to consider in studies of tolerance and sensitization?

Tolerance has a strong historical relationship to drugs of abuse, where it has been regarded widely as a key component of the dependence process.[2-4] This notion of tolerance extends back to the 19th century[5,6] or earlier and is perhaps best conceptualized as the vicious cycle in which increasing (and/or frequent) doses of drugs lead to tolerance, and tolerance, in turn, is an impetus to increase drug consumption further. As discussed below, this formulation is accepted less widely now, but it remains largely true that studies of tolerance still focus on drugs known to induce dependence, and the recent

1-8493-8311-0/95/$0.00+$.50

interest in sensitization similarly has been centered on psychostimulants such as cocaine and amphetamines. If one is interested in tolerance/sensitization to drugs such as psychostimulants, a reasonable argument can be made that such studies are more likely to provide results relevant to drug taking in humans if models are employed that have predictive validity for human drug use. In this regard, humans routinely report that drugs of abuse have desirable subjective effects. Consequently, the following discussion of tolerance/sensitization is mainly confined to results from three types of experimental paradigms: (1) drug discrimination, (2) conditioned place preference, and (3) self-administration. These procedures have reasonable face validity and predictive validity for modeling critical aspects of human response to drugs of abuse, including perception of drug effects, association of environments with desirable drug effects, and drug self-administration. Tolerance and sensitization will be evaluated using these models with special emphasis on the implications of these findings for the process of drug dependence.

A. BEHAVIORAL MODELS

Drug discrimination is well established as providing an animal model of human subjective drug phenomena.[7] In this paradigm, animals (or with increasing frequency, humans) are taught to detect the presence or absence of a drug effect using much the same methodology that one would use to train a subject to detect whether a light was on or off. Typically, subjects are placed in a chamber in which two or more response options are available. Responding is maintained by providing an appropriate reinforcer (e.g., money for humans; sugar pellets for rats). The problem for the subject is that only one of the response options will produce the reinforcer, and the option that is correct is determined by the drug given prior to starting the session. For example, using an operant chamber that has two levers and a hopper connected to a food-pellet dispenser, rats can be trained to differentiate cocaine (10 mg/kg) from saline by reinforcing lever presses (i.e., every 30 presses might result in the presentation of food). Prior to entering the chamber, the subject is injected: when injected with cocaine, only responses on the left-lever will be reinforced; when injected with saline, only responses on the right-lever will be reinforced. Initially, responses on the two levers occur randomly without reference to the presession treatment. However, over several training sessions, subjects will begin to emit all responses on the cocaine-lever when injected with cocaine, or on the saline-lever when injected with saline.

After such training, a remarkable feature of drug discrimination is revealed. When different drugs are examined for their ability to substitute for the training drug, only those drugs that humans described as having psychostimulant effects will substitute for cocaine. All other drugs engender saline-lever responding. This pattern of results has been found for essentially all classes of drugs trained and tested (e.g., if morphine is trained as a discriminative stimulus, only μ–receptor agonists will substitute). This general pattern of results is consistent with the proposal that subjects learn to use the effects of the training drug as a reference, and the effects of other drugs are compared to that reference to determine their degree of match. It also has led to the general conclusion that animals and humans classify drugs similarly. Because classes of abused drugs are defined in large measure by their distinctive subjective profiles (e.g., psychostimulant-like; sedative-hypnotic-like, etc.), drug discrimination in animals also has been proposed as a model for studying subjective effects of drugs in animals.

The drug discrimination procedure is thus particularly interesting for studying sensitization and tolerance phenomena. If either can be shown to occur in animals trained to discriminate psychostimulants, it is highly likely that a parallel phenomenon exists for the subjective effects of these drugs in humans. Certain cautions should be noted, however. First, typically 40 to 80 sessions are required to train subjects (from rats to nonhuman primates) to discriminate cocaine. It may be the case that sensitization and/or tolerance are already maximal by the time training is complete. Second, in the drug discrimination paradigm, subjects do not voluntarily take the drug. Thus, it might be possible that the effects of the drug under these circumstances are not comparable to the effects obtained when the drug is self-administered. Third, drug taking behavior in humans, particularly during "binges" of drug use, is not a simple reflection of the "pleasurable" subjective effects of a drug; indeed, drug self-administration continues at times when subjects report very little positive drug effect. Consequently, sensitization or tolerance in the drug discrimination procedure may have relevance for understanding changes in subjective effects of drugs, but might not be relevant to the core variable(s) of drug-taking behavior.

Because of this limitation of drug discrimination, it is also necessary to look at animal models with high predictive validity for drug preference and for drug use in humans. One such model is conditioned place preference. In this model, a large environment is subdivided into two smaller compartments separated by a barrier. The two compartments are distinctive (usually both visual and tactile cues are distinct). Over a series of days, animals are injected with a drug prior to being placed in one compartment, and on other days they are injected with vehicle prior to being placed in the other compartment. After several pairings, the barrier between the compartments is removed and subjects are allowed to explore both compartments without a pretest injection. Drugs with abuse potential produce positive conditioned place preference (i.e., more time is spent in the previously drug-paired compartment than in the vehicle-paired compartment), and drugs known to be aversive in humans produce negative conditioned place preference (i.e., less time is spent in the previously drug-paired environment; for example, κ-receptor agonists produce negative conditioned place preference[8]).

These advantages of conditioned place preference not withstanding, some of the caveats applied to drug discrimination also apply to conditioned place preference. Most notably, subjects do not voluntarily self-administer drugs tested in this model. Nonetheless, the concordance of positive and negative conditioned place preference data with reports of abuse potential of drugs in humans is impressive enough to conclude that the procedure has clear predictive validity in this regard. It should be noted that conditioned place preference requires exposure of the subject to the drug during training, so sensitization or tolerance may be affected by this exposure. However, as compared to drug discrimination, subjects can be trained with as few as three exposures to the drug,[9] which minimizes the likelihood of significant sensitization or tolerance occurring during the training process.

Perhaps the most critical measure for determining whether tolerance or sensitization are likely to be observed in humans is to examine whether these phenomena occur in animal models of i.v. drug self-administration. Intravenous drug self-administration in animals is correlated highly with similar behavior in humans. Thus, demonstrations of sensitization and tolerance in these paradigms almost certainly would suggest that similar effects occur in humans. In this regard, two variants of self-administration behavior must be understood. One set of data comes from fixed-ratio (FR) schedules (e.g., responses must be emitted on a lever, and an injection of drug is administered every time the specified number of responses occurs). In a general way, such schedules provide a model for a situation in which a large supply of drugs is readily available, and each individual use requires relatively little work (e.g., "hitting the crack pipe").

Do changes in the rate of cocaine self-administration under FR schedules reflect sensitization or tolerance to the reinforcing efficacy of cocaine? Unfortunately, the rate of drug taking in such schedules is at best an ambiguous measure of a drug's reinforcing efficacy; indeed, these schedules may not appropriately predict the relative reinforcing efficacy of different drugs or different doses of the same drug.[10,11] For example, the intake of cocaine typically is stable over time in FR schedules, independent of the dose that is available for self-administration;[12,13] that is, subjects increase or decrease their rate of responding to the degree that a constant quantity of cocaine is taken per unit time, independent of the dose offered for self-administration. Several hypotheses have been suggested to explain this phenomenon, including the conjecture that animals take as much drug as possible until the disruptive effects of the drug temporarily preclude further administration.[14] Alternatively, subjects may titrate their plasma concentration of drug to produce an optimal effect.[15] If the former hypothesis is correct, changes in the rate of cocaine self-administration in the FR schedule would reflect sensitization or tolerance to the disruptive effects of a drug; if the latter hypothesis is correct, these changes would reflect sensitization or tolerance to the reinforcing efficacy of a drug.

Because of this limitation of FR schedules, this review also will focus on data obtained from i.v. self-administration of drugs using a progressive-ratio (PR) schedule.[16] Under this schedule, the number of lever responses required to obtain a reinforcer is increased for successive reinforcers until the animal fails to meet the demands of the schedule; that is termed the "breaking point". The analysis of breaking point of PR responding has been proposed to offer a reliable and quantitative method for assessing changes in reinforcing efficacy of drugs;[17,18] for example, smaller doses of a drug typically produce lower breaking point values than larger doses of the same drug,[18,19] and cocaine, a drug known to have high abuse liability in humans, maintains the highest breaking point values in a variety of subjects tested under PR schedules.[20,21] Because subjects acquire stable PR responding over several sessions of drug exposure, however, significant sensitization or tolerance might already be present by the time training is complete.

II. SENSITIZATION TO PSYCHOSTIMULANTS

Sensitization to drugs of abuse, particularly to psychostimulants, is reported with increasing frequency in the experimental literature. However, most of these reports involve behavioral models that are not related readily to the motivational or subjective effects of drugs. For example, the primary literature on sensitization is based upon measures of psychostimulant-induced locomotor activity in rats. Unquestionably, significant sensitization to this behavioral effect of cocaine or amphetamines occurs when these drugs are administered repetitively.[1,22-24] However, the implication of these results for humans is less clear. Some authors contend that this phenomenon might represent an analogue of "drug-craving" in humans;[25] others relate sensitization to the propensity of psychostimulants to cause psychotic episodes in humans with increasing use;[24] and still others attribute this phenomenon to general enhancement of the psychomotor activating properties of these drugs, suggesting that enhanced locomotor activity does predict enhanced reinforcing efficacy.[15] Although all of these speculations are interesting, definitive experimental support for such contentions still is lacking. Finally, sensitization to the motor effects of psychostimulants occurs to a maximum extent after relatively few (fewer than ten) drug exposures. Perhaps species differences are involved, but this would seem to be too few drug exposures for the phenomenon of sensitization in animals to be related to drug dependence in humans.

If locomotor sensitization is in fact germane to the problem of drug dependence, the data from locomotor studies should provide interesting predictions concerning parameters of sensitization that one might expect to find in models of subjective and motivational effects of psychostimulants. First, sensitization occurs most readily when injections are intermittent. Typically, sensitization is observed following injections which are administered once daily, and sensitization frequently is more robust after a period of abstinence between the last chronic injection and the administration of a challenge dose. This observation supports the speculation that the key variable controlling the appearance of sensitization may be the time since the last drug treatment rather than the spacing of injections that are used to induce sensitization. Second, greater sensitization usually is obtained if all drug injections occur in the chamber where the final test of sensitization occurs.[26,27] Indeed, sensitization may not even be apparent if injections occur outside the environment where tests are conducted, although significant sensitization also has been reported for such "context independent" injections. Third, sensitization reaches a maximum after relatively few drug exposures. Although this variable has not been explored in detail, little additional sensitization to cocaine appears after that which is present following the first five or ten exposures. Finally, sensitization is relatively permanent. Once subjects have been exposed to several injections of psychostimulants, they rarely return to baseline activity levels characteristic of first exposure, even if long periods intervene in which subjects are drug free.

As reviewed below, it is difficult to generalize the findings from locomotor activity studies to studies of sensitization to the discriminative or motivational effects of drugs because of the limited data available. Where such comparisons can be made, the general pattern of results from locomotor activity studies appears to extend to these other models.

A. SENSITIZATION TO DISCRIMINATIVE EFFECTS OF PSYCHOSTIMULANTS

Sensitization to the discriminative effects of psychostimulants has not been reported. Once training is complete, dose-response curves are remarkably reproducible across replications. This is true even if subjects are withheld from training for periods up to a month. However, given the pattern of results described above for sensitization to locomotor effects of psychostimulants, it might not be surprising that sensitization to their discriminative effects is unknown. To discriminate a drug, subjects usually need to be trained with the drug for many more than ten exposures, and sensitization to the discriminative effect may occur to a maximum extent during this training.

A large body of evidence from discrimination studies has shown that larger doses of drug produce more intense stimulus effects than smaller doses. Overton systematically has studied the relationship between the dose of the training drug and the time required to learn a discrimination, concluding that discrimination of larger drug doses is learned more rapidly than discrimination of smaller doses.[28] Thus, an experimental paradigm that might provide an opportunity to observe sensitization to the discriminative effects of psychostimulants would be to expose subjects to the training dose of the drug once daily for several days, just prior to the start of discrimination training. If these subjects are sensitized compared to a control group, then they should learn the discrimination more rapidly. This approach has been attempted in rats with modest success,[29] but the lengthy nature of discrimination training makes it difficult to demonstrate reliable differences between control and experimental groups.

In contrast to the time required for nonhuman subjects to learn such a discrimination, humans have been trained to discriminate *d*-amphetamine in as few as ten total sessions.[30] Thus, a variant of the experiment proposed above would be to preexpose one group of human subjects to the training drug. If sensitization occurs, the preexposed group should learn the discrimination more rapidly. Such an experiment is not without problems, however. In the interest of minimizing drug exposure, it may be difficult to justify the preexposure procedure ethically. On the other hand, subjects usually receive drug during both the training and testing phases in most human drug discrimination studies, so the total number of drug exposures is usually more than the number required to test the sensitization hypothesis. Thus, by using subjects only for the duration required to train the discrimination, such an experiment could be conducted with a minimum number of drug exposures. Perhaps more troubling is the issue of whether subjects already are sensitized to a maximal extent prior to entering the experiment. This consideration raises two concerns. First, subjects would have to be naive to psychostimulants (raising issues of verification as well as ethical concerns for exposing such subjects to these drugs). Second, it appears that exposure to other drugs (including ubiquitous substances such as caffeine) may produce maximal cross-sensitization to the reinforcing effects of cocaine as described below. If the same finding extends to the discriminative effects of psychostimulants, it might not be possible to find a population of human subjects that is not sensitized already. In spite of these considerations, such an experiment is certainly feasible, and the information is likely to be very important to our understanding of how the perceived effects of psychoactive drugs change with experience. At the very least, it would be interesting to know how subjects who have a history of no psychostimulant use rate the subjective effects of the training drug across the first several sessions of exposure. If sensitization occurs, those effects might be perceived to intensify.

B. SENSITIZATION TO CONDITIONED PLACE PREFERENCE EFFECTS OF PSYCHOSTIMULANTS

Little work has been reported on sensitization to the positive conditioned place preference effects of psychostimulants. Lett originally reported sensitization to conditioned place preference effects of morphine and amphetamine following repeated pretreatment with these drugs.[31] Recently, Shippenberg and colleagues reported a shift to the left on the low-end of the cocaine dose-response curve in conditioned place preference of rats pretreated daily for 5 days with 10 mg/kg cocaine.[32] Enhanced conditioned place preference was observed most readily for doses of drug that required several sessions of conditioned place preference training to produce a significant preference under control conditions; following the pretreatment regimen with cocaine, conditioned place preference was produced after only three to four total training sessions. This finding is in agreement with those from i.v. self-administration paradigms (reviewed below) in which sensitization is seen most clearly with doses that are marginally reinforcing upon initial exposure.

Because of the rapid acquisition of conditioned place preference behavior, this model is well suited to assessing sensitization to the motivational effects of drugs. Among the issues which could be examined are: whether sensitization will modify the effects of cocaine at doses that fail to produce conditioned place preference under control conditions; and, whether the maximum drug effect increases, or only the effect of low doses. Finally, a number of variables should be investigated in a parametric manner—most notably, dose-response relationships should be determined for conditioned place preference measures using a range of sensitizing cocaine doses. Such studies are crucial to determine the extent to which sensitization is an orderly phenomenon, and to provide data on the maximal possible magnitude of the effect. These studies are unfortunately difficult to perform because they rely on acquisition measures, and each subject can only be used once.

C. SENSITIZATION TO REINFORCING EFFECTS OF PSYCHOSTIMULANTS

One of the remarkable features of i.v. cocaine self-administration is that once the behavior is acquired, cocaine administration is very stable across sessions if subjects are placed on a limited access schedule (drug is available for only a few injections or a few hours each day). Most studies report that the entire dose-response curve is highly reproducible although sensitization reportedly occurs in response to very low doses of methamphetamine in rhesus monkeys.[33] In that study, subjects eventually self-administered doses of methamphetamine that would not maintain behavior when tested initially. This report aside, the weight of the literature supports the conclusion that self-administration of psychostimulants is stable over time under limited access conditions.

Psychostimulant self-administration typically is acquired by subjects at variable rates, ranging from a few to several days (in our laboratory for example, about 85% of all subjects acquire cocaine self-administration, and 65% reach our criterion for stability after approximately ten self-administration sessions). Because subjects are exposed to drugs on each training session, it might be the case that they are maximally sensitized by the time that they reach criterion for stable responding. Thus, to determine whether sensitization occurs to the reinforcing effects of psychostimulants, the only viable strategy might be to examine the effect of various pretreatments (e.g., daily i.p. injections of cocaine for several days) on the rate of acquisition of i.v. cocaine self-administration in subjects naive to this task.

The literature concerning sensitization to the reinforcing effects of psychostimulants has come primarily from the laboratories of Piazza and Schenk, who focused on acquisition of drug-taking behavior under FR schedules in rats. Piazza and colleagues have conducted systematic studies of the effects of various stressors, including drugs, on sensitization to *d*-amphetamine as measured by the number of sessions necessary to acquire self-administration behavior.[34] However, the effect of parametric manipulations of the sensitizing regimen (i.e., dose magnitude; duration of treatment; etc.) has yet to be determined. Perhaps even more critical, the effect of pretreatments has not been investigated across a full range of amphetamine doses. These difficulties aside, the focus of this group upon stress might offer key insight into a crucial variable that unifies the sensitization literature.

Schenk and colleagues have focused more exclusively on the role of drug pretreatment in causing sensitization, and have provided substantive data defining the conditions that produce sensitization to cocaine. When rats were pretreated once daily for 9 days with cocaine (10 mg/kg, i.p.), they subsequently learned to self-administer cocaine more rapidly than did vehicle-treated controls.[35] This effect was most robust for small doses of cocaine, perhaps because self-administration of larger doses was acquired so rapidly that it would be difficult for sensitization to manifest. In subsequent experiments, a variety of drug pretreatments have been shown to enhance the acquisition of cocaine self-administration, including *d*-amphetamine,[36] caffeine,[37,38] and nicotine,[36] while other drugs prevent sensitization (e.g., MK-801 or dizocilpine[39]; cocaine-induced sensitization also was prevented when amphetamine pretreatment was combined with dizocilpine pretreatment[40]).

These results are intriguing because they provide direct evidence supporting the hypothesis that sensitization to the motivational effects of psychostimulants occurs in such a way that their efficacy as reinforcers increase. As with conditioned place preference studies of sensitization, however, this work requires significantly more exploration of fundamental variables controlling the expression of sensitization. At the very least, complete dose-response curves must be generated for acquisition of cocaine self-administration, and these determinations must follow parametric variation of pretreatment variables including dose magnitude, dose frequency, and total number of doses. To date, dose-response data for cocaine self-administration only have been published in two studies of sensitization to the reinforcing effects of cocaine,[39,41] and the effects of systematic variation of independent variables have not been determined yet. Extensive characterization will be difficult, however, because subjects can be tested only once to determine acquisition rate.

Finally, one of the critical points that must be established in this line of research is the range of drugs that are capable of producing sensitization to cocaine. If relatively ubiquitous stimuli such as caffeine truly sensitize animals to the reinforcing effects of cocaine to the same extent that cocaine itself does,[38] then the predictive validity of sensitization for understanding vulnerability to substance abuse in humans is likely to be minimal. In an attempt to differentiate the effects of amphetamine pretreatment from those of caffeine pretreatment, Schenk and colleagues recently have examined the rate of recovery from sensitization following termination of pretreatment with these agents (personal communication). They hypothesized that amphetamine would produce more enduring sensitization, which would differentiate it from the sensitization produced by caffeine. Unfortunately, this turned out not to be the case in that sensitization, measured as acquisition time required to learn an operant response for cocaine, still occurred to the same extent in both groups three weeks after discontinuing treatments. Thus, it remains to be determined whether cocaine and caffeine pretreatments produce sensitization that can be differentiated from one another. Such a demonstration will be critical to support the argument that vulnerability to the reinforcing effects of drugs of abuse is not shared to the same extent by everyone due to individual histories of caffeine use.

D. IMPLICATIONS OF SENSITIZATION FOR DRUG DEPENDENCE

Drug dependence frequently has been viewed as a phenomenon comprising of at least three distinct phases: initiation, maintenance, and relapse. Various researchers have argued that experimental sensitization is germane to each of these components. Concerning initiation, the data reviewed generally support the hypothesis that the discriminative and reinforcing/rewarding effects of psychostimulants, and perhaps also opioids, become more robust over the first several exposures to these drugs. Thus, if sensitization plays an important role in the initiation of drug self-administration in humans, one might expect to see that humans require several exposures to a drug before the drug becomes reinforcing. In fact, this might be the case for a drug such as heroin, because humans routinely (though not universally) report that the first several uses of the drug were not pleasurable.[4] Although these reports have been interpreted as showing tolerance to the dysphoric effects of opioids, they could as easily be interpreted as supporting the occurrence of sensitization to the positive reinforcing and subjective effects of opioids. Furthermore, sensitization to the psychoactive effects with marijuana also has frequently been reported, and many users report no psychoactive effects upon first exposure to this drug;[4] however, sensitization to the discriminative or reinforcing effects of marijuana has not been investigated in animal studies. On the other hand, sensitization to the subjective or reinforcing effects of psychostimulants in humans is essentially unknown. Considering that most animal studies of sensitization utilize psychostimulants, it is apparent that a discrepancy exists between animal and human literature. This discrepancy might be accounted for by cross-sensitization of drugs such as caffeine to psychostimulants.[37, 38] If so, then sensitization is unlikely to be a significant risk factor in explaining differential vulnerability to the initial use of psychostimulants, because most people would be maximally sensitized to psychostimulants by virtue of their history of caffeine use.

The evidence for the role of sensitization in maintaining drug dependence is even less well established. If sensitization continued to increase with longer periods of drug use, then one might expect that drugs abusers would report enhancement of drug effects over time. Indeed, one might expect that sensitization would lead either to the self-administration of smaller doses (if the desired effect is obtained at a lower dose) or to increasing potency of the normally administered dose. In contrast to this expectation, drug use tends to escalate over time, and many chronic drug abusers complain of decreased drug effectiveness.

Robinson and Berridge recently have suggested an intriguing hypothesis that posits a critical role for sensitization in the maintenance of drug dependence.[25] They note that although drug "liking" may be unchanged or actually diminish with frequent use, the degree to which a chronic user "wants" drugs continues to increase. Relying heavily on the motivation theory developed by Bindra,[41] they attribute this phenomenon to sensitization of the brain circuitry responsible for mediating incentive motivation. This theory has intuitive appeal; it accounts for a number of attributes seen with chronic drug use in humans. Support for the theory comes from studies of locomotor activity, which the authors argue is a reflection of the drive to obtain drugs. However, sensitization to locomotor activity reaches a peak within a few exposures to psychostimulants, whereas this theory involves processes that should occur later, and continue to escalate for much longer, in humans. In animal models where i.v. self-administration of stimulants has been maintained for long periods, most investigators report that the rate of drug self-administration is very consistent across sessions,[42] and the same is true under PR schedules.[19] Such results are inconsistent with an increasing motivation to obtain drugs. Thus, while sensitization to the incentive motivational effects of drugs is an intriguing hypothesis, data supporting it are equivocal at best.

Does sensitization play a significant role in relapse to drug use? A substantial set of experiments has documented that injection of a wide variety of drugs will reinstate responding following extinction.[43-45] Although these data may be relevant to the role of drug-related stimuli in promoting relapse, it is not clear why the concept of sensitization is invoked to describe the phenomenon. It is well known that following extinction of food-reinforced responding, presentation of a single food pellet will reinstate responding, but such subjects are not "sensitized" to the reinforcing effects of food.

Sensitization to psychostimulants is an interesting phenomenon, but its relevance to drug dependence is currently not established. Thus, sensitization may not be a significant factor either in maintenance or relapse to drugs of abuse, or in differential susceptibility to initiating drug use.

I. TOLERANCE TO PSYCHOSTIMULANTS

A. TOLERANCE TO THE DISCRIMINATIVE STIMULUS EFFECTS OF PSYCHOSTIMULANTS

When psychostimulants are trained as discriminative stimuli, tolerance to this effect can be produced such that the entire dose-response curve for the detection of the training drug shifts two or more fold to the right. The simplest way to induce such tolerance is to withhold training, and then administer the substance of interest in high doses multiple times daily. For example, in rats trained to detect 10 mg/kg cocaine from saline, then withheld from training and injected with cocaine (20 mg/kg) every 8 hr for 7 days, an approximate twofold shift to the right of the dose-response curve for the detection of cocaine is produced.[46-49] In contrast, when these subjects are injected with an equivalent volume of saline every 8 hr for a week, no change is seen from the baseline sensitivity to the stimulus. Similarly, subjects trained to detect *d*-amphetamine become tolerant to the discriminative stimulus effects of this drug when injected with high doses of amphetamine for several days, whereas vehicle treatment produces no effect on the sensitivity to amphetamine.[50,51]

The parameters of tolerance to the discriminative stimulus effects of psychostimulants have been well established. For example, the degree of tolerance increases as a function of increasing dose used during chronic treatment: no tolerance occurred to cocaine (15 mg/kg/day) administered every 8 hr for 1 week,[48,52] whereas the dose-response curve shifted nearly twofold at 30 mg/kg/day, and more than twofold tolerance occurred at 60 mg/kg/day.[48] With chronic amphetamine administration, it appears to be possible to obtain even more tolerance. For example, in pigeons trained to detect *d*-amphetamine and then given chronic *d*-amphetamine (5.6 mg/kg/12 hr for one week) a sevenfold shift to the right of the dose effect curve was obtained.[51] Chronic *d*-amphetamine also has been found to induce fourfold cross-tolerance in rats trained to detect the stimulus effects of cocaine.[48]

Many reasons might account for the greater tolerance exhibited after chronic administration of amphetamine compared to chronic treatment with cocaine, but a particularly relevant variable may be the half-life of elimination for these two drugs. It is well established that tolerance to discriminative effects of psychostimulants does not occur with once daily administration, presumably because drug-receptor interactions occur for a limited duration each day, and this duration is not sufficient to produce tolerance of 24-hr duration. Tolerance does occur following prolonged exposure to the drug with multiple daily administrations. Because amphetamine has a longer elimination half-life than does cocaine, the processes leading to adaptation are presumably engaged for longer periods each day, resulting in greater tolerance.

Beyond parameters of dosing, tolerance to the discriminative stimulus effects of either *d*-amphetamine or cocaine occur as a direct function of the duration of dosing, reaching a maximum after approximately 4 to 6 days of chronic treatment and remaining stable as chronic treatment continues thereafter. Upon discontinuation of chronic treatment, tolerance is spontaneously lost over approximately the same time course required for its acquisition.[47,48,51] In one study, 2 weeks of chronic treatment produced approximately a 2-week time course for recovery,[48] but generally, subjects recover full sensitivity to the stimulus approximately 1 week following termination of chronic administration.

Tolerance to the discriminative effects of psychostimulants is not accompanied by tolerance to the rate depressant effects of these drugs.[48,51] Thus, tolerance is not a global phenomenon as might be expected if pharmacokinetic mechanisms mediate its occurrence (i.e., tolerance in this paradigm cannot be the result of an increase in the rate of excretion of the drug because if this were the case, one would expect that all doses of amphetamine or cocaine would produce lesser effects on all dependent measures).

The profile of drugs producing tolerance and cross-tolerance to the discriminative effects of psychostimulants also is orderly. Drugs that humans describe as having psychostimulant-like subjective effects substitute for amphetamine or cocaine in animals trained to detect these drugs, whereas drugs from other classes (e.g., opioids, sedative-hypnotics) do not.[49,53,54] Tolerance to cocaine has been shown to confer cross-tolerance to a variety of other psychostimulants.[49] In contrast, chronic administration of opioids produces no cross-tolerance to the discriminative effects of either amphetamine[48] or cocaine.[51] This lack of cross-tolerance between different classes of drugs is characteristic of a variety of drug discriminations,[55] and conforms to anecdotal accounts from human polydrug users who report that cross-tolerance is confined within drug classes.[4]

If drug discrimination is a model of human subjective phenomena, then one would expect to see the development of tolerance to the subjective effects of psychostimulants. Within a single session of cocaine use, this is certainly the case.[56] When cocaine was administered to subjects i.v., a constant, high plasma concentration was associated initially with profound subjective effects, but during the course of the next few minutes, even though plasma concentration of cocaine remained stable, subjective ratings of euphoria decreased to baseline values.[57] In contrast to these acute studies, virtually no tolerance was observed when exposures were separated by one or more days. However, tolerance is seen only after several days of multiple daily administrations in the animal drug discrimination paradigm; this dosing regimen presents ethical concerns if humans are to be subjects. Because such data are potentially important for understanding whether tolerance occurs during binges of drug use, the reports of users at the end of their binge might be useful. This is based upon the observation that when cocaine is available in relatively unlimited quantities, humans who self-administer the drug typically stop after approximately 4 days of use, and usage is not resumed for roughly another 4 days.[42] It would be particularly interesting to measure subjective effects in response to cocaine both during this four-day recovery period and again several days later. If the data from animals are correct, then tolerance should be present during early withdrawal, then reduced when subjects are maintained drug free.

Tolerance in drug discrimination paradigms has been attributed to an artifact arising from retraining subjects to recognize higher doses of the drug as their training stimulus.[58]

However, the presence of pharmacodynamic tolerance provides a more parsimonious explanation for the pattern of results from drug discrimination studies of animals treated chronically with high doses of drug. For the interested reader, the argument in favor of a pharmacodynamic interpretation of tolerance has been presented in detail elsewhere.[8,48,55,59]

B. TOLERANCE TO THE CONDITIONED PLACE PREFERENCE EFFECTS OF PSYCHOSTIMULANTS

To date, no work on tolerance to the conditioned place preference effects of psychostimulants has been reported, although a positive result is expected based upon the pattern of results with opioid tolerance in conditioned place preference. Shippenberg and colleagues have reported the presence of tolerance to the conditioned place preference effects of morphine in rats.[8] Although such results are encouraging, the use of conditioned place preference has significant difficulties for the study of tolerance. First, studies of acquisition behavior require an enormous number of subjects for complete dose-response determinations. More importantly, if the data from drug discrimination and drug self-administration studies apply to conditioned place preference, one might expect that tolerance to psychostimulants would last for a maximum of 3 or 4 days after terminating the chronic dosing regimen. Thus, training would have to be complete during this period to reveal tolerance in the conditioned place preference paradigm. In view of this problem, it seems unlikely that conditioned place preference will be of significant value for investigating tolerance phenomena.

C. TOLERANCE TO THE REINFORCING EFFECTS OF PSYCHOSTIMULANTS

In a recent series of publications, we have documented that tolerance occurs to the reinforcing effects of cocaine.[60-62] Before describing these results, it may be useful to describe the historical background to these experiments. Little if any tolerance to the reinforcing effects of psychostimulants had been reported previously in various animal experiments. Indeed, increased sensitivity to the lowest doses of methamphetamine self-administered in primates was reported after a period of chronic methamphetamine administration.[33] However, no one had ever performed a self-administration experiment based upon the type of parameters established in tolerance experiments using drug discrimination procedures. Two immediate discrepancies were apparent. Under training conditions in a drug discrimination experiment, subjects can be described as "irregular users" of drug. They might receive a relatively small dose of the training drug perhaps three times per week (e.g., cocaine is administered in rats at a dose of 10 mg/kg, i.p.). In contrast, drug self-administration procedures typically permit animals access to much larger quantities of drug (e.g., 30 to 60 mg/kg per session in rats) at least five times per week. Thus, one possibility is that subjects in self-administration procedures are already maximally tolerant before an attempt is made to assess whether tolerance exists. Also, psychostimulants are administered multiple times daily in drug discrimination paradigms used to assess tolerance. Tolerance had been examined in two ways in self-administration procedures: subjects were given either daily access for a limited period of time (e.g., 3 hr), or they were allowed to binge (continuous access for 12 or more hr). In neither of these cases did animals receive drugs in patterns that resembled those producing tolerance in drug discrimination procedures.

Therefore, we trained rats to self-administer cocaine (1 mg/kg, i.v.) using an FR2 schedule of lever pressing.[60] Sessions occurred once daily and permitted a maximum of 15 injections (chronic administration of 15 mg/kg cocaine did not produce tolerance in the drug discrimination paradigm). After obtaining complete dose-response curves, training was halted, and subjects were injected with cocaine (1 mg/kg, i.v. each min for 20 min) every 8 hr for 1 week (i.e., a dosing regimen that produced approximately twofold tolerance in drug discrimination experiments). When retested over the next 4 days, the dose-response curve for self-administration of cocaine had shifted to the right approximately twofold, but returned to baseline values 1 week after terminating chronic treatment.[60]

This initial set of observations was followed by a critical experiment to determine the mechanism(s) responsible, which might include (1) pharmacokinetic mechanisms (e.g., increased metabolism) or (2) pharmacodynamic mechanisms (e.g., decreased behavior-disrupting effects or decreased reinforcing efficacy). The pharmacokinetic mechanism is plausible but unlikely because chronic administration is not known to increase cocaine metabolism in rats.[63,64] It is also unlikely that the disrupting effects of cocaine on behavior decrease. This hypothesis assumes that drug self-administration occurs as rapidly as possible, so that the time between injections under an FR2 schedule is primarily a function of behavioral intoxication; that is, subjects take as much drug as possible until they can no longer emit an operant, and the next injection occurs when subjects recover sufficiently from this disruption.[65] Thus, tolerance to the disruptive effects of cocaine would produce an increased rate of drug self-administration (i.e., subjects would take more drug per unit time simply because they were able to emit the operant sooner). However, subjects are capable of emitting operants between self-injections of cocaine,[66] suggesting that they might titrate their plasma concentration of the drug in order to obtain a "desired" effect. If this is the case, then tolerance might be expected to produce a diminished effect of all doses. We tested these alternate predictions by determining the lowest dose of cocaine that would support self-administration. If tolerance occurred to the disruptive effects of cocaine after chronic administration, then subjects should still find this dose reinforcing and would continue to self-administer this dose at least as rapidly as before. If, however, tolerance occurs to the *motivational* effects of cocaine, the lowest dose that previously was capable of supporting self-administration would now be ineffective. In other words, the lowest dose capable of supporting self-administration would shift to the right during tolerance. Such a twofold shift of the dose-response curve was observed.[60] These results were replicated and extended to reveal that chronic treatment with saline produced essentially no effect on the rate of cocaine self-administration in contrast to the results obtained with a chronic cocaine regimen.[61]

If tolerance to the reinforcing effects of psychostimulants truly is associated with tolerance to their motivational effects, then comparable tolerance would be expected in subjects administering cocaine under a PR schedule. Yanagita used a PR schedule to determine whether the intensity of drug-seeking behavior would increase following development of physical dependence on cocaine.[67] Chronic treatment for 1 month with high-dose cocaine did not increase the breaking point, and the author concluded that cocaine did not induce physical dependence. Interestingly, however, chronic treatment with cocaine produced a 50% reduction of total responses emitted in both of the nonhuman primate subjects tested, which is consistent with our data showing that treatment with high doses of cocaine decreased breaking points for cocaine in rats trained on a PR schedule;[62] thus, these data support the hypothesis that tolerance occurs to the reinforcing effects of cocaine self-administration under a PR schedule.

Following tests for tolerance, the PR schedule dose-response curves returned to baseline values within 1 week of discontinuation of chronic cocaine treatment, and chronic saline injections had no effect on the dose-response curve obtained for cocaine in the PR procedure.[62] Thus, the time course of tolerance to cocaine in this PR paradigm appears to parallel the time course of tolerance reported following comparable dosing regimens in both drug discrimination[40] and FR2 self-administration[61] paradigms.

Variability of results across studies of tolerance (or sensitization) has been attributed frequently to an interaction between drug effects and environmental factors. That is, tolerance (or sensitization) is thought to be regulated by specific environmental stimuli that are associated with previous drug injections.[1,68] Because subjects in some of our studies received chronic cocaine while in operant chambers similar to those in which they self-administered the drug (except that levers were removed), environmental stimuli might influence the expression of tolerance in those experiments. However, a similar degree of tolerance to the reinforcing effects of cocaine develops when subjects are injected with high doses of psychostimulants in their home cages,[69] suggesting that environmental context is unlikely to account for differential development of tolerance. Rather, tolerance appears to occur solely as a function of chronic drug administration per se.

A criticism of drug discrimination studies that show tolerance to the effects of psychostimulants is that the experimenter administers drug to subjects, whereas tolerance might not develop when the drug is self-administered. Therefore, we studied the time course for the development and loss of tolerance to the reinforcing effects of cocaine in two groups of rats, including one group in which the experimenter infused cocaine (1 mg/kg, i.v. each min for 20 min) every 8 hr, and another group in which rats were permitted to self-administer 20 infusions of cocaine (1 mg/kg, i.v.) every 8 hr. The degree of tolerance and its time course for onset and offset were essentially identical in the two groups.[61] Thus, tolerance is a result of high dose and frequent drug administration, and it occurs independent of contingencies of self- versus experimenter-administration.

D. IMPLICATIONS OF TOLERANCE FOR DRUG DEPENDENCE

Tolerance to either subjective/discriminative or reinforcing effects of drugs should have minimal impact upon the initiation of drug self-administration. As reviewed above, little tolerance occurs to these effects of psychostimulants unless multiple daily administration of the drug occurs, and this type of continuous drug taking is not representative of the pattern of use seen with most people upon initial involvement with cocaine or amphetamine. The remarkable stability of behavior observed when access to these drugs is limited to once daily, whether in drug discrimination or self-administration paradigms, suggests that tolerance might occur only after multiple daily drug use.

The results described above imply that tolerance is likely to be a significant factor in the maintenance of drug use. For example, one might expect that chronic, high-dose drug self-administration would decrease efficacy to the degree that more drug would be self-administered to obtain the same level of subjective or reinforcing effects. In addition, the data on acute tolerance to the subjective effects of cocaine suggest that drug use should escalate rapidly during a binge. Curiously, neither of these predictions is well supported by the literature. Humans are reported to engage in binges of i.v. drug administration (or crack smoking) in which the quantity of drug used remains fairly constant. Similarly, in animal models of such binge paradigms, the total daily intake of cocaine does not change over days (although the pattern of drug taking changes dramatically toward the end of the binge).[70]

How can these conflicting results be reconciled? The pattern of drug administration and testing used in these experiments might provide a clue. Tolerance was tested at times when subjects had been free of drug administration for at least 8 hr. As noted by Markou and Koob,[70] this seems to be a time point following binge cocaine use which contributes to a cocaine "withdrawal syndrome", including an anhedonic effect. This anhedonic effect can last for approximately 48 hr following termination of 48-hr access to cocaine. Thus, it is possible that tolerance is associated with withdrawal phenomena. Indeed, the general pattern of results is consistent with the speculation that binges of drug use actually might forestall the appearance of tolerance, which would otherwise be present for at least 4 days. Such tolerance could underlie the fact that when i.v. cocaine users terminate a binge of drug use, they typically do not engage in another binge for at least 4 days.[42]

Tolerance is unlikely to play a role in relapse to drug dependence. A key feature of the tolerance experiments described above is that the duration of tolerance is at most a few days, and baseline sensitivity spontaneously recovers upon discontinuation of chronic treatment. Thus, larger quantities of drug would be self-administered only if relapse were to occur in the few days immediately following a binge. However, normal sensitivity to cocaine should be present thereafter.

IV. CONCLUSIONS

A variety of issues concerning sensitization and tolerance remain to be resolved. The most obvious of these is why chronic administration of psychostimulants leads either to sensitization or tolerance. It has been suggested that the dosing regimen (i.e., intermittent versus continuous drug exposure) might play a crucial role.[1] The typical "intermittent" daily dosing regimen used in sensitization experiments differs from the 8-hr "continuous" regimen used in tolerance experiments; however, this hypothesis remains to be tested rigorously. One interesting question is whether the same dosing regimen might produce both tolerance and sensitization. For example, if a high-dose, chronic regimen of psychostimulant administration were given to drug-naive animals, tolerance might appear initially, but sensitization might then appear as tolerance spontaneously fades. Thus, sensitization and tolerance might not be mutually exclusive; rather they might manifest variably depending upon the dosing regimen and the time of testing.

Still, there is little evidence at this time that either sensitization or tolerance are important in drug dependence. Both phenomena can occur in animals under well-defined experimental conditions, but the paucity of corroboration that the phenomena are encountered by human drug abusers suggests that neither phenomenon may be important in regulating drug use in humans. Recent data by Schenk and Davidson suggest that children exposed to methylphenidate for treatment of hyperactive behavior go on to manifest a greater likelihood of cocaine use than children with the same disorder who were not treated with psychostimulants.[71] If additional data of this type can be documented, then sensitization might be immediately relevant to substance abuse. Similarly, tolerance might occur near the end of drug abuse binges in humans, as proposed above. In either case, it is premature to conclude that either sensitization or tolerance is a critical component of drug dependence. The emerging literature on the occurrence of these phenomena in various animal models provides insight into their relevance in humans, and with continued effort, both sensitization and tolerance will be better understood in the near future.

REFERENCES

1. Stewart, J. and Badiani, A., Tolerance and sensitization to the behavioral effects of drugs, *Behav. Pharmacol.,* 4, 289-312, 1993.
2. Kalant, H., LeBlanc, A. E., and Gibbins, R. J., Tolerance to, and dependence on, some non-opiate psychotropic drugs, *Pharmacol. Rev.,* 23, 135-191, 1971.
3. Baker, S. T. and Tiffany, S. T., Morphine tolerance as habituation, *Psychol. Rev.,* 92, 78-107, 1985.
4. Jaffe, J. H., Drug addiction and drug abuse, in *The Pharmacological Basis of Therapeutics,* Gilman, A. G., Rall, T. W., Nies, A. S., and Taylor, P., Eds., Pergamon Press, Elmsford, New York, 1990, 522-573.
5. Siegel, S., Wilkie Collins: Victorian novelist as psychopharmacologist, *J. Hist. Med. Allied Sci.,* 38, 161-175, 1983.
6. Siegel, S., Alcohol and opiate dependence. Re-evaluation of the Victorian perspective, in *Research Advances in Alcohol and Drug Dependence,* Cappell, H. D., Glaser, F. B., Israel, Y., Kalant, H., Schmidt, W., Sellers, E. M., and Smart, R. C., Eds., Vol. 9, Plenum, New York, 1986, 279-314.

7. Preston, K. L. and Bigelow, G. E., Subjective and discriminative effects of drugs, *Behav. Pharmacol.*, 2, 293-313, 1991.
8. Bals-Kubik, R., Herz, A., and Shippenberg, T. S., Evidence that the aversive effects of opioid antagonists and kappa agonists are centrally mediated, *Psychopharmacology,* 98, 203-206, 1989.
9. Shippenberg, T. S., Emmett-Oglesby, M. W., Ayesta, J., and Herz, A., Tolerance and selective cross-tolerance to the motivational effects of opioids, *Psychopharmacology,* 96, 110-115, 1988.
10. Iwamoto, M. E. and Martin, W. A., A critique of drug self-administration as a method for predicting abuse potential of drugs, in *NIDA Research Monograph,* Harris, L. S., Ed., Problems of drug dependence. Proceedings of the committee on problems of drug dependence, 1988, 457-465.
11. Katz, J. L., Models of relative reinforcing efficacy of drugs and their predictive utility, *Behav. Pharmacol.,* 1, 283-301, 1990.
12. Johanson, C. E. and Schuster, C. R., Animal models of drug self-administration, in *Advances in Substance Abuse, Behavioral and Biological Research,* Mello, N. K., Ed. Vol. II, JAI Press, Greenwich, CT, 1981, 219-297.
13. Emmett-Oglesby, M. W. and Lane, J. D., Tolerance to the reinforcing effects of cocaine, *Behav. Pharmacol.,* 3, 193-200, 1992.
14. Wilson, M. C. and Schuster, C. R., Cholinergic influence on intravenous cocaine self-administration by rhesus monkeys, *Pharmacol. Biochem. Behav.,* 1, 643-649, 1973.
15. Wise, R. A. and Bozarth, M. A., A psychomotor stimulant theory of addiction, *Psychol. Rev.,* 94, 469-492, 1987.
16. Hodos, W., Progressive ratio as a measure of reward strength, *Science,* 134, 943-944, 1961.
17. Roberts, D. C. S. and Richardson, N. R., Self-administration of psychomotor stimulants using progressive ratio schedules of reinforcement, in *Animal Models of Drug Addiction,* Boulton, A. A., Baker, G. B., and Wu, P. H., Eds., Neuromethods series: Vol. 24, Humana Press, Totowa, NJ, 1992, 233-269.
18. Roberts, D. C. S., Loh, E. A., and Vickers, G., Self-administration of cocaine on a progressive ratio schedule in rats: dose-response relationship and effect of haloperidol pretreatment, *Psychopharmacology,* 97, 535-538, 1989.
19. Depoortere, R. Y., Li, D.-H., Lane, J. D., and Emmett-Oglesby, M. W., Parameters of self-administration of cocaine in rats under a progressive-ratio schedule, *Pharmacol. Biochem. Behav.,* 45, 539-548, 1993.
20. Griffiths, R. R., Brady, J. V., and Snell, J. D., Progressive-ratio performance maintained by drug infusions: comparison of cocaine, diethylpropion, chlorphentermine, and fenfluramine, *Psychopharmacology,* 56, 5-13, 1978.
21. Risner, M. E. and Silcox, D. L., Psychostimulant self-administration by beagle dogs in a progressive-ratio paradigm, *Psychopharmacology,* 75, 25-30, 1981.
22. Wise, R. A. and Leeb, K., Psychomotor-stimulant sensitization: a unitary phenomenon?, *Behav. Pharmacol.,* 4, 339-349, 1993.
23. Kalivas, P. W., Duffy, P., Dumars, L. A., and Skinner, C., Behavioral and neurochemical effects of acute and daily cocaine administration in rats, *J. Pharmacol. Exp. Ther.,* 245, 485-492, 1988.
24. Kalivas, P. W., Sorg, B. A., and Hooks, M. S., The pharmacology and neural circuitry of sensitization to psychostimulants, *Behav. Pharmacol.,* 4, 315-333, 1993.
25. Robinson, T. E. and Berridge, K. C., The neural basis of drug craving: an incentive-sensitization theory of addiction, *Brain Res. Rev.,* 18, 247-291, 1993.
26. Stewart, J. and Vezina, P., Conditioning and behavioral sensitization, in *Sensitization in the Nervous System,* Kalivas, P. and Barnes, C. D., Eds., Telford Press, Caldwell, NJ, 1988, 207-224.
27. Stewart, J., Conditioned stimulus control of the expression of sensitization of the behavioral activating effects of opiate and stimulant drugs, in *Learning and Memory: Behavioral and Biological Substrates,* Gormezano, I. and Wasserman, E. A., Eds., Erlbaum, Hillsdale, NJ, 1992, 129-151,
28. Overton, D. A., Drug discrimination training with progressively lowered doses, *Science,* 205, 720-721, 1979.
29. Gaiardi, M., Bartoletti, M., Guvellini, C., Bacchi, A., and Babbini, M., Sensitivity to the narcotic cue in non-dependent, morphine-dependent and post-dependent rats, *Neuropharmacology,* 25, 119-123, 1986.
30. Chait, L. D., Uhlenhuth, E. H., Johanson, C. E., Individual differences in the discriminative stimulus effects of d-amphetamine in humans, *Psychopharmacology,* 96, 212-217, 1989.
31. Lett, B. T., Repeated exposures intensify rather than diminish the rewarding effects of amphetamine, morphine, and cocaine, *Psychopharmacology,* 98, 357-362, 1989.

32. Shippenberg, T. S. and Heidbreder, C. H., Kappa-opioid receptor agonists prevent sensitization to the rewarding effects of cocaine. Proceedings of the Annual Meeting of the College on Problems of Drug Dependence, *NIDA Research Monograph,* 1994, in press.
33. Woolverton, W. L., Cervo, L., Johanson, C. E., Effects of repeated methamphetamine administration on methamphetamine self-administration in rhesus monkeys, *Pharmacol. Biochem. Behav.,* 21, 737-741, 1984.
34. Piazza, P. V., Deminiere, J. M., LeMoal, M., Simon, H., Stress- and pharmacologically-induced behavioral sensitization increases vulnerability to acquisition of amphetamine self-administration, *Brain Res.*, 514, 22-26, 1990.
35. Horger, B. A., Shelton, K., Schenk, S., Preexposure sensitizes rats to the rewarding effects of cocaine, *Pharmacol. Biochem. Behav.*, 37, 707-711, 1990.
36. Schenk, S., Snow, S., and Horger, B., Pre-exposure to amphetamine but not nicotine sensitizes rats to the motor activating effect of cocaine, *Psychopharmacology,* 103, 62-66, 1991.
37. Horger, B. A., Wellman, P. J., Morien, A., Davies, B. T., and Schenk, S., Caffeine exposure sensitizes rats to the reinforcing effects of cocaine, *NeuroReport,* 2, 53-56, 1991.
38. Schenk, S., Valadez, A., Horger, B. A., Snow, S., and Wellman, P. J., Interactions between caffeine and cocaine in tests of self-administration, *Behav. Pharmacol.,* 5, 153-158, 1994.
39. Schenk, S., Valadez, A., McNamara, C., House, D., Higley, D., Bankston, M. T., Gibbs, S., and Horger, B. A., Development and expression of sensitization to cocaine's reinforcing properties: role of NMDA receptors, *Psychopharmacology,* 390-394, 1993.
40. Schenk, S., Valadez, A., Worley, C. M., McNamara, C., Blockade of the acquisition of cocaine self-administration by the NMDA antagonist MK-801 (dizocilpine), *Behav. Pharmacol.,* 4, 652-659, 1993.
41. Bindra, D., How adaptive behavior is produced: a perceptual-motivation alternative to response reinforcement, *Behav. Brain Sci.,* 1, 41-91, 1978.
42. Johanson, C.-E. and Fischman, M. W., The pharmacology of cocaine related to its abuse potential, *Phamacol. Rev.,* 41, 3-52, 1989.
43. de Wit, H. and Stewart, J., Reinstatement of cocaine-reinforced responding in the rat, *Psychopharmacology,* 75, 134-143, 1981.
44. Stewart, J. and Vezina, P., Extinction procedures abolish conditioned stimulus control but spare sensitized responding to amphetamine, *Behav. Pharmacol.,* 2, 65-71, 1989.
45. Worley, C. M., Valadez, A., Schenk, S., Reinstatement of extinguished cocaine-taking behavior by cocaine and caffeine, *Pharmacol. Biochem. Behav.,* 48, 217-221, 1994.
46. McKenna, M. and Ho, B. T., Induced tolerance to the discriminative stimulus properties of cocaine, *Pharmacol. Biochem. Behav.,* 47, 153-155, 1977.
47. Wood, D. M., Lal, H., and Emmett-Oglesby, M. W., Acquisition and recovery of tolerance to the discriminative stimulus properties of cocaine, *Neuropharmacology,* 23, 1419-1423, 1984.
48. Wood, D. M. and Emmett-Oglesby, M. W., Characteristics of tolerance, recovery from tolerance, and cross-tolerance to cocaine used as a discriminative stimulus, *J. Pharmacol. Exp. Ther.,* 237, 120-125, 1986.
49. Wood, D. M. and Emmett-Oglesby, M. W., Substitution and cross-tolerance profiles of anorectic drugs in rats trained to detect the discriminative stimulus properties of cocaine, *Psychopharmacology,* 95, 364-368, 1988.
50. Barrett, R. J. and Leith, N. J., Tolerance to the discriminative stimulus properties of *d*-amphetamine, *Neuropharmacology,* 20, 251-255, 1981.
51. Young, A. M., Walton, M. A., and Carter, T. L., Selective tolerance to discriminative stimulus effects of morphine or *d*-amphetamine, *Behav. Pharmacol.,* 3, 201-209, 1992.
52. Hutchings, G. C., Gonzales, L. P., and Altshuler, H. L., Changes in cocaine discrimination in rats during chronic cocaine exposure, *Fed. Proc.,* 37, 274, 1978.
53. Evans, S. M. and Johanson, C. E., Amphetamine-like effects of anorectics and related compounds in pigeons, *J. Pharmacol. Exp. Ther.,* 245, 17-23, 1987.
54. Järbe, T. U. C., Discriminative stimulus properties of d-amphetamine in pigeons, *Pharmacol. Biochem. Behav.,* 17, 671-675, 1982.
55. Young, A. M. and Sannerud, C. A., Tolerance to drug discriminative stimuli, in *Psychoactive Drugs: Tolerance and Sensitization,* Goudie, A. J. and Emmett-Oglesby, M. W., Eds., Humana Press, Totowa, NJ, 1989, 221-278.

56. Fischman, M. W., Schuster, C. R., Javaid, J., Hatano, Y., and Davis, J., Acute tolerance development to the cardiovascular and subjective effects of cocaine, *J. Pharmacol. Exp. Ther.*, 235, 677-682, 1985.
57. Ambre, J. J., Belknap, S. M., Nelson, J., Ruo, T. I., Shin, S., Atkinson, A. J., Acute tolerance to cocaine in humans, *Clin. Pharmacol.*, 44, 1-8, 1988.
58. Colpaert, F. C., Niemegeers, C. J. E., and Janssen, P. A. J., Studies on the regulation of sensitivity to the narcotic cue, *Neuropharmacology*, 17, 705-713, 1978.
59. Emmett-Oglesby, M. W., Shippenberg, T. S., and Herz, A., Tolerance and cross-tolerance to the discriminative stimulus properties of fentanyl and morphine, *J. Pharmacol. Exp. Ther.*, 245, 17-23, 1988.
60. Emmett-Oglesby, M. W. and Lane, J. D., Tolerance to the reinforcing effects of cocaine, *Behav. Pharmacol.*, 3, 193-200, 1992.
61. Emmett-Oglesby, M. W., Peltier, R. L., Depoortere, R. Y., Pickering, C. L., Hooper, M. L., Gong, Y. H., and Lane, J. D., Tolerance to self-administration of cocaine in rats: time course and dose-response determination using a multi-dose method, *Drug Alcohol Depend.*, 32, 247-256, 1993.
62. Li, D.-H., Depoortere, R.Y., and Emmett-Oglesby, M. W., Tolerance to the reinforcing effects of cocaine in a progressive-ratio paradigm, *Psychopharmacology*, 116, 326-332, 1994.
63. Misra, A. L., *Cocaine: Chemical, Biological, Clinical, Social, and Treatment Aspects*, Mule, S. J., Ed., CRC Press, Cleveland, 1976, 73-90.
64. Katz, J. L., Griffiths, J. W., Sharpe, L. G., De Souza, E. B., Witkin, J. M., Cocaine tolerance and cross-tolerance, *J. Pharmacol. Exp. Ther.*, 264, 183-192, 1993.
65. Katz, J. L., Drugs as reinforcers: pharmacological and behavioural factors, in *The Neuropharmacological Basis of Reward*, Liebman, J. M. and Cooper, S. J., Eds., Oxford University Press, Oxford, UK, 1989, 164-213.
66. Wise, R. A., Yokel, R. A., Hansson, P., and Gerber, G. J., Concurrent intracranial self-stimulation and amphetamine self-administration in rats, *Pharmacol. Biochem. Behav.* 7, 459-461, 1977.
67. Yanagita, T., An experimental framework for evaluation of dependence liability of various types of drugs in monkeys, *Bulletin on Narcotics*, 25, 57-64, 1973.
68. Wolgin, D. L. and Kinney, G. G., The role of instrumental learning in behavioral tolerance to drugs, in *Psychoactive Drugs: Tolerance and Sensitization*, Goudie, A. J. and Emmett-Oglesby, M. W., Eds., Humana Press, Totowa, NJ, 1989, 17-114.
69. Peltier, R. L. and Emmett-Oglesby, M. W., *d*-Amphetamine produces cross-tolerance to cocaine in a self-administration paradigm, *NIDA Research Monograph*, Proceedings of the Annual Meeting of the College on Problems of Drug Dependence, 1994.
70. Markou, A. and Koob, G. F., Postcocaine anhedonia: an animal model of cocaine withdrawal, *Neuropsychopharmacology*, 4, 17-26, 1991.
71. Schenk, S. and Davidson, E. S., Stimulant preexposure sensitizes rats and humans to the rewarding effects of cocaine, *NIDA Research Monograph*, Proceedings of the Annual Meeting of the College on Problems of Drug Dependence, 1994.

Chapter 4

Neuroendocrine Effects of Cocaine

Cynthia M. Kuhn and Patrick J. Little

CONTENTS

I. MECHANISM OF PSYCHOACTIVE DRUG ACTION ON HORMONE SECRETION

Many drugs of abuse influence endocrine function. Chronic alcohol or marijuana use can be associated with impaired reproductive function, while chronic opiate exposure disrupts reproduction, thyroid function, and glucose homeostasis.[1] There are many possible mechanisms by which such actions can occur. The most common of these is by modulation of the hypothalamo-pituitary-target organ axis which regulates many of the body's endocrine functions (Figure 1). Reproduction, adrenocortical responses to stress, thyroid function, prolactin, and growth hormone secretion all are controlled by the hypothalamo-pituitary system. Multiple neural pathways impinge upon the hypothalamic peptidergic neurons which regulate the release of hormones from the anterior and posterior pituitary gland. Modulation of these neural mechanisms represents a common mechanism by which psychoactive drugs can influence hormone secretion. Actions directly at the pituitary or target organ represent a less common site of action, although such effects have been described. The second major site of psychoactive drug modulation of endocrine function involves the autonomic nervous system which controls secretion of many peripheral hormones including pancreatic hormones, such as insulin and glucagon. Drugs which influence noradrenergic or cholinergic transmission either directly, or indirectly by influencing parasympathetic or sympathetic motor nuclei in the brain, can mimic or prevent their effects.

It is thought that cocaine exerts behavioral and physiologic effects by binding to monoaminergic transporter molecules located on the presynaptic terminal, and by subsequent inhibition of monoamine transport into the terminal following release. Through this action, cocaine increases synaptic levels of norepinephrine, dopamine and serotonin (see Chapter 12). Cocaine is not very selective in this action, as it shows nearly identical affinity for serotonergic, dopaminergic, and noradrenergic transporters.[2-4] Therefore, cocaine would be expected to affect neuroendocrine processes regulated by all of these neurotransmitters, and actions on a specific hypothalamic neuropeptide would reflect the

1-8493-8311-0/95/$0.00+$.50

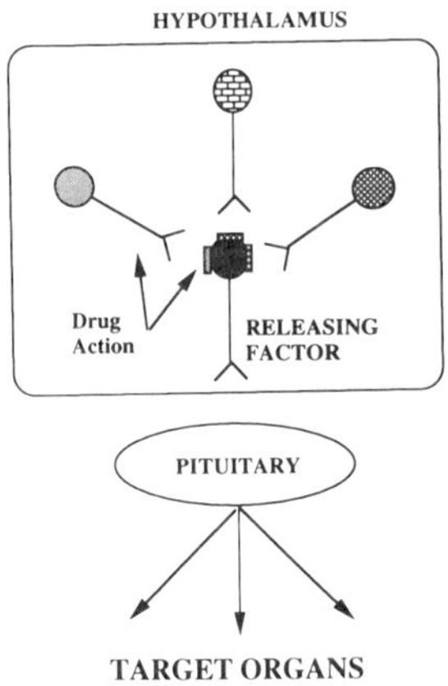

Figure 1. Diagram of hypothalamo-pituitary-target organ axis, showing the common sites of psychoactive drug action on releasing factor secretion.

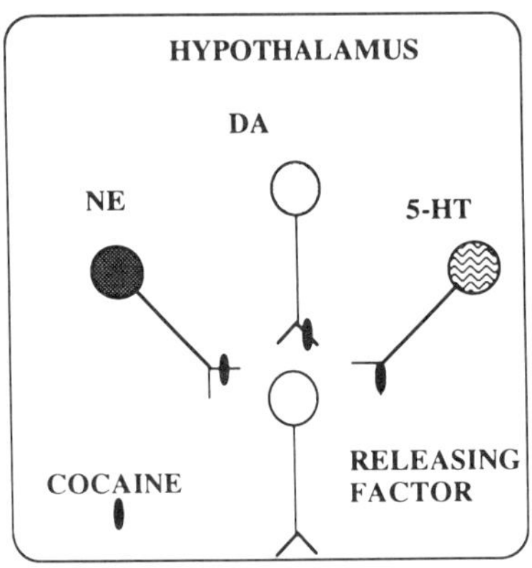

Figure 2. Diagram of hypothalamus showing the postulated site of cocaine action on releasing factor secretion.

integration of signals deriving from the elevation of serotonin, norepinephrine, and dopamine (Figure 2).

Blockade of norepinephrine uptake at peripheral sympathetic neurons provides another mechanism by which cocaine might influence endocrine function. For example, increased synaptic levels of norepinephrine in the pancreas theoretically could enhance sympathetic inhibition of insulin release and elevate blood glucose. Cocaine also acts as a local anesthetic through blockade of sodium channels.[5] However, there is less evidence to support a role for this action of cocaine in the production of neuroendocrine responses. The role of the local anesthetic properties of cocaine in behavioral responses to nontoxic blood levels of cocaine is not clear, and procaine, a drug with similar local anesthetic properties, lacks the same profile of neuroendocrine activity as cocaine.[6,7]

The pharmacokinetics of cocaine could play a significant role in the endocrine effects of cocaine. Cocaine has a fairly short half-life in the circulation, and its behavioral effects can be even shorter than the apparent half-life.[8] The repeated presentation of brief "pulses" of cocaine, as it is often self-administered, provides a pattern of stimulation similar to the normal pulsatile presentation of stimuli to the pituitary and target organs. Such episodic presentation of cocaine across days or weeks elicits many adaptive responses which contribute to the endocrine effects of cocaine.

II. NEUROENDOCRINE ACTIONS OF COCAINE

The endocrine effects of cocaine administration are described poorly in comparison to other drugs of abuse. There are few reports of the effects of cocaine on endocrine function in humans, and those that exist have been conducted in subjects who have self-administered unknown quantities of drug for varying periods of time. Furthermore, these reports reflect many different strategies, including the evaluation of acute responses to a single dose of cocaine, responses to cocaine after chronic treatment, and endocrine function during withdrawal from chronic administration. However, a growing body of more systematic research in animals and humans has provided data consistent with the model of action proposed above, and provided suggestions regarding the potential consequences of drug exposure in humans. Effects on all the major pituitary hormones except growth hormone have been described, and preliminary data indicate an involvement of peripheral sympathomimetic effects of cocaine in the regulation of peripheral endocrine functions such as glucose homeostasis.

III. HYPOTHALAMO-PITUITARY-ADRENAL AXIS

The secretion of cortisol by the adrenal gland is controlled by the release of adrenocorticotropic hormone (ACTH) from the pituitary in response to corticotropin releasing hormone (CRF) secretion from the hypothalamus. Norepinephrine, serotonin, and dopamine all provide stimulatory input to regulate the secretion of CRF into the hypophyseal-portal circulation.[9-13] Therefore, it is not surprising that cocaine produces a marked increase of ACTH secretion in both rats and humans.[14-17] This stimulation of ACTH can be blocked by pretreatment with an antibody to CRF or a CRF receptor antagonist, suggesting that the direct stimulation of CRF release is the primary mechanism.[18,19] For reasons that are unclear, this increased ACTH secretion is extremely brief, significantly shorter than either the half-life or behavioral activation of cocaine.[6,16]

Specific neurotransmitter antagonists as well as uptake blockers for specific neurotransmitter systems have been used to explore the relative contribution of noradrenergic, dopaminergic, and serotonergic neurons to the elevation of ACTH induced by cocaine. Both dopaminergic (DA_2) and serotonergic ($5HT_2$) antagonists diminish the effect of cocaine, while the uptake inhibitors GBR 12909 and fluoxetine, which are selective for dopaminergic and serotonergic transporters respectively, stimulate ACTH secretion.[6,10,12,20,21] These pharmacologic studies support a role for both neurotransmitters in the actions of cocaine. Although a significant stimulatory role for norepinephrine in ACTH secretion has been demonstrated,[13] the contribution of this neurotransmitter to the actions of cocaine is less well established. Neurotoxic destruction of hypothalamic norepinephrine terminals produces only small decrements in cocaine-induced ACTH release, and there are conflicting data about the ability of alpha-1 and beta antagonists to block cocaine-induced hypothalamo-pituitary-adrenal (HPA) activation.[6,20] However, a recent study utilizing central administration of antagonists to avoid peripheral cardiovascular side effects suggests that noradrenergic neurons also contribute to cocaine-induced HPA activation.[21]

In fact, one puzzling aspect of these studies is that antagonists which selectively affect different neurotransmitter receptors each completely prevent cocaine-induced stimulation of ACTH secretion. If the actions of cocaine simplistically represented the summation of actions at several sites, this would be accomplished only with combined administration of selective noradrenergic, dopaminergic, and serotonergic receptor antagonists. However, this is not the case, which suggests the possibility that other actions of cocaine might contribute to the stimulation of CRF.

There are mixed reports of HPA reactivity to cocaine and other agents following chronic administration. Early studies[22,23] observed no significant alteration of the HPA response to cocaine after a daily, intermittent treatment regimen which produced an augmented behavioral response to cocaine. Testing the ability of specific monoaminergic pathways to modulate HPA reactivity has revealed effects that were not obvious after challenge using a nonselective uptake inhibitor such as cocaine. Chronic cocaine administration causes a specific suppression of ACTH secretion by the serotonin releasing drug parachloroamphetamine, enhanced responses to the $5HT_2$ agonist DOI, but no change in response to the receptor agonists mCPP, RU24969 and MK212.[24-26] While not comprehensive, these results suggest that the balance of different noradrenergic, dopaminergic, and serotonergic influences modulating ACTH secretion might change during chronic cocaine administration.

IV. PROLACTIN SECRETION

Prolactin secretion is inhibited tonically by dopamine released into the hypophyseal portal circulation by tuberoinfundibular dopamine (TIDA) neurons.[27] These neurons are thought to possess a very low density of dopamine transporters, and therefore are resistant to the actions of cocaine.[28,29] Early negative studies of cocaine-induced changes in prolactin secretion were consistent with these neurochemical findings.[30] However, neuroendocrine effects of chronic cocaine exposure in humans include the development of hyperprolactinemia during withdrawal,[31-35] although normal prolactin levels have been reported following cocaine withdrawal.[36,37] Subsequent studies have shown that cocaine can inhibit prolactin secretion in rats and humans significantly.[6,17,34,38,39] The discrepancies among these different studies could reflect the timing and duration of cocaine administration, the duration of abstinence, and the basal level of prolactin secretion. Thus, if prolactin secretion already is inhibited maximally, cocaine-induced elevation of dopamine in portal blood would produce no further decrease of prolactin secretion. However, if dopamine levels are suppressed even slightly, an enhanced dopaminergic inhibition of prolactin secretion by cocaine would be detected. The consistent reports of elevated prolactin following chronic cocaine use suggest that an acute inhibition of prolactin secretion is sufficient to activate adaptive mechanisms during chronic exposure. Both decreased turnover in TIDA neurons and decreased pituitary D_2 receptor sensitivity have been invoked to explain these findings, although two reports of normal prolactin responses to apomorphine challenge the latter conclusion.[40,41] It is possible that the negative findings reflect brief chronic treatment regimens which did not induce this adaptation, or long withdrawal times which permitted recovery.

While most studies have focused on dopaminergic regulation of prolactin secretion, cocaine also influences serotonergic regulation. Chronic cocaine administration blunts prolactin responses to the serotonin agonist MK212, but enhances responses to the $5HT_{1c}$/$5HT_2$ agonist DOI.[25,26] As nonserotonergic mechanisms might contribute to the MK212 responses, these authors conclude that prolactin responses to $5HT_2$ agonists are enhanced by chronic cocaine treatment.

V. COCAINE EFFECTS ON REPRODUCTIVE FUNCTION

Disruption of reproduction represents the hallmark neuroendocrine consequence of exposure to drugs of abuse. Cocaine is no exception in this regard, as some of the earliest reports of clinical problems in cocaine users included altered reproductive function, such as disruption or cessation of menstrual cycles and decreased libido in women, and impotence and gynecomastia in men.[42,43] Similar findings have emerged from animal studies in which rats exhibit interruption of normal estrous cyclicity,[44-46] and rabbits display suppression of ovulation during chronic cocaine administration.[47,48]

Although impaired reproduction is observed consistently in humans, primates, and rodents after chronic cocaine treatment, opposite effects of acute cocaine have been reported. Acute administration of cocaine reportedly increases luteinizing hormone (LH) secretion in male or intact cycling female primates and in humans.[17,38,39] However, decreased LH secretion has been reported in ovariectomized monkeys, and after high doses of cocaine in ovariectomized rats.[49,50] Similarly, biphasic effects of cocaine on testosterone have been reported in male rats, with early stimulation and later inhibition of secretion.[51] Many of the apparent inconsistencies are reconciled when treatment paradigms are considered. High doses of cocaine inhibit LH secretion and eventually inhibit production of gonadal steroids in both males and females,[49-51] while low doses stimulate LH secretion.

The stimulatory effects of cocaine on LH secretion are best explained by altered gonadotropin releasing hormone (GnRH) secretion. Both norepinephrine and dopamine are stimulatory to LH secretion in most mammals,[52] which suggests that a single cocaine dose stimulates LH secretion. The finding that cocaine administration does not affect LH responses to GnRH in either rabbits or monkeys suggests a hypothalamic site for inhibitory effects as well.[47,53] Finally, cocaine-induced suppression of noradrenergic stimulation of GnRH release has been associated with impaired estrous cyclicity in females.[54] However, increased testosterone production has been observed in the absence of changes in LH,[51] and the possibility of direct cocaine effects on the gonads has not been investigated thoroughly. It is possible that elevation of synaptic norepinephrine at peripheral sites could modulate gonadal steroid production directly, as sympathetic innervation of the uterus, ovaries, and testes has been described. Finally, the role of serotonergic neurons in cocaine-induced reproductive dysfunction has not been studied.

Dose-dependent effects of cocaine on reproductive function also might be explained by the recruitment of additional mechanisms of cocaine action. Although cocaine inhibits monoamine uptake potently, blockade of sodium channels can be achieved at plasma levels observed in humans during self-administration.[7,8,55] Recently, it has been shown that cocaine inhibits GnRH release triggered by phorbol esters, an effect that cannot be explained easily by inhibition of monoamine uptake.[56] Furthermore, this effect is mimicked by tetrodotoxin, another sodium channel blocker. This finding suggests that inhibition of LH release at high doses of cocaine, or following prolonged, high-dose chronic regimens might be the result of sodium channel blockade by cocaine.

Any pharmacologic treatment which influences food intake theoretically could suppress reproductive function secondary to malnutrition. However, most of the published studies report endocrine effects after treatment which only influence weight gain at the beginning of chronic treatment. The studies of King et al.[45] are convincing in this regard, as they demonstrate that animals whose food intake was matched to that of cocaine-treated animals maintained normal estrous cyclicity.

In summary, the effects of cocaine on reproductive hormone secretion are complex. While a single dose can trigger a surge of LH release, diminished LH release and impaired reproductive function are associated more commonly with chronic treatment. The role of specific neural mechanisms in these actions of cocaine on LH have not been established yet, but both inhibition of monoamine uptake and blockade of sodium channels might be important.

VI. THYROID FUNCTION

Although noradrenergic and dopaminergic stimulation of thyroid stimulating hormone (TSH) secretion are well-described in both humans and animals,[57] there has been little study of cocaine-induced changes in thyroid function. Two early reports of blunted TSH responses to exogenous thyrotropin releasing hormone (TRH)[58,59] suggest that regular cocaine use produced repeated stimulation of TSH secretion with subsequent downregulation of pituitary TRH receptors. However, no significant effects of cocaine were reported in a later study in which a more thorough evaluation of thyroid function in cocaine users was conducted.[60] It is possible that the transient alteration of TSH produced by cocaine is compensated easily by thyroid hormone feedback, resulting in stable thyroid status despite transient changes of hormone secretion. However, this speculation remains to be investigated.

VII. POSTERIOR PITUITARY HORMONE SECRETION

Although monoaminergic regulation of both vasopressin and oxytocin secretion have been described, there are few direct reports of cocaine effects on these posterior pituitary hormones. Oxytocin secretion is of particular interest, as it could contribute to premature labor that is a common result of cocaine use during pregnancy.[61] One study of acute cocaine effects on maternal oxytocin concentration in the pregnant ewe reported that cocaine did not affect oxytocin levels after a single dose.[62] However, diminished oxytocin plasma levels have been observed after chronic cocaine treatment in adult rats.[63] Furthermore, a daily cocaine regimen which altered serotonergic responses to ACTH and prolactin also suppressed $5HT_2$-mediated oxytocin secretion.[25,26] The former results suggest that more careful study of posterior pituitary function during cocaine use is required.

VIII. COCAINE EFFECTS ON FETAL NEUROENDOCRINE ACTIVITY

Numerous reports have demonstrated that cocaine adversely affects fetal development.[61,64] However, there are multiple mechanisms by which cocaine potentially could influence fetal neuroendocrine function, few of which have been studied directly. These include effects on maternal/placental hormone secretion, indirect effects through cocaine-induced impairment of uterine blood flow, or direct effects of cocaine on neuroendocrine tissues in the fetus. One report observed normal maternal progesterone level after chronic cocaine treatment, which suggests that maternal/placental hormone secretion is not impaired by cocaine.[47] Furthermore, as described above, maternal oxytocin levels are not affected by a dose of cocaine large enough to increase maternal blood pressure. However, there are no reports of hormone levels in pregnant humans or in animal models, nor have any studies examined the effect of cocaine on the progression of endocrine changes that occur at parturition.

There are very few direct studies of fetal neuroendocrine development during prenatal cocaine exposure. The data that have been published employ an animal model for studying neuroendocrine ontogeny: the pregnant ewe. The ewe and its fetus are large enough that both the maternal and the fetal circulation can be sampled during drug infusion into the ewe. Furthermore, the neuroendocrine ontogeny of the sheep is extremely well described.[65] Maternal cocaine administration produces a small but significant increase in fetal ACTH secretion, but direct fetal administration of cocaine which bypasses indirect effects mediated by changes in maternal physiology does not affect fetal ACTH levels.[66,67] In addition, maternal cocaine administration increased fetal glucose and lactate concentrations,[68] which could reflect fetal hypoxia following uterine vasoconstriction or perhaps a direct effect on glucose production or utilization in the fetus. These results suggest that the disruption of maternal physiology and the production of fetal hypoxia contribute significantly to fetal endocrine dysfunction.

There have been few studies of direct cocaine action on neuroendocrine function in the developing animal. The most likely target of cocaine action in the fetus is the same as in adults: neuroendocrine secretion that is under monoaminergic control. Preliminary results from our laboratory suggest that significant neuroendocrine effects of cocaine occur in developing animals. Administration of cocaine to neonatal rat pups on postnatal day 10

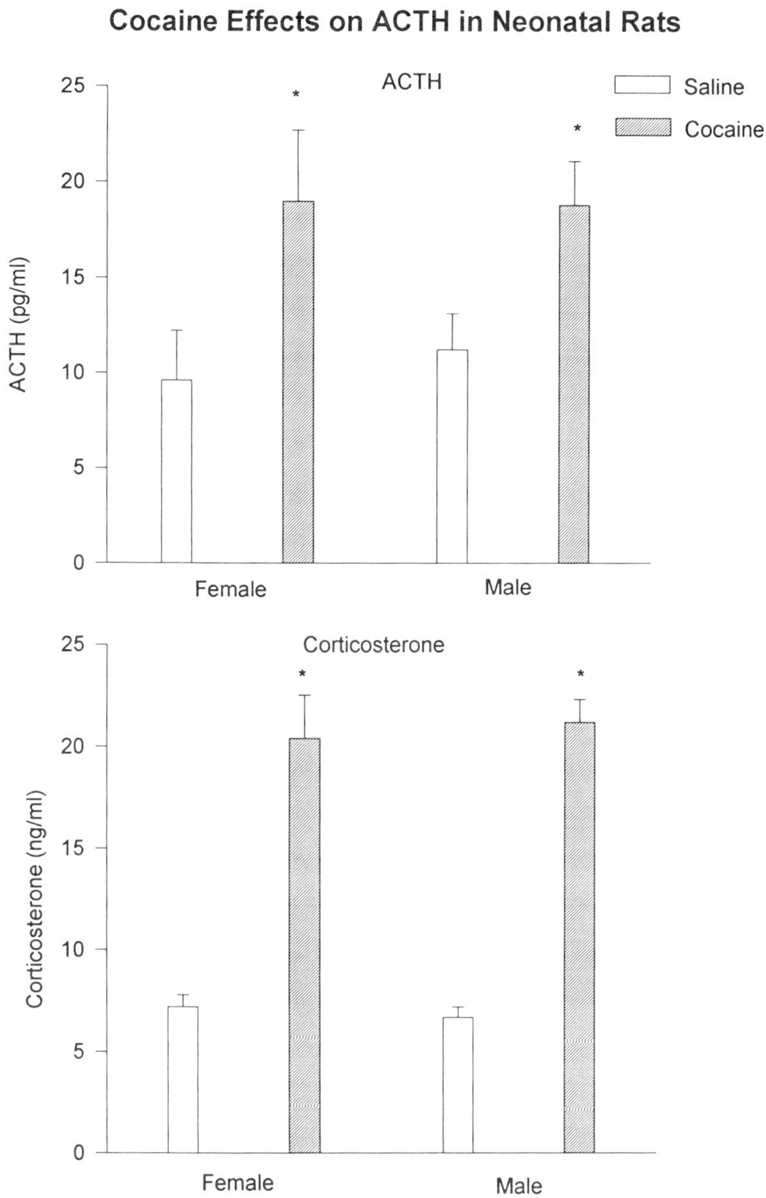

Figure 3. Plasma ACTH levels in neonatal (day 10) male and female rats treated with saline or cocaine, 10 mg/kg i.p. 15 min before kill. N = 8/group. * indicates p <0.05 or better relative to saline-injected control.

produces a substantial elevation of ACTH and corticosterone secretion (Figure 3), similar to that observed in adults. This suggests that cocaine could influence hormone secretion as soon as monoaminergic neurons innervate the hypothalamus.

Finally, the effects of cocaine on the development of arousal and responsivity to environmental stimuli might influence neuroendocrine reactivity substantially. It has been reported that salivary cortisol levels in cocaine-exposed infants are normal under baseline conditions, but react significantly less to stressful stimuli than those of normal infants.[69] This demonstrates that cocaine can influence neuroendocrine function indirectly, but persistently, long after the drug has been eliminated.

In summary, several mechanisms for cocaine-induced neuroendocrine function in developing animals have been demonstrated to exist, including influences on maternal physiology, direct interaction with monoaminergic neurons in the developing animal, and effects mediated by altered arousal or stress/responsivity. As endocrine function is critical for normal fetal development, these effects represent an important potential mediator of the developmental effects of cocaine.

IX. ROLE OF NEUROENDOCRINE RESPONSES IN THE BEHAVIORAL ACTIONS OF COCAINE

Even the most severe endocrine consequences of psychoactive drug administration often are viewed as minor side effects in comparison to the life-threatening cardiovascular complications and seizures that accompany cocaine self-administration in humans. Endocrine responses can be useful as indices of specific neuronal adaptations to chronic cocaine administration. However, endocrine responses rarely have been viewed as central to the actions of drug. The behavioral actions of cocaine represent an important exception to this assumption. A growing body of evidence suggests that endocrine responses to cocaine can contribute to one of the most important behavioral consequences of chronic cocaine administration.

In most animal models, repeated administration of cocaine produces a progressive augmentation or "sensitization" of locomotor activity[70,71] (see Chapter 6). This phenomenon is thought to contribute to cocaine dependence and craving.[70-72] Several endocrine responses to cocaine have been reported to influence this process. First, it has been suggested that vasopressin is important for the development of sensitization. This assertion is based on a finding that the Brattleboro rat, which lacks biologically active vasopressin, does not exhibit sensitization in response to chronic cocaine treatment. Furthermore, it has been suggested that vasopressin antagonists reduce the acquisition of cocaine (and heroin) self-administration.[74,75] However, there are few reports regarding cocaine effects on vasopressin secretion. Therefore, it is not clear whether cocaine-induced vasopressin secretion is crucial for the development of sensitization, or whether the presence of normal physiologic secretion of vasopressin is permissive in this process. Similarly, conflicting reports of oxytocin effects on behavioral responses to cocaine exist, including blockade of locomotor activation, enhancement of cocaine-induced stereotypy, and blockade of "tolerance" to cocaine-induced stereotypy following daily cocaine administration.[63,76] However, it is not clear whether cocaine-induced vasopressin and oxytocin secretion from the posterior pituitary, or their release from other neuronal populations, might mediate the observed behavioral effects.

The role of activation of the HPA axis in cocaine sensitization has been the focus of considerable research, which has produced more experimental support. A growing body of evidence suggests that repeated stimulant-induced HPA activation might contribute to the development of sensitized behavioral responses to locomotor stimulants and stress. The HPA axis has been implicated through the actions of the hypothalamic releasing peptide CRF and glucocorticoid hormone secreted by the adrenal cortex. Many of the studies which address this issue have utilized amphetamine rather than cocaine, and it should be noted that the prolonged effects of amphetamine might produce more robust effects. Nevertheless, these studies are quite relevant to the actions of cocaine.

The results of pharmacologic studies have implicated central CRF receptors in the mechanism underlying stimulant sensitization. For example, administration of CRF into the brain enhances the locomotor response to a single stimulant (amphetamine) dose.[77] Furthermore, repeated CRF administration sensitizes animals to a subsequent dose of amphetamine.[78] Finally, pretreatment of animals with a CRF antagonist before each daily dose of CRF retards the development of stress-induced sensitization to amphetamine administration.[79] These effects may occur independently of the ability of CRF to stimulate ACTH release, as peripheral administration of CRF does not produce the same effects.

Additional evidence suggests that the endocrine response to CRF neuron activation also contributes to stimulant-induced sensitization and perhaps self-administration. Corticosterone administration itself also sensitizes the locomotor response to amphetamine,[80] and blockade of glucocorticoid receptor activation prevents this response.[81] Furthermore, suppression of glucocorticoid secretion impairs amphetamine sensitization, and prevents the development of a sensitized behavioral response to alpha-1 noradrenergic drugs.[82,83] In fact, the individual "vulnerability" to amphetamine self-administration can be predicted by the magnitude of the HPA response to a novel stressor.[84]

The ability of HPA activation to enhance stimulant sensitization and of stress to predict vulnerability to stimulant self-administration are consistent with numerous studies showing that stress and stimulants produce "cross-sensitization"[85] (see Chapter 6). These findings have led to the hypothesis that reactivity of central CRF neurons to external events represents the critical neurobiologic factor in stimulant sensitization.[85] This hypothesis suggests that responsivity to external stimuli may dictate sensitivity to locomotor stimulants, perhaps through activation of the HPA axis. Therefore, this neuroendocrine response to cocaine could represent a critical factor in the development of stimulant self-administration and addiction.

Cocaine-induced suppression of reproductive function also might play a significant role in determining the behavioral responses to cocaine. There are prominent gender differences in behavioral activation following either amphetamine or cocaine treatment in rodents. Females show enhanced behavioral responses to cocaine and amphetamine relative to males and more profound sensitization during repeated cocaine treatment than males.[86-91] Evaluation of stimulant responses in gonadectomized animals, hormone replacement studies, and investigations of estrous cycle differences support an important role for estrogen in facilitating responses to both amphetamine and cocaine.[92-96] This facilitation could reflect direct actions of estrogen to increase both dopamine release and dopamine receptor sensitivity.[97-105] Regardless of mechanism, these findings suggest that there may be substantial gender differences in human responses to cocaine, and that cocaine-induced reproductive dysfunction might suppress the effects of cocaine slightly in females but not in males.

X. SUMMARY OF COCAINE ACTIONS ON NEUROENDOCRINE FUNCTION

In summary, the studies reviewed above demonstrate that cocaine alters the secretion of pituitary and peripheral hormones through elevation of synaptic concentrations of dopamine, norepinephrine, and serotonin. A single dose of cocaine initiates changes in most anterior pituitary hormones which reflect the elevation of synaptic monoamine concentrations. With chronic exposure to cocaine, adaptations in these neuroendocrine responses occur, depending upon the monoamine system involved, the postsynaptic receptor mediating the response, and the endocrine axis under investigation. Enhanced, diminished, and unchanged responses all have been observed. However, no clear sensitization to cocaine-induced endocrine responses develops during chronic cocaine exposure which parallels the sensitized behavioral responses.

Pharmacokinetic considerations also influence the neuroendocrine response to cocaine. The dose of cocaine determines both the magnitude, and in the case of LH secretion, even the direction of change in hormone secretion. While low doses of cocaine can produce effects which probably are regulated by changes in extracellular monoamine concentration, the extremely high doses experienced by human cocaine users produce sodium channel blockade and other adaptations which can reverse the initial endocrine response. Furthermore, the fairly short half-life of cocaine leads to large but transient alterations of hormone secretion.[6]

Finally, some of the endocrine responses to cocaine may be secondary to nonendocrine effects of the drug. For example, clinically depressed cocaine users might exhibit perturbed endocrine responses as a result of their depression rather than cocaine use. Similarly, stress-mediated endocrine responses probably contribute to the pattern of response both during acute cocaine administration and withdrawal.

XI. CLINICAL SIGNIFICANCE OF ENDOCRINE RESPONSE TO COCAINE

There are several important clinical ramifications to the endocrine responses described above. First, some effects such as hyperprolactinemia and impaired fertility in males and females represent significant clinical side effects. In addition, a recent report suggests that the HPA response might well provide an index of successful blockade of cocaine-induced euphoria, and so provide an adjunct while monitoring the early phases of treatment. The partial opiate agonist buprenorphine blocks both the cocaine-induced high, and the ACTH response to cocaine in human subjects.[106] This suggests that monitoring the ACTH response to cocaine might provide an index of individual responsivity to cocaine and to subsequent treatment. This speculation certainly is consistent with the postulated role of CRF neurons in cocaine-induced behavioral action reviewed above. Finally, a growing body of information suggests that endocrine state influences behavioral responsivity to cocaine. If so, the administration of cocaine potentially could affect endocrine state, and so alter behavioral sensitivity to subsequent doses of cocaine. Although complete examination of these possibilities is beyond the scope of the current review, the literature reported above relating to cocaine effects on adrenocortical function, reproductive state, and posterior pituitary hormones suggests that these effects could be involved in such adaptations. Such a phenomenon could have important clinical import in determining individual vulnerability to cocaine effects. A recent report that the progesterone antagonist mifepristone deceases the cardiotoxicity of cocaine[107] supports this possibility and suggests an important future area of research.

REFERENCES

1. Smith, C. G. and Asch, R. H., Drug abuse and reproduction, *Fertil. Steril.*, 48, 355-373, 1987.
2. Ritz, M. C., Lamb, R. J., Goldberg, S. R., and Kuhar, M., Cocaine receptors on dopamine transporters are related to self administration of cocaine, *Science*, 237, 1219-1222, 1987.
3. Ross, S. B. and Renyi, A. L., Uptake of some tritiated sympathomimetic amines by mouse brain cortex in vitro, *Acta Pharmacol. Toxicol.*, 24, 297-309, 1966.
4. Ritz, M. C., Cone, E. J., and Kuhar, Cocaine inhibition of ligand binding at dopamine, norepinephrine, and serotonin transporters: a structure activity study, *Life Sci.*, 46, 635-645, 1990.
5. Reith, M. E. A., Kim, S. S., and Lajtha, A., Structural requirements for cocaine congeners to interact with [^{3}H]batrachotoxinin A 20-α-benzoate binding sites on sodium channels in mouse brain synaptosomes, *J. Biol. Chem.*, 261, 7300-7305, 1986.
6. Borowsky, B. and Kuhn, C. M., Monoamine mediation of cocaine-induced HPA Activation, *J. Pharmacol. Exp. Ther.*, 256, 201-210, 1991.
7. Crumb, W. J. and Clarkson, C. W., Characterization of cocaine-induced block of cardiac sodium channels, *Biophys. J.*, 57, 589-599, 1990.
8. Jeffcoat, A. R., Perez-Reyes, M., Hill, J. M., Sadler, B. M., and Cook, C., Cocaine disposition in humans after intravenous injection, nasal insufflation (snorting) or smoking, *Drug Metabol. Dispos.*, 17, 153-159, 1989.
9. Grossman, A., Costa, A., Navarra, P., and Tsagarakis, S., The regulation of hypothalamic corticotropin-releasing factor release: in vitro studies, in *Corticotropin Releasing Factor*, CIBA Foundation Symposium, John Wiley, New York, 172, 129-143, 1993.

10. Fuller, R. W., Serotonergic stimulation of pituitary-adreno-cortical function in rats, *Neuroendocrinology,* 32, 118-127, 1981.
11. Van de Kar, L. D., Neuroendocrine pharmacology of serotonergic (5-HT) neurons, *Annu. Rev. Pharmacol. Toxicol.,* 31, 289-320, 1991.
12. Borowsky, B. and Kuhn, C. M., GBR 12909 stimulates hypothalamo-pituitary-adrenal activity by inhibition of uptake at hypothalamic dopamine neurons, *Brain Res.,* 613, 251-258, 1993.
13. Plotsky, P. M., Cunningham, E. T., and Widmaier, E. P., Catecholaminergic modulation of corticotropin-releasing factor and adrenocorticotropin secretion, *Endocr. Rev.,* 10, 437-458, 1989.
14. Moldow, R. L. and Fischman, A. J., Cocaine-induced secretion of ACTH, beta endorphin and corticosterone, *Peptides,* 8, 819-822, 1987.
15. Calogero, A. E., Gallucci, W. T., Kling, M. A., Chrousos, G. P., and Gold, P. W., Cocaine stimulates rat hypothalamic corticotropin releasing hormone secretion in vitro, *Brain Res.,* 505, 7-11, 1989.
16. Torres, G. and Rivier, C., Cocaine-induced ACTH secretion: dependence of plasma levels of the drug and mode of exposure, *Brain Res. Bull.,* 29, 51-56, 1992.
17. Mendelson, J. H., Teoh, S. K., Mello, N. K., Ellingboe, J., and Rhoades, E., Acute effects of cocaine on plasma adrenocorticotropic hormone, luteinizing hormone and prolactin levels in cocaine-dependent men, *J. Pharmacol. Exp. Ther.* 263, 505-509, 1992.
18. Rivier, C. and Vale, W., Cocaine stimulates adrenocorticotropin (ACTH) secretion through a corticotropin-releasing factor (CRF) mediated mechanism, *Brain Res.* 422, 403-406, 1987.
19. Sarnyai, Z., Biro, E., Penke, B., and Telegdy, G., The cocaine-induced elevation of plasma corticosterone is mediated by endogenous corticotropin-releasing factor (CRF) in rats, *Brain Res.,* 589, 154-156, 1992.
20. Levy, A. D., Li, Q., Kerr, J. E., Rittenhouse, P. A., Milnoas, G., Cabrera, T. M., Batttaglia, G., Alvarez Sanz, M. C., and Van de Kar, L. D., Cocaine-induced elevation of plasma adrenocorticotropin hormone and corticosterone is mediated by serotonergic neurons, *J. Pharmacol. Exp. Ther.,* 259, 495-500, 1991.
21. Sarnyai, Z., Biro, E., and Telegdy, G., Cocaine-induced elevation of plasma corticosterone is mediated by different neurotransmitter systems in rats, *Pharmacol. Biochem. Behav.,* 45, 209-2145, 1993.
22. Borowsky, B. and Kuhn, C. M., Chronic cocaine administration sensitizes behavioral but not neuroendocrine responses, *Brain Res.,* 543, 301-306, 1991.
23. Torres, G. and Rivier C., Differential effects of intermittent or continuous exposure to cocaine on the hypothalamic-pituitary-adrenal axis and c-fos expression, *Brain Res.,* 571, 204-211, 1992.
24. Van de Kar, L. D., Bonadonna, A. M., Rittenhouse, P. A., Kerr, J. E., Levym, A. D., Iyer, L., Herbert, G. B., Alvarez Sanz, M. C., Lent, S. J., and Carnes, M. Prior chronic exposure to cocaine inhibits the serotonergic stimulation of ACTH and secretion of corticosterone, *Neuropharmacology,* 31, 169-175, 1992.
25. Van de Kar, L. D., Rittenhouse, P. A., O'Connor, P., Palionis, Brownfield, M. S., Lent, S. J., Carnes, M., and Bethea, C. L., Effect of cocaine injections on the neuroendocrine response to the serotonin agonist MK-212, *Biol. Psychiatr.,* 32, 258-269, 1992.
26. Levy, A. D., Li, Q., Alvarez-Sanz, M. C., Ritenhouse, P. A., Brownfield, M. S., and Van de Kar, L. D., Repeated cocaine modifies the neuroendocrine responses to the 5HT1a/5HT2 receptor agonist DOI, *Eur. J. Pharmacol.,* 221, 121-127, 1992.
27. Ben Jonathan, N., Dopamine: a prolactin inhibiting hormone, *Endocr. Rev.,* 6, 564-589, 1985.
28. Demarest, K. T. and Moore, K. E., Lack of a high affinity transport system for dopamine in the median eminence and posterior pituitary, *Brain Res.,* 171, 545-551, 1979.
29. Janowsky, A., Berger, P., Vocci, F., Labarca, R., Skolnick, P., and Paul, S. M., Characterization of sodium-dependent [^{3}H]GBR-12935 binding in brain: a radioligand for selective labelling of the dopamine transport complex, *J. Neurochem.,* 46, 1272-1276, 1986.
30. Ravitz, A. J. and Moore, K. E., Effects of amphetamine, methylphenidate and cocaine on serum prolactin concentrations in the male rat, *Life Sci.,* 21, 267-272, 1977.
31. Cocores, J. A., Dackis, C. A., and Gold, M. S., Sexual dysfunction secondary to cocaine abuse in two patients, *J. Clin. Psychiatr.,* 47, 384-385, 1986.
32. Dackis, C. A. and Gold, M. S., New concepts in cocaine addiction: the dopamine depletion hypothesis, *Neurosci. Biobehav. Rev.,* 9, 469-477, 1985.
33. Kranziler, H. R. and Wallington, D. J., Serum prolactin level, craving and early discharge from treatment in cocaine-dependent patients, *Am. J. Drug Alcohol Abuse,* 18, 187-195, 1992.

34. Pilotte, N. S., Sharpe, L. G., and Dax, E. M., Multiple but not acute infusions of cocaine alter the release of prolactin in male rats, *Brain Res.*, 512, 107-112, 1990.
35. Mendelson, J. H., Teoh, S. K., Lange, U., Mello, N. K., Weiss, R., Skupny, A., and Ellingboe, J., Anterior pituitary, adrenal and gonadal hormones during cocaine withdrawal, *Am. J. Psychiatr.* 145, 1094-1098, 1988.
36. Gawin, F. H. and Kleber, H. D., Neuroendocrine findings in chronic cocaine abusers: a preliminary report, *Br. J. Psychiatr.*, 147, 469-477, 1985.
37. Swartz, C. M., Breen, K., and Leone, F., Serum prolactin levels during extended cocaine abstinence, *Am. J. Psychiatr.*, 147, 777-779, 1990.
38. Mello, N. K., Mendelson, J. H., Drieze, J., and Kelly, M., Acute effects of cocaine on prolactin and gonadotropins in female rhesus monkey during the follicular phase of the menstrual cycle, *J. Pharmacol. Exp. Ther.*, 254, 815-823, 1990.
39. Mendelson, J. H., Mello, N. K., Teoh, S. K., Ellingboe, J., and Cochin, J., Cocaine effects on pulsatile secretion of anterior pituitary, gonadal and adrenal hormones, *J. Clin. Endocrinol. Metabol.*, 69, 1256-1260, 1989.
40. Hollander, E., Nunes, E., De Caria, C. M., Quitkin, F. M., Cooper, T., Wager, S., and Klein, D. F., Dopaminergic sensitivity and cocaine abuse: response to apomorphine, *Psychiatr. Res.*, 33, 161-169, 1990.
41. Lee, M. A., Bowers, M. M, Nash, J. F., and Meltzer, H. Y., Neuroendocrine measures of dopaminergic function in chronic cocaine users, *Psychiatr. Res.*, 33, 151-159, 1990.
42. Cregler, L. L. and Mark, H., Medical complications of cocaine abuse, *N. Engl. J. Med.*, 315, 1495-1500, 1986.
43. Siegel, R. K., Cocaine and sexual dysfunction: curse of mama coca, *J. Psychoactive Drugs,* 14, 71-74, 1982.
44. King, T. S., Schenken, R. S., Kang, I. S., Javors, M. A., and Riehl, R. M., Cocaine disrupts estrous cyclicity and alters the reproductive neuroendocrine axis in the rat, *Neuroendocrinology,* 51, 15-22, 1990.
45. King, T. S., Canez, M. S., Gaskill, S., Javors, M. A., and Schenken, R. S., Chronic cocaine disruption of estrous cyclicity in the rat: dose dependent effects, *J. Pharmacol. Exp. Ther.*, 264, 29-34, 1993.
46. Roberts, D. C. S., Bennett, S. A. L., and Vickers, G., The estrous cycle affects cocaine self-administration on a progressive ratio schedule in rats, *Psychopharmacology,* 98, 408-411, 1989.
47. Atlas, S. J. and Wallach, E. E., Effects of intravenous cocaine on reproductive function in the mated rabbit, *Am. J. Obst. Gynecol.* 165, 1785-1790, 1991.
48. Kaufman, R. A., Savoy-Moore, R. T., Sacco, A. G., and Subramanian, M. G., The effect of cocaine on oocyte development and the follicular microenvironment in the rabbit, *Fertil. Steril.*, 54, 921-926, 1990.
49. Canez, M. S., Samuels, M. H., Luther, M. F., King, T. S., and Schenken, R. S., Cocaine impairs gonadotropin secretion in oophorectomized monkeys, *Am. J Obst. Gynecol.*, 167, 1785-1793, 1992.
50. Steger, R. A., Silverman, A. J. and Asch, R. H., Interactions of cocaine and delta-9 tetrahydrocannabinol with the hypothalamic-hypophysial axis of the female rats, *Fertil. Steril.*, 35, 567-572, 1981.
51. Berul, C. I. and Harclerode, J. E., Effects of cocaine hydrochloride on the male reproductive system, *Life Sci.*, 45, 91-95, 1989.
52. Kalra, S. P. and Kalra, P. S., Neural regulation of luteinizing hormone secretion in the rat, *Endocr. Rev.*, 4, 311-351, 1983.
53. Mello, N. K., Mendelson, J. H., Drieze, J., and Kelly, M., Cocaine effects on luteinizing hormone-releasing hormone-stimulated anterior pituitary hormones in female rhesus monkeys, *J. Clin. Endocrinol. Metabol.*, 71, 1434-1441, 1990.
54. King, T. S., Schenken, R. S., Kang, I. S., Javors, M. A., and Riehl, R. M., Cocaine disrupts estrous cyclicity through interactions with the reproductive neuroendocrine axis in the rat, *Neuroendocrinology,* 51, 15-22, 1990.
55. Spiehler, V. R. and Reed, D., Brain concentrations of cocaine and benzoylecgonine in fatal cases, *J. Forensic. Sci.*, 30, 1003-1011, 1985.
56. King, T. S., Kang, I. S., Javors, M. A., and Schenken, R. S., Phorbol esters stimulate in vitro GnRH release from the hypothalamus of the estrogen-primed, ovariectomized rat: inhibitory effects of cocaine, *Neuroendocrinology,* 56, 526-532, 1992.

57. Tuomisto, J. and Manisto, P., Neurotransmitter regulation of anterior pituitary hormones, *Pharmacol. Rev.*, 37, 249-332, 1985.
58. Dackis, C. A., Estroff, T. W., Sweeney, D. R., Pottash, A. L. C., and Gold, M. S., Specificity of the TRH test for major depression in patients with serious cocaine abuse, *Am. J. Psychiatr.*, 142, 1097-1099, 1985.
59. Gianini, A. J., Malone, D. A., Loiselle, R. H., and Price, W. A., Blunting of TSH response to TRH in chronic cocaine and phencyclidine abusers, *J. Clin. Psychiatr.*, 48, 25-26, 1987.
60. Dhopesh, V. P., Burke, W. M., Maany, I., and Ravi, N. V., Effect of cocaine on thyroid function, *Am. J. Drug Alcohol Use*, 17, 423-427, 1991.
61. Neerhof, M. G., MacGregor, S. N., Retzky, S. S., and Sullivan, T. P., Cocaine abuse during pregnancy: peripartum prevalence and perinatal outcome, *Am. J. Obst. Gynecol.*, 161, 633-638, 1989.
62. Owiny, J. R., Myers, T., Massmann, G. A., Sadowsky, D. W., Jenkins, S., and Nathanielsz, P. W., Lack of effect of maternal cocaine administration on myometrial electromyogram and maternal plasma oxytocin concentrations in pregnant sheep at 124-146 days gestational age, *Obst. Gynecol.*, 79, 81-84, 1992.
63. Sarnyai, Z., Biro, E., Babarczy, E., Vecsernyes, M., Laczi, F., Szabo, G., Krivan, M., Kovacs, G. L., and Telegdy, G., Oxytocin modulates behavioral adaptation to repeated treatment with cocaine in rats, *Neuropharmacology*, 31, 593-598, 1992.
64. Zuckerman, B. and Bresnahan, K., Developmental and behavioral consequences of prenatal drug and alcohol exposure, *Pediatr. Clin. North Amer.*, 38, 1387-1406, 1991.
65. Gluckman, P. D., The hypothalamic-pituitary unit: the maturation of the neuroendocrine system in the fetus, in *Clinical Neuroendocrinology*, Besser, G. M. and Martini, L., Eds., Academic Press, 1982, 31-66.
66. Owiny, J. R., Jones, M. T., Sadowsky, D., Myers, T., Massmann, A., and Nathanielsz, P. W., Cocaine in pregnancy: the effect of maternal administration of cocaine on the maternal and fetal pituitary-adrenal axis, *Am. J. Obstetr. Gynecol.*, 164, 658, 1991.
67. Owiny, J. R., Jones, M. T., Sadowsky, D., Massmannm, A, Ding, X. Y., and Nathanielsz, P. W., Lack of effect of fetal administration of cocaine on maternal and fetal plasma adrenocorticotropin, cortisol and lactate concentrations at 127-138 days gestational age, *Gynecol. Obst. Invest.*, 32, 196-199, 1991.
68. Owiny, J. R., Sadowsky, D., Jones, M. T., Reimers, T. J., and Nathanielsz, P. W., Effect of maternal cocaine administration on maternal and fetal glucose, lactate and insulin in sheep, *Obst. Gynecol.*, 77, 901-904, 1991.
69. Magnano, C. L., Gardner, J. M., and Karmel, B. Z., Differences in salivary cortisol levels in cocaine-exposed and noncocaine-exposed NICU infants, *Develop. Psychobiol.*, 25, 93-103, 1992.
70. Kalivas, P. W. and Stewart, J., Dopamine transmission in the initiation and expression of drug- and stress-induced sensitization of motor activity, *Brain Res. Rev.*, 16, 223-244, 1991.
71. Post, R. M. and Weiss, S. R. B., Chronic cocaine administration: sensitization and kindling effects, in *Cocaine: Clinical and Behavioral Aspects*, Risher, S., Raskin, A., and Uhlenhuths, E. H., Eds., Oxford University Press, New York, 1987, 109-173.
72. Robinson, T. E. and Barridge, K. C., The neural basis of drug craving: an incentive-sensitization theory of addiction, *Brain Res. Rev.*, 18, 247-291, 1993.
73. Post, R. M., Contel, N., and Gold, P., Impaired behavioral sensitization to cocaine in vasopressin deficient rats, *Life Sci.*, 31, 2745-2750, 1982.
74. Van Ree, J. M., Burbach-Bloemarts, M. M., and Wallace, M., Vasopressin neuropeptides and the acquisition of heroin and cocaine self-administration, *Life Sci.*, 42, 1091-1099, 1988.
75. de Vry, J., Donselaar, I., and van Ree, J. M., Effects of desglycinamide^9 (Arg^8)vasopressin and vasopressin antiserum on the acquisition of intravenous cocaine self-administration in the rat, *Life Sci.*, 42, 2709-2715, 1988.
76. Sarnyai, Z., Babarczy, E., Krivan, M., Szabo, G., Kovacs, G. L., Barth, T., and Telegdy, G., Selective attenuation of cocaine-induced stereotyped behavior by oxytocin: putative role of basal forebrain target sites, *Neuropeptides*, 19, 51-56, 1991.
77. Cole, B. J. and Koob, G. F., Low doses of corticotropin releasing factor potentiate amphetamine-induced stereotyped behavior, *Psychopharmacology*, 99, 27-33, 1989.
78. Cador, M., Cole, B. J., Koob, G. F., Stinus, L., and Le Moal, M., Central administration of corticotropin releasing factor induces long-term sensitization to d-amphetamine, *Brain Res.*, 606, 181-186, 1993.

79. Cole, B. J., Cador, M., Stionu, L., Rivier, J., Vale, W., Koob, G. F., and Le Moal, M., Central administration of a CRF antagonist blocks the development of stress-induced behavioral sensitization, *Brain Res.*, 512, 343-346, 1990.
80. Derouche, V., Piazza, P. V., Maccari, S., Le Moal, M., and Simon, H., Repeated corticosterone administration sensitizes the locomotor response to AMPH, *Brain Res.*, 584, 309-313, 1992.
81. Rivet, J. M., Stinus, L., Le Moal, M., and Mormede, P., Behavioral sensitization to AMPH is dependent on corticosteroid receptor activation, *Brain Res.*, 498, 149-153, 1989.
82. Cools, A. R., Differential role of mineralocorticoid and glucocorticoid receptors in the genesis of dexamphetamine-induced sensitization of mesolimbic alpha1 adrenergic receptors in the ventral striatum, *Neuroscience*, 43, 419-428, 1991.
83. Cole, B. J., Cador, M., Stinus, L., Rivier, C., Rivier, J., Vale, W., Le Moal, M. and Koob, G. F., Critical role of the hypothalamic pituitary adrenal axis in AMPH-induced sensitization of behavior, *Life Sci.*, 47, 1715-1720, 1990.
84. Piazza, P., Maccari, S., Deminere, J. M., Le Moal, M., Mormede, P., and Simon, H., Corticosterone levels determine individual vulnerability to amphetamine self-administration, *Proc. Natl. Acad. Sci. USA*, 88, 2088-2092, 1991.
85. Antelman, S. M. and Chiodo, L. A., Amphetamine as a stressor, in *Stimulants: Neurochemical, Behavioral, and Clinical Perspectives*, Creese, I., Ed., Raven Press, New York, 1983, 269-299.
86. Beatty, W. W. and Holzer, G. A., Sex differences in stereotyped behavior in the rat, *Pharmacol. Biochem. Behav.*, 9, 777-873, 1978.
87. Glick, S. D. and Hinds, P. A., Sex differences in sensitization to cocaine-induced rotation, *Eur. J. Pharmacol.*, 99, 119-121, 1984.
88. Van Haaren, F. and Meyer, M. E., Sex differences in locomotor activity after acute and chronic cocaine administration, *Pharmacol. Biochem. Behav.*, 39, 923-927, 1991.
89. Camp, D. M. and Robinson, T. E., Susceptibility to sensitization. I. Sex differences in the enduring effects of chronic D-amphetamine treatment on locomotion, stereotyped behavior and brain monoamines, *Behav. Brain Res.*, 30, 55-68, 1988.
90. Savageau, M. M. and Beatty, W. W., Gonadectomy and sex differences in hyperactivity produced by amphetamine in rats, *Physiol. Behav.*, 22, 47-51, 1979.
91. Schneider, B. F. and Norton, S., Circadian and sex differences in hyperactivity produced by amphetamine in rats, *Physiol. Behav.*, 22, 47-51, 1979.
92. Robinson, T. E., Camp, D. J., Jacknow, D. S., and Becker, J. B., Sex differences and estrous cycle dependent variation in rotational behavior elicited by electrical stimulation of the mesostriatal dopamine system, *Behav. Brain Res.*, 6, 273-287, 1982.
93. Camp, D. M. and Robinson, T. E., Susceptibility to sensitization. II. The influence of gonadal hormones on enduring changes in brain monoamines and behavior produced by the repeated administration of d-amphetamine or restraint stress, *Behav. Brain Res.*, 30, 69-88, 1988.
94. Van Hartesveldt, C. and Joyce, J. N., Effects of estrogen on the basal ganglia, *Neurosci. Biobehav. Rev.*, 10, 1-14, 1986.
95. Camp, D. M., Becker, J. B., and Robinson, T. E., Sex differences in the effects of gonadectomy on amphetamine-induced rotational behavior in rats, *Behav. Neural Biol.*, 46, 491-495, 1986.
96. Roberts, D. C. S., Bennett, S. A. L., and Vickers, G. J., The estrous cycle affected cocaine self-administration on a progressive ratio schedule in rats, *Psychopharmacology*, 98, 408-411, 1989.
97. Peris, J., Decambre, N., Coleman-Hardee, M. L., and Simpkins, J. W., Estradiol enhances behavioral sensitization to cocaine and amphetamine-stimulated striatal [^{3}H]dopamine release, *Brain Res.*, 566, 255-264, 1991.
98. Becker, J. B. and Ramirez, V. D., Sex differences in the amphetamine-induced release of catecholamines from rat striatal tissue in vitro, *Brain Res.*, 204, 361-372, 1980.
99. Castner, S. A., Xiao, L., and Becker, J. B., Sex differences in striatal dopamine: in vivo microdialysis and behavioral studies, *Brain Res.*, 610, 127-134, 1993.
100. Becker, J. B., Estrogen rapidly potentiates amphetamine-induced striatal dopamine release and rotational behavior during microdialysis, *Neurosci. Lett.*, 118, 169-171, 1990.
101. Becker, J. B. and Cha, J. H., Estrous cycle-dependent variation in amphetamine-induced behaviors and striatal dopamine release assessed with microdialysis, *Behav. Brain Res.*, 35, 117-125, 1989.

102. Maus, M., Bertrand, P., Drouva, S., Rasolonjanahary, R., Kordon, C., Glowinski, J., Premont, J., and Enjalbert, A., Differential modulation of D1 and D2 dopamine sensitive adenylate cyclases by 17ß estradiol in cultured striatal neurons and anterior pituitary cells, *J. Neurochem.*, 52, 410-418, 1989.
103. Tonnaer, J. A. D., Leinders, T., and Van Delft, A. M. L., Ovariectomy and subchronic estradiol 17ß administration decrease dopamine D_1 and D_2 receptors in rat striatum, *Psychoneuroendocrinology,* 14, 469-476, 1989.
104. Hruska, R. E., Elevation of striatal dopamine receptors by estrogen: dose and time studies, *J. Neurochem.*, 47, 1908-1915, 1986.
105. Levesque, D. and Di Paolo, T., Dopamine receptor reappearance after irreversible receptor blockade: effect of chronic estradiol treatment of ovariectomized rats, *Mol. Pharmacol.*, 39, 659-665, 1991.
106. Mendelson, J. H., Teoh, S. K., Mello, N. K., and Ellingboe, J., Buprenorphine attenuates the effects of cocaine on adrenocorticotropin (ACTH) secretion and mood states in man, *Neuropsychopharmacology,* 7, 157-162, 1992.
107. Sharma, A., Plessinger, M. A., Miller, R. K., and Woods, J. R., Progesterone antagonist mifepristone (RU486) decreases cardiotoxicity of cocaine, *Proc. Soc. Exp. Biol. Med.*, 202, 279-287, 1993.

Chapter 5

Brain Imaging Studies of the Cocaine Addict: Implications for Reinforcement and Addiction

Nora D. Volkow and Joanna S. Fowler

CONTENTS

I. INTRODUCTION

The highly reinforcing and addictive properties of cocaine are manifested to the extreme in laboratory animals who, given free access to cocaine, will self-administer until death.[1] This sets cocaine apart from other drugs of abuse which are not self-administered at the expense of life-prolonging and life-preserving behaviors. Animal studies have shown that the brain dopamine (DA) systems appear central to the reinforcing and addictive properties of cocaine.[2-7]

Studies in humans evaluating the effects of chronic cocaine on dopamine brain function have been marked by inconsistencies. Because of the inaccessibility of the human brain for investigation, these studies were limited until recently to the evaluation of indirect parameters of dopamine activity and to measurements in postmortem brains. Studies measuring prolactin and growth hormone as indirect indices of central DA activity have reported either increases[8-11] or no changes[12-14] in cocaine abusers. Studies on plasma HVA concentration in cocaine abusers also have been unsuccessful in delineating a consistent pattern of abnormalities.[14-16] Postmortem studies have reported that the brains of cocaine abusers exhibit decreased DA concentration,[17,18] decreased[19] or increased[20] DA transporter sites, and decreased DA D_1[21] and D_2[22] receptor mRNA. Pharmacological studies report findings suggestive of decreased and/or abnormal function of DA receptors in cocaine abusers: blunted response to DA agonists[23,24] and increased sensitivity to DA antagonists.[25-27]

With positron emission tomography (PET), an imaging method used to track the regional distribution and kinetics of chemical compounds labeled with short-lived positron-emitting isotopes in the living body,[28] it is now possible to directly evaluate the involvement of the dopamine system in the addictive and reinforcing properties of cocaine in human subjects. Different aspects of the DA synapse have been investigated using PET. The presynaptic elements (DA neurons) have been examined using tracers to measure DA metabolism, DA transporter sites, and vesicular transporters.[29] The postsynaptic elements have been investigated with tracers that measure DA D_1 and D_2 receptor availability. There are currently no PET ligands which may be used to differentially evaluate DA D_3, D_4, and D_5 receptors. The use of 2-deoxy-2-[^{18}F]-fluoro-D-glucose (FDG) has made

1-8493-8311-0/95/$0.00+$.50

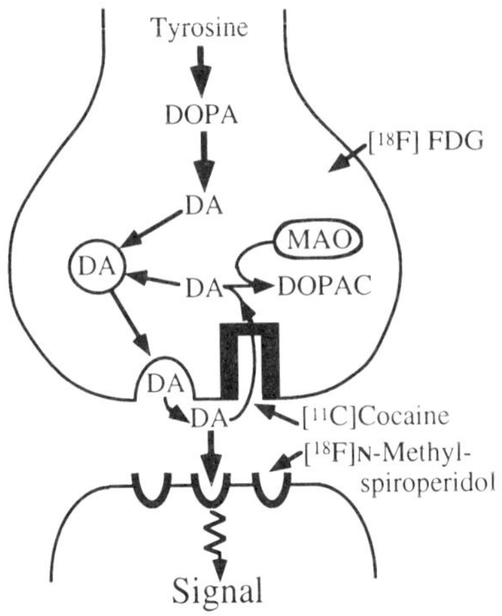

Figure 1. Schematic diagram of a DA synapse illustrating the different tracers used to evaluate the various components of the DA system: ^{18}F-N-Methylspiroperidol to measure DA D_2 receptors; [^{11}C]cocaine to measure DA transporters; and ^{18}F-FDG to measure regional cerebral metabolic rate for glucose.

possible the measurement of regional brain glucose metabolism in brain areas neuroanatomically connected with DA systems. Since positron emitters can be used to label compounds without affecting their pharmacological characteristics, PET has been used to measure the pharmacokinetics and binding pattern of cocaine in DA neurons in the human brain.[29]

PET also has been used to evaluate the effects of acute cocaine on brain glucose metabolism[30] and the effects of chronic cocaine on cerebral blood flow,[31] glucose metabolism during early[32] and late[33] cocaine withdrawal, DA metabolism,[34] and DA D_2 receptor availability.[35] The pharmacokinetics and distribution of cocaine and cocaethylene, a metabolite of cocaine and ethanol, have been compared in baboon brain[36] and the pharmacokinetics of cocaine in the presence or absence of alcohol intoxication has been examined.[37] In addition, the distribution of [^{11}C]cocaine in the human body has been examined.[38] This chapter summarizes the results of studies which examined different components of the DA system in cocaine abusers, the consequences of cocaine-induced changes on brain function, and their relationship to addictive behaviors. The effects of cocaine on the various elements of the DA synapse (Figure 1) are described separately; then correlational studies between these variables are described. This chapter also summarizes pharmacokinetic studies of cocaine, since they are relevant to our understanding of the uniquely reinforcing properties of this drug.

II. DOPAMINE D_2 RECEPTOR AVAILABILITY

[^{18}F]N-Methylspiroperidol (NMS) was used to measure DA D_2 receptor availability in cocaine abusers (n = 7) tested within one week of last cocaine use.[35] PET scans were performed using the PETT VI tomograph in the high-resolution mode [in-plane resolution of 9 mm, full width half maximum (FWHM)]. Dynamic scans were obtained after injection of 4 to 6 mCi [^{18}F]NMS for a total of 4 hr. Values for DA D_2 receptor availability were obtained using the ratio index (RI).[39] RI values were significantly lower in cocaine abusers (C) than in normal control subjects (N) ($C = 2.6 \pm 0.67$, $N = 3.98 \pm 0.92$; $t = 3.2$, $p \leq 0.01$) (Figure 2). DA D_2 receptor availability values obtained by Patlak's graphical analyses[40] revealed a significantly lower K_i value for cocaine abusers than for normal controls ($C = 0.081 \pm 0.04$, $N = 0.122 \pm 0.02$; $t = 2.7$, $p \leq 0.05$). Three cocaine abusers tested two weeks after last cocaine use displayed D_2 receptor availability values similar to those of normal control subjects.

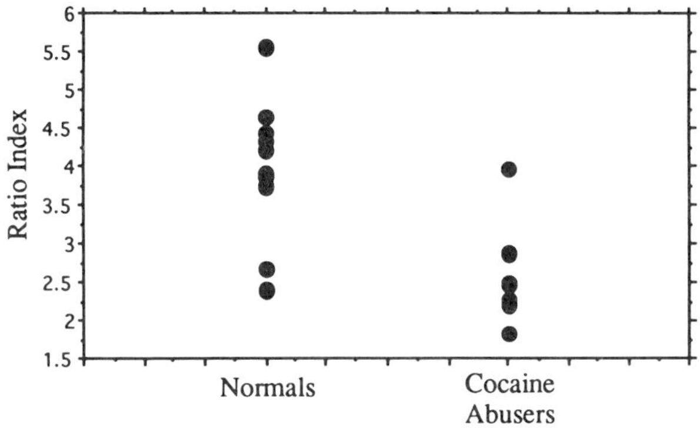

Figure 2. Ratio index in normal control subjects (n = 10) and in cocaine abusers scanned within one week of last cocaine use (n = 7).

To evaluate whether these changes were temporary or long lasting, a second study was undertaken in another group of 20 cocaine abusers who were tested 1 to 7 weeks after last use of cocaine and were compared to a group of 20 normal control subjects.[41] In contrast to the previous study, the cocaine abusers were maintained in the hospital to assure the absence of psychoactive substances. The PET studies were done using a whole-body, high-resolution tomograph (Computer Technologies, Incorporated, CTI 931; 6.5mm × 5.9mm, FWHM at the center of the field of view, interslice distance 5.9mm, 15 slices). Dynamic scans were obtained after injection of 4 to 6 mCi [^{18}F]NMS for a total of 3 hr. RI values for DA D_2 receptor availability were significantly lower in cocaine abusers than those obtained in normal control subjects (C = 2.65 ± 0.46, N = 3.08 ± 0.45; t = 2.9, $p \leq 0.01$) (Figure 3). Values of DA D_2 receptor availability obtained with Patlak's graphical analyses revealed a significantly lower K_i value for cocaine abusers than for normal control subjects (C = 0.093 ± 0.02, N = 0.119 ± 0.02; t = 3.2, $p \leq 0.005$). The 7 subjects who remained in the rehabilitation program displayed RI values similar to those observed initially (Initial RI = 2.5 ± 0.4; RI after 3 months of detoxification = 2.6 ± 0.3) (Figure 4). RI values for DA D_2 receptor availability were correlated negatively with scores on the Beck Depression Inventory ($r = -0.70$, $p < 0.001$) (Figure 5), but no correlation was observed between DA D_2 receptor availability and cocaine craving in detoxified cocaine abusers. In addition, RI values decreased as a function of age in both normal control subjects and cocaine abusers (N: $r = 0.72$, $p \leq 0.0005$; C: $r = 0.58$, $p \leq 0.01$).

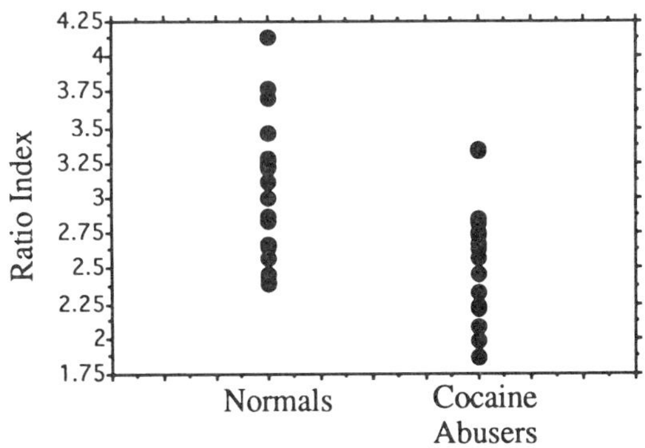

Figure 3. Ratio index in normal control subjects (n = 20) and in cocaine abusers scanned 1 to 7 weeks after last cocaine use (n = 20).

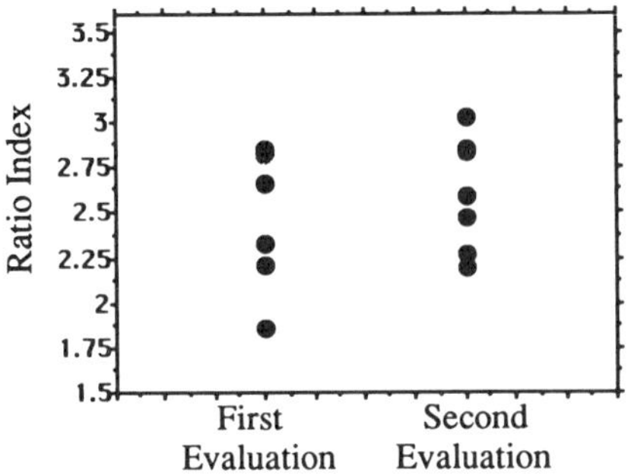

Figure 4. Ratio index in cocaine abusers who completed a 3-month detoxification period. The first evaluation was done 1 to 7 weeks after last cocaine use, while the second evaluation was done 3 months later after completion of an inpatient rehabilitation program.

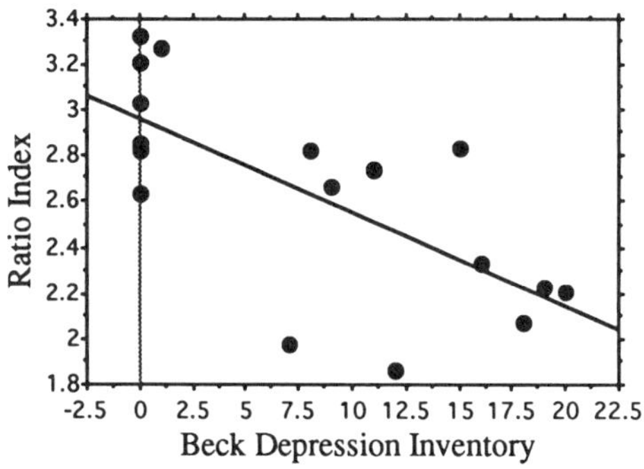

Figure 5. Correlation between the ratio index of DA receptor availability and scores on the Beck Depression Inventory ($r = -0.70$, $p \leq 0.001$).

III. PRESYNAPTIC DOPAMINE TERMINALS

Because [^{11}C]cocaine has been shown to bind predominantly to the DA transporter in the basal ganglia[29], this ligand was used to evaluate presynaptic DA terminals. The results compiled from 15 cocaine abusers and 12 normal control subjects reveal no difference in the pharmacokinetics or metabolism of [^{11}C]cocaine.[42] The uptake of [^{11}C]cocaine decreased in the basal ganglia of cocaine abusers when compared with normal control subjects as measured by the K_1 obtained using Logan Plots[43] (C = 0.403, N = 0.521; $p \leq 0.01$). The distribution volume for [^{11}C]cocaine also decreased in the basal ganglia of cocaine abusers compared to normal control subjects (C = 3.25 ± 0.645, N = 4.82 ± 18 ml/min; t = 3.43, $p \leq 0.01$). Decreased [1^{1}C]cocaine binding in the basal ganglia could represent either downregulation of the DA transporter or degeneration of presynaptic DA terminals. In fact, a preliminary evaluation of DA metabolism reported decreased uptake of [^{18}F]l-Dopa in cocaine abusers (n = 2).[34] However, because [^{11}C]cocaine binding also was observed to be lower in the thalamus and cerebellum, we cannot exclude the possibility that the apparent reduction of binding could reflect decreased cerebral blood flow (CBF), a decrease of binding sites other than the DA transporter, and/or a decrease in nonspecific binding. Therefore, the extent to which presynaptic DA neurons are affected in cocaine abusers is still unclear and requires further investigation.

IV. REGIONAL BRAIN GLUCOSE METABOLISM

It has been postulated that cocaine addiction develops due to a depletion of brain DA secondary to chronic cocaine administration.[9] However, studies evaluating the effects of chronic cocaine administration on the DA system have yielded contradictory results,[44] and the use of DA agonists has not been uniformly successful in the long-term treatment of the cocaine addict.[12] This suggests that DA could act as a functional modulator in brain regions that subserve addictive behaviors rather than directly encoding these behaviors.[45]

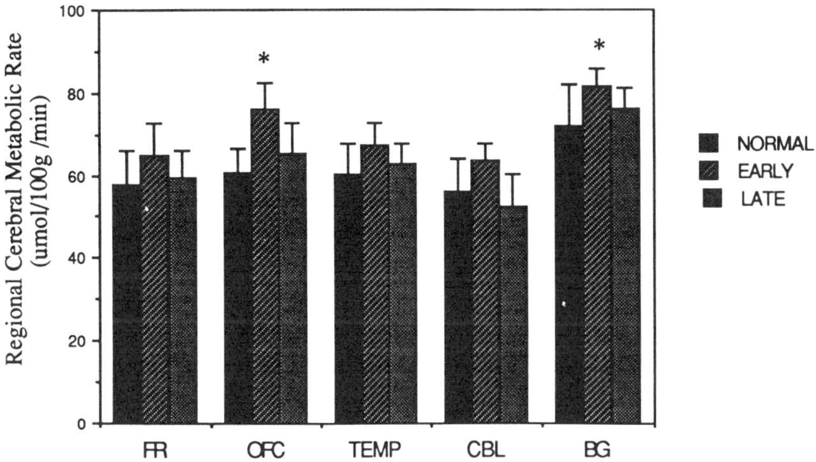

Figure 6. Regional cerebral metabolic rate for glucose (μmol/100 g/min) in cocaine abusers tested within 1 week of cocaine use (early) and cocaine abusers tested 1 to 4 weeks after last use (late). * – significantly ($p \leq 0.01$) different from normal control values.

Thus, chronic cocaine use could lead to addiction through consequences on the regional modulation of brain activity by DA. Alternatively, abnormalities in these brain regions prior to drug exposure could be associated with a higher vulnerability to drug addiction. To assess whether the brain regions innervated by the DA system might be involved in the addictive behavior of cocaine abusers, we used FDG to evaluate regional cerebral metabolic rate for glucose ($rCMR_{glc}$) in brain regions receiving projections from both the nigrostriatal and the mesocortical dopamine system. Because brain glucose metabolism reflects brain activity,[46] its measurement in conjunction with that of specific neurotransmitter ligands enables an assessment of their functional significance. Furthermore, because these studies were carried out in human subjects, neurochemical and metabolic brain changes can be linked to the clinical symptomatology.

Because cocaine withdrawal and detoxification may entail changes in dopaminergic activity, we evaluated patients both during early[32] and late[33] cocaine withdrawal. $rCMR_{glc}$ was significantly higher in the orbitofrontal cortex (OFC) and basal ganglia in cocaine abusers (n = 10) during early cocaine withdrawal (<1 week) than 1 to 4 weeks later (n = 5) or in normal control subjects (n = 15) (Figure 6). Also, $rCMR_{glc}$ in these brain regions correlated significantly with time since last cocaine use. Metabolic activity in the OFC ($r = 0.63$, $p \leq 0.02$) and prefrontal cortex (PFC) ($r = 0.70$, $p \leq 0.003$) were significantly correlated with intensity of cocaine craving (Figure 7). Cocaine abusers did not show this hypermetabolic pattern after 1 week of detoxification.

A second study examined a group of 20 cocaine abusers tested 1 to 7 weeks after last use of cocaine, compared with 24 normal control subjects.[33] Cocaine abusers were maintained in the hospital to assure the absence of psychoactive substances. Global cerebral metabolic rate for glucose did not differ significantly between normal control subjects and cocaine abusers (N = 38.4 ± 3, C = 36.5 ± 5 μmol/100 g/min, NS). However, $rCMR_{glc}$ in frontal cortical areas was significantly decreased ($p \leq 0.05$) in cocaine abusers, who exhibited lower $rCMR_{glc}$ in 16 of 21 frontal regions in the left hemisphere and 8 of 21 frontal regions on the right. This effect persisted after 3 to 4 months of detoxification.

Regional metabolic data were normalized with regard to whole brain metabolism in order to assess whether decreased frontal metabolism was related to an alteration of global brain metabolism. This relative measure revealed a significant diminution of frontal metabolism which persisted 3 months after detoxification. In contrast, the relative decrease

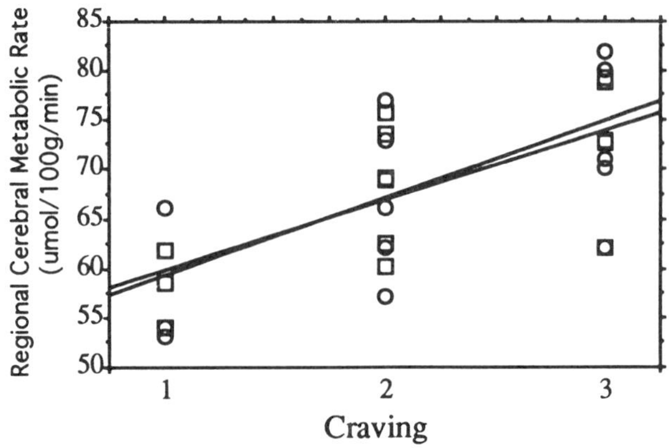

Figure 7. Regional cerebral metabolic rate for glucose (μmol/100 g/min) in prefrontal (squares) and orbitofrontal cortex (circles) correlated with the subjective sense for drug craving (1 = mild, 2 = moderate, 3 = severe).

of parietal metabolism demonstrated during the first 7 weeks of detoxification was no longer evident 3 months later (Figure 8). Frontal metabolism was correlated negatively with the dose of cocaine ($r = -0.47$, $p \leq 0.01$) and with the length of time the drug was used ($r = -0.42$, $p \leq 0.05$). Interestingly, there were no significant differences in striatal metabolism of cocaine abusers tested during late withdrawal.

The hypofrontality observed in detoxified cocaine abusers was interpreted as a reflection of decreased DA activity in mesocortical projections, since hypofrontality has been associated with conditions which decrease DA activity such as Parkinson's disease[47-51] and supranuclear palsy.[52-56] Furthermore, the strong correlation observed between frontal metabolism and D_2 receptor availability in cocaine abusers (see Correlational Studies) supports this hypothesis.

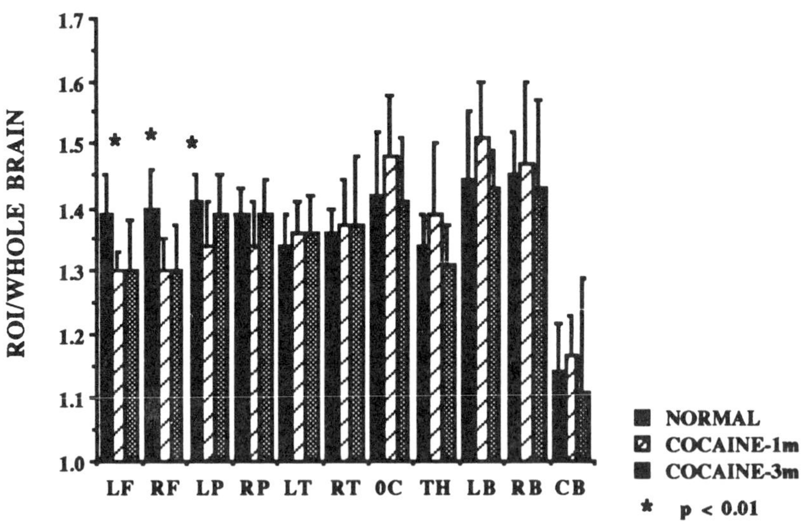

Figure 8. Relative metabolism in 20 normal control subjects compared to 20 cocaine abusers tested 1 to 7 weeks after detoxification (cocaine-1m), and seven of these patients who were tested 3 months later (cocaine-3m). * – Significantly ($p \leq 0.01$) different from normal control subjects. Abbreviations: L = left, R = right, F = frontal, P = parietal, T = temporal, OC = occipital, TH = thalamus, B = basal ganglia, CB = cerebellum.

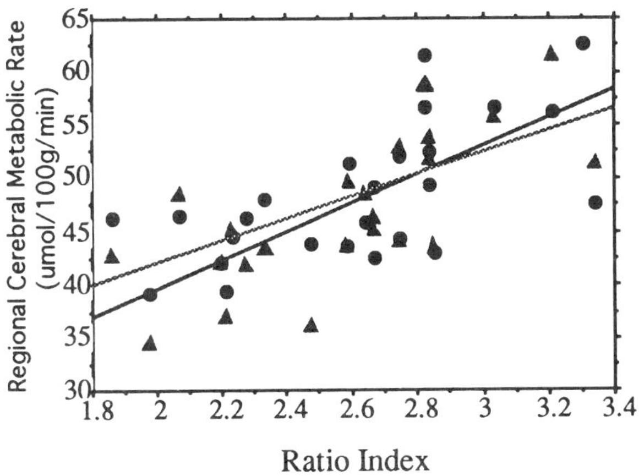

Figure 9. Correlation between the ratio index of DA D_2 receptor availability and regional cerebral metabolic rate for glucose (μmol/100 g/min) in orbitofrontal cortex (circles) and in cingulate gyrus (triangles).

V. CORRELATIONAL STUDIES

Because brain metabolism is coupled tightly with brain function, measurement of brain metabolism in addition to the various indices of DA activity allows the assessment of functional changes associated with changes in brain DA activity. The following data summarize the correlational analyses between DA D_2 receptor availability and regional brain glucose metabolism in 20 cocaine abusers[41] and in 4 of 7 subjects who were retested 3 months after detoxification. Several regions of the frontal cortex, most markedly OFC ($r = 0.70$, $p \leq 0.0001$), PFC ($r = 0.56$, $p \leq 0.005$), and cingulate cortex ($r = 0.63$, $p \leq 0.0005$) exhibited significant correlations (Figure 9). Comparison of [^{11}C] cocaine binding in the basal ganglia and regional glucose metabolism revealed a similar pattern to that observed above with significant correlations limited predominantly to frontal cortex ($r = 0.67$, $p \leq 0.01$).

Because regional brain glucose metabolism primarily represents activity in nerve terminals,[57] this association could reflect cellular activity in the ventral tegmental area, which directly innervates the OFC, PFC, and cingulate cortex.[58,59] These cortical regions, in turn, could modulate striatal DA receptor availability via the corticostriatal pathway.[45] This association also could reflect striatal modulation of frontal metabolism via the striato-pallidal-thalamic cortical pathway[60] (Figure 10).

The mesolimbic DA system appears to be critical in mediating the reinforcing properties of cocaine and its addiction liability.[61,62] Disruption of the mesocortical DA system by chronic cocaine use could lead to abnormal function in projection areas (i.e., OFC, PFC, cingulate cortex, nucleus accumbens, hippocampus, and amygdala[45]), which could then give rise to the complex pattern of behaviors associated with addiction. In fact, the OFC, PFC, and cingulate cortex have been shown to be involved in processes mediating positive reinforcement in animal studies.[63] The initial process by which the intake of the drug is experienced as pleasurable, or the *intrinsic reinforcing drug effect,* is associated with increased DA in nucleus accumbens and PFC.[61,63,64] The memory of the drug experience and of the circumstances and the behaviors associated with this experience have also been shown to contribute to repeated cocaine intake.[65] With repeated administration, the ability of this *memory* to elicit a desire or "craving" for cocaine becomes more frequent and serves to perpetuate the use of cocaine.[7] The neurochemical and neuroanatomical substrates for consolidation of this memory and for eliciting cocaine craving are not well understood, but probably involve the hippocampus among other brain regions.

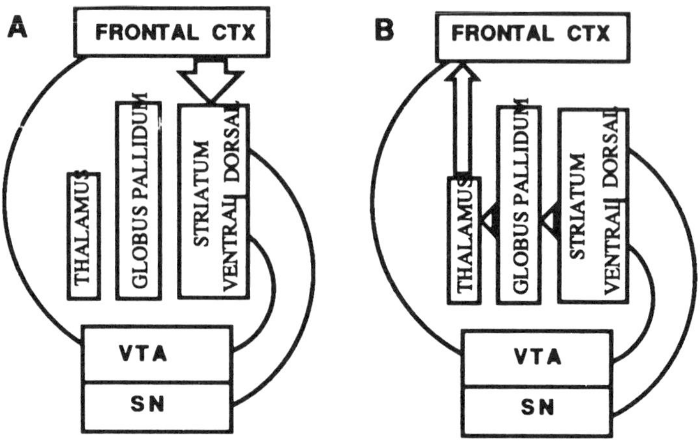

Figure 10. Schematic diagrams illustrating the mesocortical DA pathway with efferent projections from the ventral tegmental area (VTA) to the frontal cortex and ventral striatum and the nigrostriatal pathway with efferent projections from the substantia nigra (SN) to the dorsal striatum. Interactions between frontal cortex and the striatum could occur via the cortico-striatal pathway (A) or indirectly via the striato-pallidal-thalamo-cortical pathway (B).

While the memory and intrinsic reinforcing properties of cocaine are important, we hypothesize that other processes are also involved, since compulsive cocaine administration in the addicted individuals occurs despite rapid tolerance to the subjective effects of cocaine[66] and even in the presence of adverse physical reactions. The drive and *loss of control* leading to compulsive self-administration of cocaine probably is regulated both by DA and serotonin[67,68] and may involve the OFC, PFC, and cingulate cortex. Other processes, such as *sensitization*, have also been reported to occur with repeated cocaine administration[44] and may also participate in triggering and/or perpetuating compulsive drug self-administration. Abnormalities in the amygdala could facilitate kindling and sensitization. Another factor which could facilitate repeated cocaine use is the *emotional reaction* of the individual to the losses experienced from his/her cocaine addiction. In particular, dysphoria during withdrawal has been associated with a higher relapse rate in the cocaine abuser.[9] One could postulate that because the mesolimbic DA system is involved in the reward processes, its dysfunction in the cocaine abuser could intensify depressive symptoms such as anhedonia and loss of drive.[69]

We postulate that DA abnormalities in the cocaine abuser lead to disregulation of frontal regions favoring the emergence of behaviors associated with addiction, such as impulsivity, compulsion to self-administer cocaine,[70] dysphoria, and the inability to restrain from using cocaine. Animal studies have documented a central role of these frontal regions in the reinforcing properties of drugs.[63] Abnormalities in other projection areas of the mesocortical DA system such as hippocampus and amygdala probably also contribute to these addictive behaviors. Because of the limited spatial resolution of the PET tomograph, however, function in these areas could not be evaluated.

VI. STUDIES OF COCAINE PHARMACOKINETICS

The absolute uptake, regional distribution, and pharmacokinetics of cocaine binding in the human and baboon brain have been examined using PET and [^{11}C]cocaine.[29,36] These studies have revealed an heterogeneous distribution of [^{11}C]cocaine binding in the human brain. In studies of cocaine binding in the baboon brain, values of cocaine uptake were

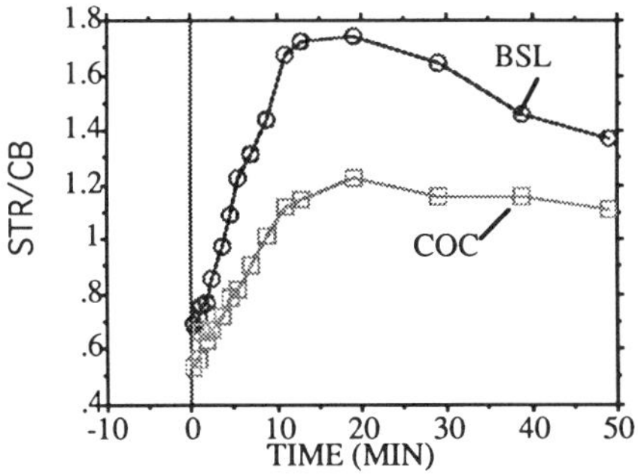

Figure 11. Effects of cocaine pretreatment on the ratio of [^{11}C]cocaine in the striatum vs. the cerebellum (STR/CB). Cocaine (COC) pretreatment (2 mg/kg, i.v.) significantly decreased the STR/CB ratio compared to baseline (BSL) over time.

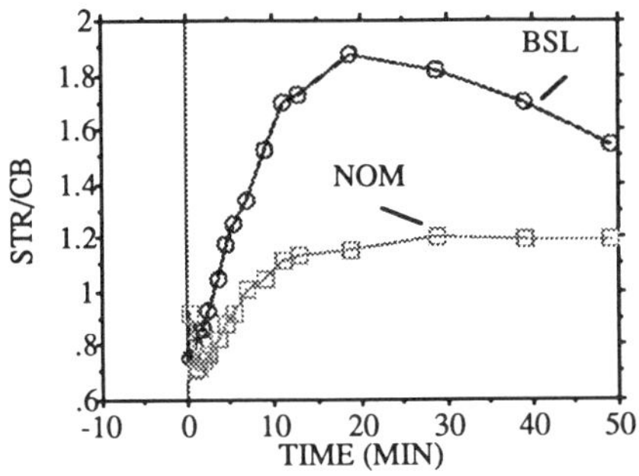

Figure 12. Effects of nomifensine (NOM) on the ratio of [^{11}C]cocaine in the striatum vs. the cerebellum (STR/CB). NOM significantly decreased the STR/CB compared to baseline (BSL) over time.

normalized by computing a ratio of concentration in the striatum to that of the cerebellum, a region which contains no dopamine transporters. The uptake of [^{11}C]cocaine was maximal in the basal ganglia, and the striatal-to-cerebellar ratio was reduced from 1.75 to 1.17 by preadministration of cocaine (Figure 11), suggesting that these data reflect specific binding. Preadministration of nomifensine, a drug that binds selectively to the dopamine transporter, also decreased striatal [^{11}C]cocaine binding (Figure 12) to the same extent as observed following cocaine preadministration, which suggests that cocaine binds predominantly to the dopamine transporter in the striatum.

Pharmacokinetic studies of [^{11}C]cocaine in human brain revealed a wide variability among individuals. Although the maximal uptake differed markedly among subjects, the shape of the time-activity curves for the uptake of [^{11}C]cocaine was remarkably similar in each of eight normal volunteers investigated (Figure 13). Cocaine uptake was very fast, and peak activity was reached in most subjects between 4 to 8 min after injection. The clearance of cocaine from the brain was slower in the basal ganglia than in other subcortical regions, cortical areas, or cerebellum. Half-peak clearance of [^{11}C]cocaine was reached approximately 20 min after injection in the basal ganglia and in approxi-

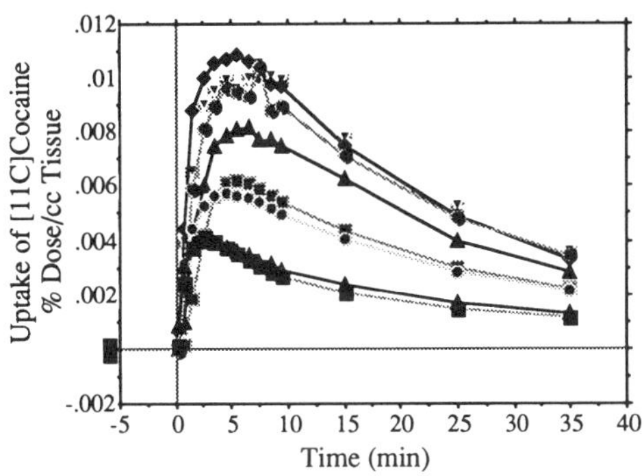

Figure 13. Uptake of [^{11}C]cocaine (percent of dose per cc of tissue) measured over time in the basal ganglia of eight normal volunteers. Notice the variability of peak uptake.

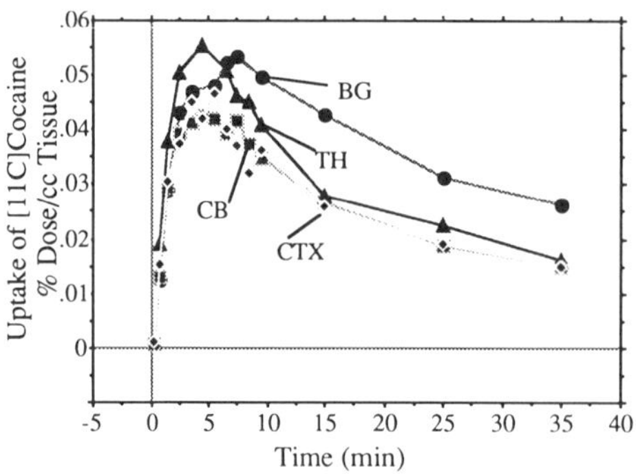

Figure 14. [^{11}C]Cocaine uptake (percent of dose per cc of tissue) in various brain regions. Abbreviations: BG = basal ganglia, TH = thalamus, CTX = cortex, CB = cerebellum.

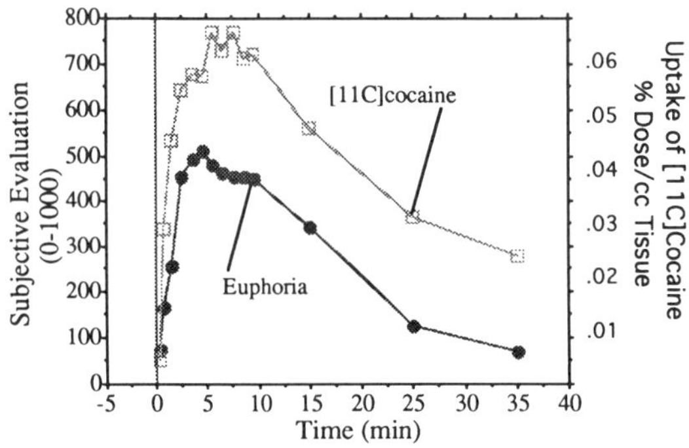

Figure 15. Mean data from ten subjects showing [^{11}C]cocaine uptake in the basal ganglia ([^{11}C]cocaine) and the subjective rating of euphoria experienced by subjects after intravenous cocaine (Euphoria) plotted over time. Values for uptake of [^{11}C]cocaine were rescaled so as to correspond to the scale for euphoria (0-1000). Notice the similarity between the curves.

mately 15 min in thalamus, cortex, and cerebellum (Figure 14). These pharmacokinetic characteristics parallel the time course of intensity of euphoria experienced after intravenous cocaine administration, as is illustrated by the similarity of [^{11}C]cocaine uptake in the basal ganglia plotted along with values indicating euphoria experienced after intravenous cocaine administration over time, which were measured in another study[71] (Figure 15).

VII. SUMMARY AND CONCLUSION

These studies document dopaminergic abnormalities in cocaine abusers, and suggest a regulatory role for DA in metabolism of frontal brain regions. The correlation of striatal D_2 receptor availability with regional metabolism was greatest in OFC, PFC, and cingulate cortex. In cocaine abusers tested during early withdrawal (<1 week), the OFC was found to be hypermetabolic and metabolism in OFC and PFC was observed to be associated significantly with cocaine craving.[32] Thus, we postulate that repeated and intermittent DA stimulation such as that which occurs during a cocaine binge activates the OFC and PFC, increasing the drive to compulsively self-administer cocaine. These frontal cortical regions become hypometabolic after cocaine discontinuation and during protracted withdrawal. Dopaminergic stimulation by a DA-enhancing drug and/or environmental conditioning could reactivate these frontal regions resetting the compulsion to self-administer cocaine and the inability to terminate this behavior.

The pharmacokinetic data from studies using [^{11}C]cocaine are consistent with behavioral and pharmacological studies in animals as well as *in vitro* studies which have revealed that the mechanisms underlying the reinforcing properties of cocaine probably involve the dopamine system of the brain. These data also highlight the close relationship between the pharmacokinetic characteristics and the unique reinforcing properties of cocaine.

ACKNOWLEDGMENT

Supported by NIDA R01 DA06278, and DOE DE-ACO2-76CH00016.

REFERENCES

1. Bozarth, M. S. and Wise, R. A., Toxicity associated with long-term intravenous heroin and cocaine self-administration in rat, *J. Am. Med. Assoc.*, 254, 81, 1985.
2. Ritz, M. C., Lamb, R. J., Goldberg, S. R., and Kuhar, M. J., Cocaine receptors on dopamine transporters are related to self-administration of cocaine, *Science*, 237, 1219, 1987.
3. Roberts, D. C. S., Corcoran, M. E., and Fibiger, H. C., On the role of ascending catecholaminergic systems in intravenous self-administration of cocaine, *Pharmacol. Biochem. Behav.*, 6, 615, 1977.
4. Woolverton, W. L. and Johnson, K. M., Neurobiology of cocaine abuse, *Trends Pharmacol. Sci.*, 13, 193, 1992.
5. DeWit, H. and Wise, R. A., Blockade of cocaine reinforcement in rats with the dopamine receptor blocker pimozide but not with the noradrenergic blockers phentolamine or phenoxybenzamine, *Can. J. Psychol.*, 31, 195, 1977.
6. Galloway, M. P., Neurochemical interactions of cocaine with dopaminergic systems, *Trends Pharmacol. Sci.*, 9, 45, 1988.
7. Johanson, C. E. and Fischman, M. W., The pharmacology of cocaine related to its abuse, *Pharm. Rev.*, 41, 3, 1989.
8. Cocares, J. A., Dackis, C. A, and Gold, M. S., Sexual dysfunction secondary to cocaine abuse in two patients, *J. Clin. Psychiatry*, 47, 384, 1986.
9. Dackis, C. A. and Gold, M. S., New concepts in cocaine addiction. The dopamine depletion hypothesis, *Neurosci. Biobehav. Rev.*, 9, 469, 1985.
10. Kranzler, H. R. and Wallington, D. J., Prolactin, cocaine dependence and treatment, *Am. Psychiatr. Assoc. New Res. Abstr.*, 199, 1989.
11. Mendelson, J. H., Tesh, S. K., Lange, U., Mello, N. K., Weiss, R., Skupny, A., and Ellingboe, J., Anterior pituitary, adrenal, and gonadal hormones during cocaine withdrawal, *Am. J. Psychiatry*, 145, 1094, 1988.
12. Gawin, F. H. and Ellinwood, E. H., Cocaine and other stimulants, *N. Engl. J. Med.*, 318, 1173, 1988.
13. Swartz, C. M., Breen, K., and Leone, F., Serum prolactin levels during extended cocaine abstinence, *Am. J. Psychiatry*, 147, 777, 1990
14. Satel, S. L., Price, L. H, Palumbo, J. M., McDougle, C. J., Krystal, J. H., Gawin, F., Charney, D. S., Heninger, G. R., and Klebe, H. D., Clinical phenomenology and neurobiology of cocaine abstinence, *Am. J. Psychiatry*, 148, 1712, 1991.
15. Martin, S. D., Yeragani, V. K., Lodhi, R., and Galloway, M. P., Clinical ratings and plasma HVA during cocaine abstinence, *Biol. Psychiatry*, 26, 356, 1989.
16. Extein, I., Potter, W. E. Z., Gold, M. S., Andre, P., Rafuls, W. A., and Gross, D. A., Persistent neurochemical deficit in cocaine abuser, *Am. Psychiatr. Assoc. New Res. Abstr.*, 51, 1989.
17. Wilson, R. J., Deck, J., Shannak, K., Chang, L. J., DiStefano, L. M., and Kish, S. J., Markedly reduced striatal dopamine levels in brain of a chronic cocaine abuser, *Soc. Neurosci. Abstr.*, 16, 252, 1990.
18. Wyatt, R. J., Karoum, F., Suddath, R., and Fawcette, R., Persistently decreased brain dopamine levels and cocaine, *J. Am. Med. Assoc.*, 27, 2996, 1988.
19. Staley, J., Toiba, R., Ruttenber, A. J., Wetli, C. V., Lee-Hearn, W., Flynn, D. D., and Mash, D. C., ^{125}I RTI binding to the dopamine transporter in cocaine overdose deaths, *Soc. Neurosci. Abstr.*, 18, 542, 1992.
20. Little, K. Y., Effects of cocaine on the dopamine transporter, *Am. Psychiatr. Assoc. New Res. Abstr.*, 125, 1992.
21. Toiba, R., Ruttenber, A., Wetli, C. V., Lee-Hearn, W., Staley, J., Mash, D. C., Dopaminergic receptor subtype regulation in cocaine induced psychosis and sudden death: an autoradiographic study, *Soc. Neurosci. Abstr.*, 18, 542, 1992.
22. Meador-Woodruff, J. H., Dopamine receptor mRNA's in the brain: effects of cocaine, *Am. Psychiatr. Assoc. New Res. Abstr.*, 125, 1992.
23. Hollander, E., Nunes, E., DeCaria, C., Quitkin, F. M., Cooper, T., Wager, S., and Klein, D. F., Dopaminergic sensitivity and cocaine abuse: response to apomorphine, *Psychiatry Res.*, 33, 161, 1990.

24. Hitzemann, R., Burr, G., Piscani, K., Hazan, J., Krishnamoorthy, G., Cushman, P., Baldwin, C. H., Carrion, R., Volkow, N. D., Hirschowitz, J., Handelsman, L., Chiaramonte, J., and Angrist, B., Neuroendocrine and clinical features of cocaine withdrawal, *Psychiatry Res.*, in press.
25. Kumor, K., Sherer, M., and Jaffe, J., Haloperidol-induced dystonia in cocaine addicts, *Lancet*, 2, 1341, 1987.
26. Choy-Kwang, M. and Lipton, R. B., Dystonia related to cocaine withdrawal: a case report and pathogenic hypotheses, *Neurology*, 39, 996, 1989.
27. Hegarty, A., Lipton, R. B, and Merriam, A., Cocaine as a risk factor for acute dystonic reaction, *Neurology*, 40, 146, 1990.
28. Fowler, J. S. and Wolf, A. P., New directions in positron emission tomography, *Annu. Rep. Med. Chem.*, 24, 277, 1989.
29. Fowler, J. S., Volkow, N. D., Wolf, A. P., Dewey, S. L., Schlyer, D. J., MacGregor, R., Hitzemann, R., Logan, J., Bendriem, B., and Christman, D., Mapping cocaine-binding sites in human and baboon brain *in vivo*, *Synapse*, 4, 371, 1989.
30. London, E. D., Cascella, N. G., Wong, D. F., Phillips, R. L., Sannals, R. F., Links, J. M., Herning, R., Grayson, R., and Jaffe, J. H., Cocaine-induced reduction of glucose utilization in human brain. A study using positron emission tomography and [fluorine-18]-fluorodeoxyglucose, *Arch. Gen. Psychiatry*, 47, 567, 1990.
31. Volkow, N. D., Mullani, N., Gould, L., Krajewski, K., and Adler, S., Cerebral blood flow in chronic cocaine users, *Br. J. Psychiatry*, 152, 641, 1988.
32. Volkow, N. D., Fowler, J. S., Wolf, A. P., Hitzemann, R., Dewey, S., Bendriem, B., Alpert, R., and Hoff, A., Changes in brain glucose metabolism in cocaine dependence and withdrawal, *Am. J. Psychiatry*, 148, 621, 1991.
33. Volkow, N. D., Hitzemann, R., Wang, G. J., Fowler, J. S., Wolf, A. P., and Dewey, S. L., Long-term frontal brain metabolic changes in cocaine abusers, *Synapse*, 11, 184, 1992.
34. Baxter, L. R, Schwartz, J. M, Phelps, M., Mazziota, J. C, Barrio, J. , Rawson, R. A, Engel, J., Guze, B. H., Selin, C., and Sumida, R., Localization of neurochemical effects of cocaine and other stimulants in the human brain, *J. Clin. Psychiatry*, 4, 923, 1988.
35. Volkow, N. D., Fowler, J., Wolf, A., Schlyer, D., Shiue, C., Albert, R., Dewey, S., Logan, J., Bendriem, B., Christman, D., Hitzemann, R., and Henn, F., Effects of chronic cocaine abuse on postsynaptic dopamine receptors, *Am. J. Psychiatry*, 147, 719, 1990.
36. Fowler, J. S., Volkow, N. D., MacGregor, R. R., Logan, J., Dewey, S. L., Gartley, S. J., and Wolf, A. P., Comparative PET studies of the kinetics and distribution of cocaine and cocaethylene in baboon brain, *Synapse*, 12, 220, 1992.
37. Fowler, J. S., Volkow, N. D., Logan, J., MacGregor, R. R., Wang, G.-J., and Wolf, A. P., Alcohol intoxication does not change cocaine pharmacokinetics in human brain and heart, *Synapse*, 12, 228, 1992.
38. Volkow, N. D., Fowler, J. S., Wolf, A. P., Wang, G. J., Logan, J., MacGregor, R., Dewey, S. L, Schlyer, D. J, and Hitzemann, R., Distribution of ^{11}C-cocaine in human heart, lungs, liver and adrenals. A dynamic PET study, *J. Nucl. Med.*, 33, 521, 1992.
39. Smith, M., Wolf, A. P., Brodie, J. F., Arnett, C. D., Barouche, F., Shiue, C.-Y., Fowler, J. S., Russell, J. A. G., MacGregor, R. R., Wolkin, A., Angrist, B., Rotrosen, J., and Peselow, E., Serial ^{18}F N-methylspiroperidol PET studies to measure changes in antipsychotic drugs D_2 receptor occupancy in schizophrenic patients, *Biol. Psychiatry*, 23, 653, 1988.
40. Patlak, C. S, Patlak, C., and Blasberg, R. G., Graphical evaluation of blood to brain transfer constants from multiple-time activity data. Generalizations, *J. Cereb. Blood Flow Metab.*, 5, 584, 1983.
41. Volkow, N. D., Fowler, J. S., Wang, G.-J., Hitzemann, R., Logan, J., Schlyer, D., Dewey, S., and Wolf, A. P., Dopaminergic dysregulation of frontal metabolism may contribute to cocaine addiction, *Synapse*, 14, 169, 1993.
42. Volkow, N. D., Fowler, J. S., Logan, J., Wang, G. J., Hitzemann, R., MacGregor, R., and Dewey, S. L., Decreased binding of ^{11}C-cocaine in the brain of cocaine addicts, *J. Nucl. Med.*, 33, 888, 1992.
43. Logan, J., Fowler., J. S., Volkow, N. D., Wolf, A. P., Dewey, S. L., Schlyer, D., MacGregor, R. R., Hitzemann, R., Bendriem, B., Gatley, S. J., and Christman, D. R., Graphical analysis of reversible binding from time activity measurements, *J. Cereb. Blood Flow Metab.*, 10, 740, 1990.

44. Post, R., Weiss, S. R., Pert, A., and Uhde, T., Chronic cocaine administration: sensitization and kindling effects, in *Cocaine: Clinical and Biobehavioral Aspects,* Fischer, S. and Maskin, A., Eds., Oxford, New York, 1987, 109.
45. Le Moal, M. and Simon, H., Mesocorticolimbic dopaminergic network: functional and regulatory notes, *Physiol Rev.,* 71, 155, 1991.
46. Sokoloff, L., Reivich, M., Kennedy, C., Des Rosiers, M. H., Patlak, C. S., Pettigrew, K. D., Sakurada, O., and Shinohara, M., The [^{14}C] deoxyglucose method for the measurement of local cerebral glucose utilization, *J. Neurochem.,* 28, 897, 1977.
47. Perlmutter, J. S., New insights into the pathophysiology of Parkinson's disease: the challenge of positron emission tomography, *Trends Neurosci.,* 11, 203, 1988.
48. Mayberg, H. S., Starkstein, S. E., Sadzot, B., Preziosi, T., Andrezejewski, P. L., Dannals, R. F., Wagner, H. N., and Robinson, R., Selective hypometabolism in the inferior frontal lobe in depressed patients with Parkinson's disease, *Annu. Neurol.,* 28, 57, 1990.
49. Bes, A., Guell, A., and Fabre, N., Cerebral blood flow studied by xenon-133 inhalation technique in parkinsonism: loss of hyperfrontal pattern, *J. Cereb. Blood Flow Metab.,* 3, 33, 1983.
50. Wolfson, L. I., Leenders, K. L., Brown, L. L., and Jones, T., Alterations of regional cerebral blood flow and oxygen metabolism in Parkinson's disease, *Neurology,* 35, 1399, 1985.
51. Perlmutter, J. S. and Raichle, M. E., Regional blood flow in hemiparkinsonism, *Neurology,* 35, 1127, 1985.
52. Blin, J., Baron, J. C., Dubois, B., Pillon, B., Cambon, H., Cambier, J., and Agid, Y., Positron emission tomography study in progressive supranuclear palsy brain hypometabolic pattern and clinical metabolic correlations, *Arch. Neurol.,* 47, 747, 1990.
53. Leenders, K. L., Frackowiak, S. J., and Lees, A. J., Steele-Richardson-Olszewski syndrome, *Brain* III, 615, 1988.
54. Karbe, H., Grond, M., Huber, M., Herholz, K., Kessler, J., and Heiss, W. D., Subcortical damage and cortical dysfunction in progressive supranuclear palsy demonstrated by positron emission tomography, *J. Neurol.,* 239, 98, 1992.
55. Foster, N. L., Gilman, S., Berenet, S., Morin, E. M., Brown, M. B., and Koeppe, R. A., Cerebral hypometabolism in progressive supranuclear palsy studies with positron emission tomography, *Annu. Neurol.,* 24, 399, 1988.
56. Goffinet, A. M., Volder, A. G, de Gillain, C., Rectem, D., Bol, A., Michel, C., Cogneau, M., Labar, D., and Lattere, C., Positron tomography demonstrates frontal lobe hypometabolism in progressive supranuclear palsy, *Annu. Neurol.,* 25, 131, 1989.
57. Schwartz, W. J., Smith, C. B., Davidsen, L., Savaki, L., Sokoloff, M., Mata, Fink, D. J., and Gainer H., Metabolic mapping of functional activity in the hypothalamo-neurohypophysial system of the rat, *Science,* 205, 723, 1979.
58. Glowinski, J., Tassin, J. P., and Thierry, A. M., The mesocortical-prefrontal dopaminergic neurons, *Trends Neurosci.,* 7, 451-418, 1984.
59. Stuss, D. T. and Benson, D. F., *The Frontal Lobes,* Raven Press, New York, 1986, 12.
60. Heimer, L, Alheid, G. F, and Zaborzky, L., The basal ganglia, in *The Rat Nervous System,* Paxinos, G., Ed., Academic Press, New York, 1985, 37.
61. Goeders, N. E. and Smith, J. E., Reinforcing properties of cocaine in the medial prefrontal cortex: primary action on presynaptic dopaminergic terminals, *Pharmacol. Biochem. Behav.,* 25, 191, 1986.
62. Wise, R. A. and Bozarth, M. A., A psychosomatic stimulant theory of addiction, *Psychol. Rev.,* 94, 469, 1987.
63. Dworkin, S. I. and Smith, J. E., Cortical regulation of self-administration, in *Neurobiology of Drug and Alcohol Addiction,* Kalivas, P. W. and Samson, H. A., Eds., New York Academy of Sciences, 1992, 274.
64. Hurd, Y. L. and Ungerstedt, J., Cocaine: an in vivo microdialysis evaluation of its acute action on dopamine transmission in rat striatum, *Synapse,* 3, 48, 1989.
65. Wise, R. A., Neural mechanisms in the reinforcing actions of cocaine, in *Cocaine in the Brain,* Volkow, N. D and Swann, A. D., Eds., New Brunswick, Rutgers Press, 1990, 42.
66. Fischman, M. W., Schuster, C. R., Javaid, J., Hatano, Y., and Davis, J., Acute tolerance development to the cardiovascular and subjective effects of cocaine, *J. Pharmacol. Exp. Ther.,* 235, 677, 1985.

67. Di Chiara, G., Acquas, E., and Carboni, E., Role of mesolimbic dopamine in the motivational effects of drugs: brain dialysis and place preference studies, in *The Mesolimbic Dopamine System: From Motivation to Action,* Willner, P. and Scheel-Krüger, J., Eds., Wiley, New York, 1991, 367.
68. Loh, E. A. and Roberts, D. C. S., Increased motivation to administer intravenous cocaine following 5,7-dihydroxytryptamine lesions of the media forebrain bundle in the rat, *Soc. Neurosci. Abstr.,* 14, 659, 1988.
69. Willner, P., Muscat, R., Papp, M., and Sampson, D., Dopamine, depression, and anti-depressant drugs, in *The Mesolimbic Dopamine System: From Motivation to Action,* Willner, P. and Scheel-Krüger, J., Eds., Wiley, New York, 1991, 387.
70. Modell, J. G., Mountz, J. M., Curtis, G. J., and Greden, J., Neurophysiologic dysfunction in basal ganglia, limbic, striatal and thalamocortical circuits as a pathogenetic mechanism of obsessive-compulsive disorder, *Neuropsychiatry,* 1, 27, 1989.
71. Cook, C. E., Jeffcoat, A. R., and Perez-Reys, M., Pharmacokinetic studies of cocaine and phencyclidine in man, in *Pharmacokinetics and Pharmacodynamics of Psychoactive Drugs,* Barnett, G. and Chiang, C. N., Eds., Biomedical Publications, Foster City, 1985, 48.

Section II

CELLULAR EFFECTS OF COCAINE

Chapter 6

Neural Basis of Behavioral Sensitization to Cocaine

Peter W. Kalivas

CONTENTS

I. INTRODUCTION

Shortly after the introduction of cocaine and other amphetamine-like psychostimulants into European and North American societies, the presence of a medically significant sequelae of behavioral alterations resulting from continued use of these drugs was identified.[1,2] These alterations consist of progressive increases in certain psychopathological behaviors, including anxiety disorders, panic attack, and paranoid psychosis.[2-4] With chronic psychostimulant use, these behaviors supersede the euphoric effects of the drug and become the primary behavioral and interoceptive characteristic of drug use.[5,6] This progressive augmentation of negative behavioral responses to psychostimulant use has been termed behavioral sensitization.

A particularly insidious aspect of behavioral sensitization is that it is relatively enduring, even in the absence of continued drug use. Thus, while psychostimulant-induced behaviors generally dissipate when drug use is discontinued, the sensitized behavioral pathologies can be reinstated by an acute drug challenge up to a decade later.[7,8] Furthermore, spontaneous psychiatric disturbances can be precipitated in ex-psychostimulant users that are unrelated to a drug challenge, but perhaps are in response to environmental stressors.[5] The long-lasting characteristic of behavioral sensitization is likely the result of a temporally equivalent alteration in brain function. Thus, it can be hypothesized that psychostimulant-sensitized individuals sustain enduring alterations in central nervous system function which predispose them to exacerbated responses to a psychostimulant or stressor challenge. The modern prevalence of psychostimulant abuse and the resulting social and medical manifestations of behavioral sensitization underscore the need to identify the neural alterations mediating sensitization, and to determine

1-8493-8311-0/95/$0.00+$.50

appropriate treatments for reversing the changes and/or blunting their behavioral consequences.

This chapter explores (1) how chronic psychostimulant administration might produce behavioral sensitization, (2) where in the brain neuronal function is affected, and (3) what neural changes occur during sensitization. While functional imaging studies and analysis of postmortem tissue in human psychostimulant addicts have been informative, they suffer from problems associated with relatively poor technical resolution. These include a lack of anatomical definition using current imaging technologies, and difficulties of interpretation using postmortem measures of tissue content to estimate neurotransmitter function.[9-11] To some extent these difficulties are resolved by employing experimental animals where more invasive procedures are possible. The animal model most frequently employed to study behavioral sensitization utilizes repeated injection of psychostimulants in rats, which produces a progressive increase in motor activity.[12-14] The motor responses typically measured are locomotor activity when low doses are employed or stereotyped behaviors when higher doses of drug are used.[15] Similar to drug-induced psychopathologies in humans, the augmented motor responses in rats are relatively enduring.[12,16] While a single injection of psychostimulant produces behavioral augmentation lasting only a few days, more aggressive treatment regimes using repeated injections for many days or weeks can produce behavioral sensitization persisting for months.[12,16,17] Also similar to sensitization in humans, cross-sensitization between psychostimulants and environmental stressors can be demonstrated in the rat model of behavioral sensitization.[16,18]

A plethora of data has been generated over the last two decades as a result of the rodent model of behavioral sensitization to psychostimulants. Numerous contradictions have emerged from these data which only recently have begun to coalesce into reasonably consistent models of the neural changes produced by repeated psychostimulants.[14] There are two primary reasons for the long gestation period required for the evolution of such consistent, albeit incomplete models. First, the development of behavioral sensitization is dynamic. Many changes are produced by chronic psychostimulant administration, only some of which are relevant to behavioral sensitization. Furthermore, some alterations are present for only a few hours or days after discontinuing repeated drug administration while other alterations appear only after a period of withdrawal and may persist for many days or weeks. Over the last 15 years, different laboratories have examined the effects of different psychostimulant drugs, administration regimes, and withdrawal periods. Thus, it has been necessary to accumulate many bits of information before constructing a consistent image of the sequelae of neuronal alterations underlying behavioral sensitization. Second, certain neuronal systems are impacted directly by psychostimulants due to a specific molecular site of action at monoamine transport proteins, notably the dopamine transporter. However, behavioral sensitization results from a multitude of more subtle alterations of neuronal function that are distributed throughout the circuitry in which dopamine neurons are embedded. Thus, the intense focus of research on dopamine transmission has yielded a skewed perception of the neuronal mechanisms mediating sensitization which only now is being addressed by studies showing an important role of excitatory amino acids (EAA), GABA, and serotonin.

II. INITIATION AND EXPRESSION OF BEHAVIORAL SENSITIZATION

Given the caveats outlined above which continue to impede an accurate understanding of the neural mechanisms mediating sensitization to psychostimulants, this chapter will discuss two interrelated models of neuronal alteration that appear to play an important role in behavioral sensitization. These models correspond to temporally and spatially distinct phases of sensitization, termed initiation and expression.[16] Expression can be defined as the changes in neuronal function that are associated with and causally related to behavioral sensitization. Accordingly, these alterations are relatively permanent. Initiation can be defined as the cellular and molecular alterations in neuronal function that precede the relatively permanent changes mediating the expression of behavioral sensitization. The changes associated with initiation also can be relatively permanent or may be components of transitory neuronal responses. This chapter will focus on the role played by dopaminergic neurons projecting from the ventral mesencephalon (including the ventral tegmental area; VTA) to the striatum (including the nucleus accumbens; NAc) and prefrontal cortex. The relative importance of afferent connections to the VTA, NAc, and prefrontal cortex in the initiation and expression of behavioral sensitization also will be explored, including the roles of EAA, GABA, and serotonin transmission.

III. BINDING TO THE DOPAMINE TRANSPORTER

It has been known for many years that the acute motor stimulant effect of cocaine depends upon dopamine transmission, especially in the NAc and striatum. Thus, dopamine lesions in these structures or the local microinjection of dopamine blockers prevents behavioral stimulation following a systemic cocaine challenge.[19-21] The mechanism for this dependence upon dopamine transmission was revealed by studies showing a correlation between binding to the dopamine transport protein and the behavioral effects of cocaine.[22,23] Thus, by preventing the elimination of dopamine from the synaptic cleft, dopamine transmission is enhanced which results in behavioral excitation. Considering that increased dopamine transmission is a primary site of action in mediating many of the acute behavioral effects of cocaine, it was logical to postulate that an augmentation in this action on dopamine neurons could mediate behavioral sensitization to cocaine. With the advent of technologies permitting the *in vivo* estimation of dopamine release, this hypothesis generally has been borne out. Thus, using microdialysis or voltammetry to measure extracellular dopamine content, animals pretreated with repeated injections of cocaine or other psychostimulants show a greater release of dopamine in the NAc or striatum in response to an acute drug, stress, or electrical stimulation challenge than do control animals.[14,24-35] While this general hypothesis has been verified many times, some laboratories observed that during the first few days of withdrawal from cocaine or amphetamine there exists no significant increase in extracellular dopamine content in the NAc or striatum.[33,34,36,37]

While many studies have evaluated the temporal dynamics of changes in axonal dopamine transmission, relatively few investigations of sensitization-related effects on somatodendritic dopamine have been conducted. However, investigators measuring the extracellular content of dopamine in the VTA found an inverse temporal sequence in somatodendritic dopamine release compared to axonal release. In contrast to accumbal release, the increase of extracellular dopamine content in the VTA is augmented during early withdrawal in animals behaviorally sensitized to cocaine (Figure 1), and by 2 weeks of withdrawal the levels returned to normal or were subnormal.[26,38]

The temporal dissociation of the effect of cocaine on somatodendritic versus axonal dopamine transmission indicates that the long-term expression of behavioral sensitization may be mediated in part by dopamine transmission in the NAc and striatum, while initiation may involve an action by cocaine on somatodendritic dopamine transport. Indeed, numerous studies attempting to sensitize rats to the intracranial administration of psychostimulants have found that while repeated microinjection of amphetamine into the NAc, striatum, or prefrontal cortex is ineffective, daily injection into the VTA results in behavioral sensitization to a systemic psychostimulant challenge.[39-43] Moreover, recent papers by Vezina and co-workers demonstrate that behavioral sensitization produced by repeated amphetamine administration into the VTA is associated with an augmentation in the increase in extracellular dopamine content in the NAc normally elicited by a

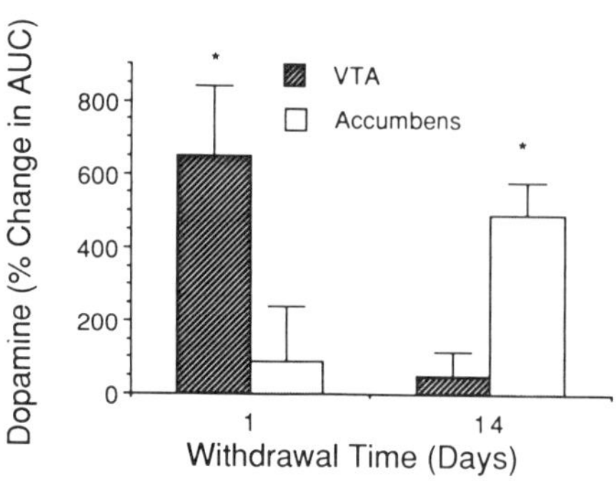

Figure 1. Opposite temporal sequence of changes in axonal versus somatodendritic dopamine release after sensitization to repeated cocaine. At one day of withdrawal, extracellular dopamine content was augmented in cocaine-sensitized rats in the VTA, but not in the nucleus accumbens; the opposite changes were present at 14 days of withdrawal. Data are derived from Kalivas and Duffy[33,38] and shown as the total change in extracellular dopamine for 2 hr following acute cocaine challenge (15 mg/kg, i.p.). AUC—area under the curve for the change in extracellular dopamine from cocaine-sensitized rats minus the cocaine-induced change in control rats.

systemic drug challenge and with an increase in the behavioral response to amphetamine microinjected into the NAc.[44,45] Thus, psychostimulants inhibit dopamine uptake in the VTA to initiate behavioral sensitization (Figure 2). However, once initiated, psychostimulant actions in the NAc to increase dopamine release at least partly mediate the expression of behavioral sensitization.

IV. A MODEL FOR INITIATION OF BEHAVIORAL SENSITIZATION IN THE VTA

Figure 3 shows the components of neurotransmission in the VTA that have been identified as potentially important in the initiation of behavioral sensitization. Each process is indicated by a number, and the evidence supporting or contradicting its role in initiation is discussed below.

A. ENHANCED SOMATODENDRITIC DOPAMINE RELEASE

When microinjected into the VTA a number of compounds can initiate behavioral sensitization to a subsequent psychostimulant challenge, including amphetamine,[39,41,43,44] μ-opioid receptor agonists,[46-50] neurotensin,[46,51,52] a neurokinin agonist (DiMeC7),[16] a calcium-dependent potassium channel antagonist (apamin),[53] and pertussis toxin.[54] Similarly, repeated electrical stimulation in the VTA produces behavioral sensitization to amphetamine.[55] This diverse array of molecules act at different binding sites and on different neuronal elements within the VTA. However, a common characteristic among these compounds is that they increase somatodendritic dopamine transmission.[53,56,57]

B. DESENSITIZATION OF D_2 AUTORECEPTORS

Stimulation of D_2 receptors by somatodendritically released dopamine hyperpolarizes the dopamine cell membrane due to a G protein-dependent increase in potassium conductance.[58] Acute presentation of amphetamine to dopamine cells *in vitro* decreases the capacity of D_2 receptor stimulation to elicit membrane hyperpolarization.[59] A similar

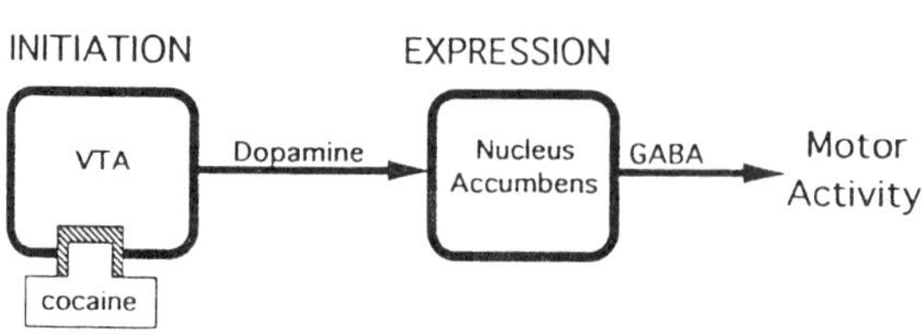

Figure 2. Repeated cocaine administration acts in the VTA to initiate behavioral sensitization, but enhanced dopamine release in the nucleus accumbens and other dopamine terminal fields mediates the expression of behavioral sensitization.

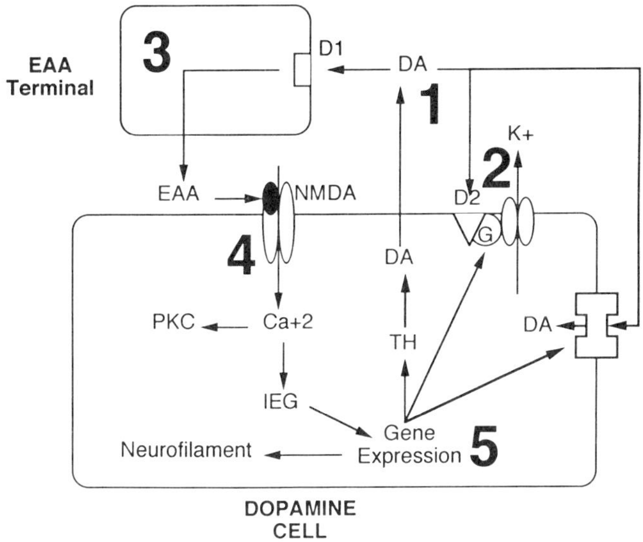

Figure 3. A model of the sequence of events involved in the initiation of behavioral sensitization to repeated cocaine. The numbers indicate steps in the sequelae of cellular events proposed to cause the initiation of behavioral sensitization and are explained in the text. DA—dopamine, EAA—excitatory amino acid, G—G protein, IEG—immediate early gene, NMDA—N-methyl-D-aspartate, PKC—protein kinase C, TH—tyrosine hydroxylase.

desensitization of the D_2 receptor is observed *in vivo* followed repeated administration of either amphetamine or cocaine. Thus, the ability of iontophoretic or systemic D_2 agonists to decrease dopamine cell firing frequency is reduced in animals sensitized to repeated psychostimulants.[34,60-63] This desensitization is not associated with a reduction in D_2 receptor density or D_2 mRNA content in the VTA or adjacent substantia nigra (unpublished observation).[64] The lack of alteration on D_2 receptor mRNA or expression suggests that psychostimulant-induced desensitization may result from a reduction of receptor coupling to G proteins. It is known that somatodendritic D_2 receptor transduction occurs via a pertussis toxin-sensitive G protein and that repeated cocaine administration elicits a short term (less than 24 hr) reduction of the *in vitro* pertussis toxin-catalyzed ADP-ribosylation of G proteins in the VTA but not in the adjacent substantia nigra.[65,66] This indicates a reduction of the heterotrimeric state of G proteins which would result in a decrease of receptor coupling. Furthermore, the reduction of *in vitro* ADP-ribosylation was correlated negatively with the extent of behavioral sensitization,[66] and, as indicated above, a single microinjection of pertussis toxin into the VTA produces behavioral sensitization to cocaine that is associated with augmented extracellular dopamine in the NAc.[54] Unfortunately, it has not been verified that the alteration of G protein function in the VTA is associated specifically with D_2 receptors, so it may represent altered transduction of other pertussis toxin-sensitive G protein coupled receptors.

Other evidence supporting a role for impaired transduction of D_2 receptors in behavioral sensitization comes from studies of the actions of neurotensin in the VTA. Neurotensin increases dopamine cell firing frequency by initiating protein kinase-dependent phosphorylation and desensitization of D_2 autoreceptors.[67] Repeated neurotensin microinjection into the VTA produces long-lasting sensitization to a subsequent amphetamine challenge, and it has been shown that inhibiting protein kinases in the VTA prevents the initiation of behavioral sensitization by repeated neurotensin administration.[68]

C. STIMULATION OF D_1 RECEPTORS AND GLUTAMATE RELEASE

The desensitization of D_2 receptors following repeated psychostimulants further increases the somatodendritic release of dopamine since this important mechanism of membrane hyperpolarization is blunted. The increased extracellular concentration of dopamine more effectively stimulates D_1 receptors, the importance of which in behavioral sensitization is indicated by the finding that microinjection of SCH 23390, a D_1 receptor antagonist, into the VTA prevents the initiation of behavioral sensitization to systemic amphetamine.[69]

D_1 receptors are located primarily on descending afferent fibers innervating the VTA, as indicated by the sparse or absent expression of D_1 receptor mRNA in the VTA[70] and by a >70% reduction of D_1 receptor density in the substantia nigra following transection of the striatonigral pathway.[17,71] It is generally assumed that D_1 receptors in the substantia nigra and VTA reside primarily on GABAergic terminals, whereupon receptor activation promotes the release of GABA to hyperpolarize dopaminergic cells.[72,73] However, at least in the VTA, stimulation of D_1 receptors more effectively releases the EAA, glutamate (minimum effective dose determined by dialysis = 3 μM), than GABA (minimum effective dose = 300 μM).[74] This suggests that stimlation of D_1 receptors by physiological levels of extracellular dopamine provides positive rather than negative feedback to dopamine cells.

D. STIMULATION OF NMDA RECEPTORS

Many laboratories have found that the systemic administration of NMDA receptor antagonists prevents the initiation of behavioral sensitization to cocaine or amphetamine.[75-77] Considering the apparent role of the VTA in initiating behavioral sensitization, it is possible that the systemically administered antagonist may be acting at least in part to block NMDA receptors in the VTA. In fact, cocaine-induced sensitization is blocked by pretreating the VTA with either a noncompetitive (MK-801) or competitive [3-(2-carboxypiperazine-4-yl)propyl-1-phosphonic acid] NMDA antagonist.[78] Thus, it is proposed that activation of D_1 receptors by somatodendritically released dopamine stimulates EAA release which in turn stimulates NMDA receptors on dopamine cells (Figure 3).

E. ALTERING GENE EXPRESSION

A number of studies indicate that an alteration of gene expression may be important in the initiation of behavioral sensitization to psychostimulants. However, no specific change has been identified as critical. Inhibition of protein synthesis prevents behavioral sensitization to repeated amphetamine,[79] and direct application of the protein synthesis inhibitor, anisomycin, into the VTA prevents behavioral sensitization to systemic cocaine injections (personal communication; B. Sorg, Washington State University). Changes in the tissue content of six different proteins have been identified in the VTA following repeated cocaine administration. One study reported a reduction of both G_i and G_o,[65] while another did not.[80] Such a reduction of G protein content could limit the coupling of D_2 receptors to potassium channels, thereby contributing to the desensitization of D_2 receptors produced during the initiation of behavioral sensitization. A short-term elevation in the content of the rate limiting enzyme in dopamine synthesis, tyrosine hydroxylase, has been demonstrated.[81-83] This increase of tyrosine hydroxylase is probably a compensatory change to accommodate the marked elevation of somatodendritic dopamine turnover associated with the initiation of sensitization. A reduction of neurofilament protein content associated with decreased axonal transport has been observed in the VTA,[84] which may account for the delayed appearance of enhanced dopamine release in the NAc and striatum of behaviorally sensitized animals.[33,34,85] Finally, a short-term

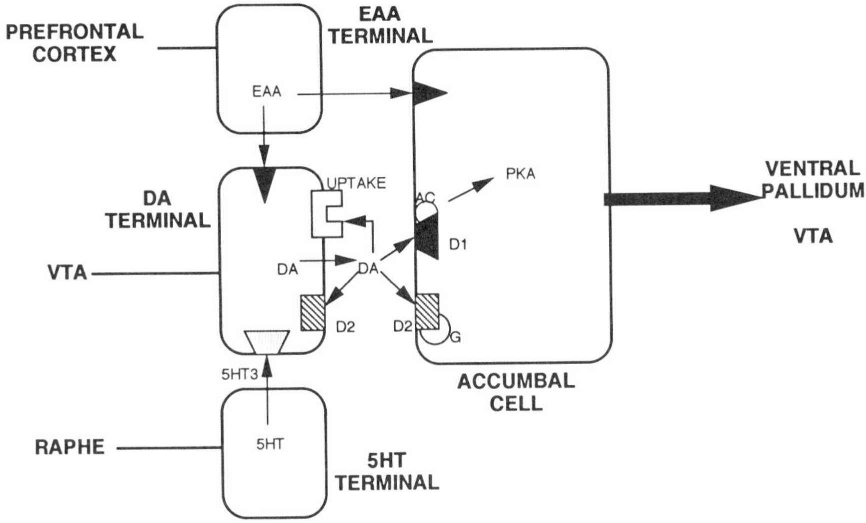

Figure 4. A model of the neural elements that may be involved in the expression of behavioral sensitization in the nucleus accumbens. Afferent projections to the nucleus accumbens arise from the prefrontal cortex (EAA), VTA (DA), and raphe nuclei (5HT) and efferent projections extend to the ventral pallidum and VTA. See text for detailed description. AC—adenylyl cyclase, DA—dopamine, G—G protein, 5HT—serotonin, PKA—protein kinase A.

decrease in mRNA encoding the dopamine transporter was observed.[86] Assuming that this reduction of mRNA decreases the expression of the dopamine transporter, clearance of dopamine would be reduced which might underlie the elevation of extracellular dopamine content associated with behavioral sensitization.

V. A MODEL FOR THE EXPRESSION OF BEHAVIORAL SENSITIZATION

As mentioned above, many studies have demonstrated that the expression of behavioral sensitization is associated with an augmentation in the increase of accumbal extracellular dopamine content following an acute psychostimulant challenge. The enhanced dopamine transmission probably results from alterations present in the dopamine terminals since augmented dopamine release can be shown in tissue slices or synaptosomes obtained from the NAc or striatum of animals sensitized to psychostimulants.[85,87-89] The cellular alterations mediating the enhanced releaseability of dopamine remain unknown. Figure 4 illustrates various aspects of the synaptic organization of the NAc that may be involved in augmenting dopamine release, including (1) alteration of presynaptic release mechanisms, (2) alteration of afferent regulation of release, (3) augmentation of postsynaptic responsiveness to dopamine, and (4) involvement of GABAergic efferent projections from the NAc.

A. ALTERATION OF PRESYNAPTIC RELEASE MECHANISMS

Enhanced dopamine release in accumbal or striatal tissue slices from sensitized rats can be revealed using either amphetamine or potassium to increase release.[90] It is generally believed that these releasing agents function by two distinct mechanisms. Amphetamine increases dopamine release by promoting the efflux of dopamine through the uptake carrier.[91] This dopamine arises directly from cytosolic pools and probably to some extent as a result of amphetamine-induced dopamine efflux from synaptic vesicles into the cytoplasm.[91,92] In either event, amphetamine-induced release is calcium-independent. In contrast, potassium depolarizes the synaptic terminal producing an increase of calcium influx via voltage-dependent calcium channels, thereby promoting dopamine release from synaptic vesicles.[93,94] Thus, it appears that both calcium-dependent and independent dopamine release is augmented in sensitized rats. However, it has been shown that the portion of amphetamine-induced dopamine release that is augmented in the NAc of behaviorally sensitized rats is calcium-dependent.[95] The apparent calcium-dependence of sensitized dopamine release strongly implicates a role for the regulation of intracellular calcium levels in the enhanced releaseability of dopamine. The regulation of intracellular calcium may be at the level of calcium release from intracellular storage pools or calcium influx from the extracellular space. At present, no data exist to determine which pool(s) may be involved. However, consistent with the importance of extracellular calcium, systemic administration of calcium channel blockers prevents the expression of cocaine-induced behavioral sensitization.[96]

In addition to increased release, a reduction of presynaptic dopamine uptake could significantly enhance the extracellular concentration of dopamine. Although a reduction of the tissue content of dopamine transporter mRNA has been reported in the ventral mesencephalon after repeated cocaine,[86] the data regarding dopamine transporter expression in axon terminals are inconsistent. Estimates of the maximal binding concentration and affinity for the dopamine transporter obtained in accumbal or striatal homogenate binding assays have revealed a reduction,[97,98] increase,[99] or no change[64,100-105] after repeated cocaine. Examining the *in vitro* uptake of [^{3}H]dopamine has revealed similar inconsistencies with one group reporting a reduction of uptake affinity in the NAc, but not the striatum,[101] while another observed an increase of striatal dopamine after an acute cocaine challenge in rats pretreated with repeated cocaine.[100] Using *in vivo* voltammetry to measure the clearance of electrically stimulated dopamine from the NAc after repeated cocaine treatment, dopamine reuptake was increased was increased at 1, but not at 10, days of withdrawal.[27] This finding has been replicated using microdialysis to demonstrate increased elimination of dopamine from the NAc at 1, but not 10 or 14, days of withdrawal following repeated cocaine.[33,106] In contrast, using *in vitro* rotating disk voltammetry, Schenk and co-workers found that the rate of dopamine uptake in striatal tissue slices increased at 1, 14, or 21 days of withdrawal following either repeated noncontingent or self-administered cocaine.[107,108] Finally, *in vivo* voltammetry has revealed a reduction of dopamine elimination in the NAc, but not the striatum after 7 days of withdrawal from repeated cocaine.[109] Based upon the data outlined above, it is possible that alterations of dopamine uptake might influence the expression of behavioral sensitization. However, the contradictory nature of the findings to date do not present a consistent image of whether these changes would promote or inhibit the augmentation of extracellular dopamine which is proposed to be critical to the expression of behavioral sensitization.

Another presynaptic mechanism whereby dopamine release could be altered is via the D_2 autoreceptor. The stimulation of presynaptic D_2 autoreceptors reduces the release of dopamine in response to a depolarizing stimulus.[110,111] Although a consistent alteration of D_2 receptor density has not been reported to be associated with behavioral sensitization,[12,16] the capacity of D_2 agonists to modulate dopamine release is altered in rats sensitized to repeated injections of amphetamine-like psychostimulants. Unfortunately, the current literature is inconsistent in this regard with reports of sensitization,[112] desensitization,[113] or no change[114] of D_2 receptor-mediated inhibition of *in vitro* dopamine release. However, daily cocaine treatment produced a significant blunting in the capacity of the D_2/D_3 receptor agonist, quinpirole, perfused through a dialysis probe placed in the NAc to inhibit dopamine release during the first 48 hr of drug withdrawal, although this effect had subsided by 3 weeks of withdrawal (unpublished observation).

B. ALTERATION OF THE AFFERENT REGULATION OF DOPAMINE RELEASE

Dopamine release in the NAc can be increased by either EAA or serotonin. Serotonin stimulation of dopamine release arises from a selective action on the $5HT_3$ receptor subtype while EAA stimulation of both NMDA and non-NMDA receptors releases dopamine.[115-119] An augmentation in the tone of either of these afferent modulators of dopamine release could be involved in the expression of behavioral sensitization. A role for serotonin is plausible since cocaine inhibits serotonin reuptake.[22,120] Furthermore, there is an augmentation in extracellular serotonin levels in the NAc of animals sensitized to repeated cocaine.[26]

It is also possible that EAA transmission may be altered in the NAc of rats sensitized to cocaine. While a change in accumbal EAA levels remains to be evaluated in sensitized animals, exposure to acute stress is known to elevate extracellular glutamate levels.[121] Furthermore, an increase in accumbal glutamate release is consistent with the effect of cocaine sensitization on extracellular dopamine in the rat prefrontal cortex. A variety of studies have revealed an inverse relationship between dopamine release in the NAc and prefrontal cortex.[122] Thus, enhanced release of dopamine in the prefrontal cortex inhibits dopamine release in the NAc and vice versa.[123] This is thought to result from the inhibitory regulation by dopamine of EAA projections from the prefrontal cortex to the NAc and VTA.[122] Cocaine-induced release of dopamine in the prefrontal cortex is blunted significantly in rats behaviorally sensitized to repeated stress or cocaine administration,[124,125] which would thereby increase EAA release in the NAc of sensitized animals resulting in enhanced dopamine release. Whether this transynaptic regulation of accumbal dopamine release by prefrontal dopamine and EAA is a factor in the expression of sensitization to cocaine remains to be evaluated directly. However, it should be noted that the blunting of prefrontal dopamine transmission in animals sensitized to cocaine or stress is not manifested by an amphetamine challenge.[126] Since behavioral cross-sensitization can be shown between amphetamine and cocaine,[5] this EAA-dependent mechanism for augmenting accumbal dopamine release does not appear to be necessary to induce expression of behavioral sensitization.

C. ENHANCED POSTSYNAPTIC RESPONSIVENESS TO DOPAMINE

When animals are behaviorally sensitized to cocaine or amphetamine, the capacity of dopamine to inhibit the firing of neurons in the NAc increases.[127,128] This augmentation of postsynaptic responsiveness is mediated by D_1, but not D_2, receptors, and persists for at least 30 days after discontinuing repeated cocaine injections.[127] Enhanced postsynaptic sensitivity to dopamine also has been shown in the NAc of animals sensitized to self-administered cocaine (personal communication: S. White, Washington State University). D_1 receptors are coupled positively to adenylyl cyclase and, consistent with an augmentation of D_1 receptor responsiveness, cocaine sensitization increases adenylyl cyclase and protein kinase activity in the NAc.[81,129] However, it is not known if these cellular correlates persist longer than 24 hr after discontinuing repeated cocaine. Also, attempts to demonstrate an increase in D_1-stimulated adenylyl cyclase in the NAc or striatum of animals sensitized to repeated cocaine have not been successful.[130] Finally, preliminary data indicate that the motor stimulant effect elicited by microinjection of the D_1 agonist, SKF 28395, into the NAc is not altered in rats pretreated with daily cocaine at either 1 or 21 days of withdrawal (unpublished observation).

D. GABAERGIC EFFERENT PROJECTIONS FROM THE NUCLEUS ACCUMBENS

The spiny NAc neurons receiving dopaminergic input are GABAergic and project to the ventral pallidum and the VTA.[131,132] Extracellular GABA content in the ventral pallidum

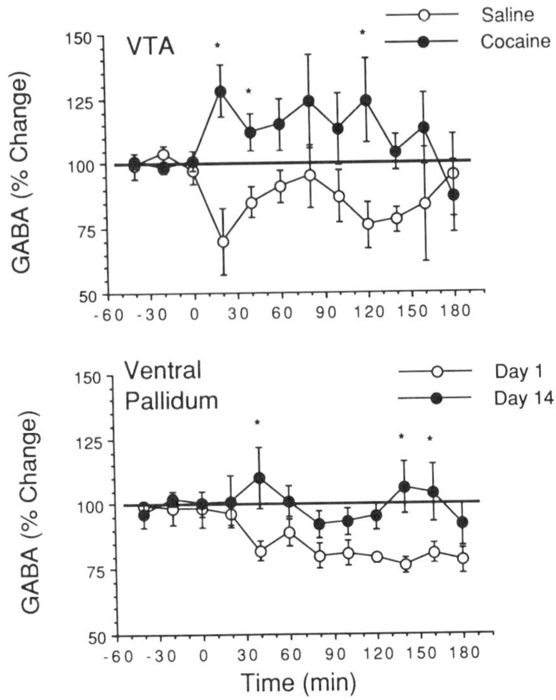

Figure 5. Changes in extracellular GABA concentrations in the ventral pallidum and VTA in rats sensitized to repeated cocaine. GABA was measured using microdialysis in the VP or VTA as described elsewhere.[137] Cocaine (30 mg/kg, i.p./day) or saline was administered for 7 days for the VTA study and GABA was measured in response to an acute cocaine challenge (15 mg/kg, i.p.) on day 21 of withdrawal. For the measurement of pallidal GABA, rats were treated for 5 days with cocaine (15 mg/kg, i.p.) and GABA measured on the first day of cocaine administration and at day 9 of withdrawal (day 14). Extracellular GABA was normalized to the average of three baseline samples. The data are shown as mean ± SEM, the N was >7 in each group and the arrow indicates the injection of cocaine at time = 0. The data in the ventral pallidum were adapted from Bourdelais.[138] *—$p \leq 0.05$—significant ($p \leq 0.05$) difference between the two treatment groups at each time using a two-way repeated measures ANOVA. A significant interaction F score was found in both experiments.

and VTA is altered in response to an acute cocaine challenge 1 or 2 weeks after discontinuing repeated cocaine (Figure 5). These data indicate that GABA transmission is elevated in response to a cocaine challenge in sensitized compared to control rats. Since increased GABAergic tone in the ventral pallidum and VTA inhibits spontaneous and psychostimulant-induced motor activity, the apparent increase in GABA release in sensitized rats may actually constitute an effort by the brain to inhibit the pharmacologically stimulated motor response.[56,133] However, if the increase of extracellular GABA content in the VTA results selectively from GABAergic input to GABA interneurons, then the apparent increase in GABA release would disinhibit dopamine cells and promote motor activity.[56]

VI. CONCLUSIONS

The models described above constitute an attempt to integrate many of the findings made over the last decade regarding the cellular mechanisms mediating the initiation and

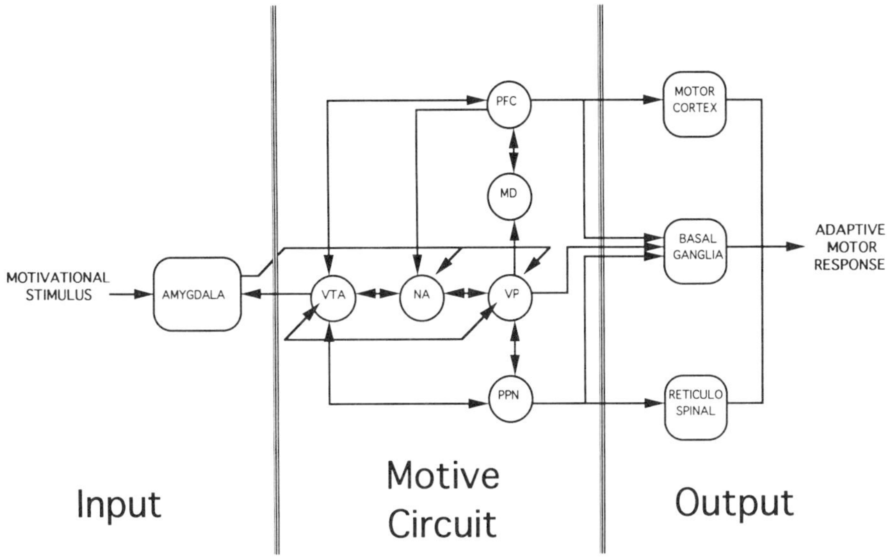

Figure 6. The circuitry of mesoaccumbens dopamine system that may be involved in the expression of behavioral sensitization to cocaine. See Kalivas et al.[14] for more details. MD—mediodorsal thalamus, NA—nucleus accumbens, PFC—prefrontal cortex, PPN—pedunculopontine region, VP—ventral pallidum.

expression of behavioral sensitization to cocaine. It should be noted that these models are incomplete, and some of the identified or postulated cellular mechanisms may be revealed in future studies to be epiphenomena which are not related casually to behavioral sensitization. Nonetheless, they provide constructs upon which to design and test hypotheses.

The thrust of experimentation to date necessarily assumes that the causal mechanisms of behavioral sensitization to cocaine ultimately can be reduced to a distinct region of the brain and/or a precise sequence of cellular and molecular alterations. This assumption has been permissive and, indeed, essential to our emerging understanding of the neural underpinnings of behavioral sensitization to cocaine and other psychostimulants. However, any given cellular alteration functionally manifests itself through the neural circuitry in which it is embedded. Thus, a full comprehension of behavioral sensitization will arise only by understanding how cellular alterations modify the functional status of the circuitry located distal to the molecular site of action of cocaine. As outlined above, an important molecular site of action for cocaine is the dopamine transport protein. As a result, alterations in dopamine transmission have been identified most readily as causal in the initiation and expression of behavioral sensitization to cocaine. Only recently have studies begun to consider how this impact on dopamine neurons may affect the circuitry in which the dopamine projections are embedded. Figure 6 illustrates a simplified version of the circuitry adjacent to the mesoaccumbens dopamine projection. This circuitry allows extensive positive and negative feedback relationships and numerous studies indicate that it plays a role in translating motivationally relevant stimuli into adaptive behavioral responses.[134] These include not only exploratory motor behaviors but also behavioral responses to conditioned stimuli and to stressors.[135,136] Therefore, in addition to future studies identifying in greater detail the cellular and molecular alterations associated with behavioral sensitization, it is important to consider how the functional status of these circuits may be altered in the initiation and expression of behavioral sensitization to repeated psychostimulants. The value of this approach is clarified by the improved understanding of behavioral sensitization that has arisen by considering the role of EAA afferents to the mesoaccumbens dopamine projection. It is likely that future consideration of both afferent and efferent mesoaccumbal interconnectivity will yield similar important insights.

REFERENCES

1. Angrist, B. M. and Sudilovsky, A., Central nervous system stimulants: historical aspects and clinical effects, in *Handbook of Psychopharmacology, Vol. 11, Stimulants,* Iversen, L. L., Iversen, S. D., and Snyder, S. H., Eds., Plenum Press, New York, 1978, 99.
2. Connell, P. H., *Amphetamine Psychosis,* Chapman and Hill, London, 1958.
3. Ellinwood, E. H., Sudilovsky, A., and Nelson, L. M., Evolving behavior in the clinical and experimental amphetamine model psychosis, *Am. J. Psychiatry,* 130, 1088, 1973.
4. Post, R. M. and Kopanda, R. T., Cocaine, kindling, and psychosis, *Am. J. Psychiatry,* 133, 627, 1976.
5. Post, R. M. and Weiss, S. R. B., Sensitization and kindling: implications for the evolution of psychiatric symptomatology, in *Sensitization in the Nervous System,* Kalivas, P. W. and Barnes, C. D., Eds., Telford Press, Caldwell, NJ, 1988, 257.
6. Gawin, F. H., Cocaine addiction: psychology and neurophysiology, *Science,* 251, 1580, 1991.
7. Sato, M., Chen, C.-C., Akiyama, K., and Otsuki, S., Acute exacerbation of paranoid psychotic state after long-term abstinence in patients with previous methamphetamine psychosis, *Biol. Psychiatry,* 18, 429, 1983.
8. Sato, M., A lasting vulnerability to psychosis in patients with previous methamphetamine psychosis, *Annu. N.Y. Acad. Sci.,* 654, 160, 1992.
9. Townsend, D. W., Optimization of signal in positron emission tomography scans: present and future developments, in *Exploring Brain Functional Anatomy with Positron Tomography,* Porter, R., Ed., John Wiley and Sons, New York, 1991, 57.
10. Volkow, N. D., Fowler, J. S., Wang, G. J., Hitzemann, R., Logan, J., Schyler, D. J., Dewey, S. L., and Wolf, A. P., Decreased dopamine D_2 receptor availability is associated with reduced frontal metabolism in cocaine abusers, *Synapse,* 14, 167, 1993.
11. Hurd, Y. L. and Herkenham, M., Molecular alterations in the neostriatum of human cocaine addicts, *Synapse,* 13, 357, 1993.
12. Robinson, T. E. and Becker, J. B., Enduring changes in brain and behavior produced by chronic amphetamine administration: a review and evaluation of animal models of amphetamine psychosis, *Brain Res. Rev.,* 11, 157, 1986.
13. Segal, D. S. and Schuckit, M. A., Animal models of stimulant-induced psychosis, in *Stimulants: Neurochemical, Behavioral, and Clinical Perspectives,* Creese, I., Ed., Raven Press, New York, 1983, 131.
14. Kalivas, P. W., Hooks, M. S., and Sorg, B., The pharmacology and neural circuitry of sensitization to psychostimulants, *Behav. Pharmacol.,* 4, 315, 1993.
15. Segal, D. S. and Kuczenski, R., Individual differences in responsiveness to single and repeated amphetamine administration: behavioral characteristics and neurochemical correlates, *J. Pharmacol. Exp. Ther.,* 242, 917, 1987.
16. Kalivas, P. W. and Stewart, J., Dopamine transmission in the initiation and expression of drug- and stress-induced sensitization of motor activity, *Brain Res. Rev.,* 16, 223, 1991.
17. Weiss, S. R. B., Post, R. M., Pert, A., Woodward, R., and Murman, D., Context-dependent cocaine sensitization: differential effect of haloperidol on development versus expression, *Pharmacol. Biochem. Behav.,* 34, 655, 1989.
18. Antelman, S. M., Stressor-induced sensitization to subsequent stress: implications for the development and treatment of clinical disorders, in *Sensitization in the Nervous System,* Kalivas, P. W. and Barnes, C. D., Eds., Telford Press, Caldwell, NJ, 1988, 227.
19. Kelly, P. H. and Iversen, S. D., Selective 6-OHDA-induced destruction of mesolimbic dopamine neurons: Abolition of psychostimulant-induced locomotor activity in rats, *Eur. J. Pharmacol.,* 40, 45, 1975.
20. Koob, G. F., Stinus, L., and le Moal, M., Hyperactivity and hypoactivity produced by lesions to the mesolimbic dopamine system, *Behav. Brain Res.,* 3, 341, 1981.

21. Clarke, P. B. S., Jakubovic, A., and Fibiger, H. C., Anatomical analysis of the involvement of mesolimbocortical dopamine in the locomotor stimulant actions of d-amphetamine and apomorphine, *Psychopharmacology,* 96, 511, 1988.
22. Reith, M. E. A., Meisler, B. E., Sershen, H., and Lajtha, A., Structural requirements for cocaine congeners to interact with dopamine and serotonin uptake sites in mouse brain and to induce stereotyped behavior, *Biochem. Pharmacol.,* 35, 1123, 1986.
23. Ritz, M. C., Lamb, R. J., Goldberg, S. R., and Kuhar, M. J., Cocaine receptors on dopamine transporters are related to self-administration of cocaine, *Science,* 237, 1219, 1987.
24. Patrick, S. L., Thompson, T. L., Walker, J. M., and Patrick, R. L., Concomitant sensitization of amphetamine-induced behavioral stimulation and *in vivo* dopamine release from rat caudate nucleus, *Brain Res.,* 538, 343, 1991.
25. Pettit, H. O., Pan, H.- T., Parsons, L. H., and Justice, J. B., Jr., Extracellular concentrations of cocaine and dopamine are enhanced during chronic cocaine administration, *J. Neurochem.,* 55, 798, 1990.
26. Parsons, L. H. and Justice, J. B., Jr., Serotonin and dopamine sensitization in the nucleus accumbens, ventral tegmental area and dorsal raphe nucleus following repeated cocaine administration, *J. Neurochem.,* 61, 1611, 1993.
27. Ng, J. P., Hubert, G. W., and Justice, J. B., Jr., Increased stimulated release and uptake of dopamine in the nucleus accumbens after repeated cocaine administration as measured by *in vivo* voltammetry, *J. Neurochem.,* 56, 1485, 1991.
28. Sorg, B. A. and Kalivas, P. W., Behavioral and neurochemical cross-sensitization between footshock stress and cocaine, *Brain Res.,* 559, 29, 1991.
29. Akimoto, K., Hamamura, T., Kazahaya, Y., Akiyama, K., and Otsuki, S., Enhanced extracellular dopamine level may be the fundamental neuropharmacological basis of crossbehavioral sensitization between methamphetamine and cocaine—an *in vivo* dialysis study in freely moving rats, *Brain Res.,* 507, 344, 1990.
30. Akimoto, K., Hamamura, T., and Otsuki, S., Subchronic cocaine treatment enhances cocaine-induced dopamine efflux, studies by *in vivo* intracerebral dialysis, *Brain Res.,* 490, 339, 1989.
31. Kazahaya, Y., Akimoto, K., and Saburo, O., Subchronic methamphetamine treatment enhances methamphetamine- or cocaine-induced dopamine efflux *in vivo, Biol. Psychiatr.,* 25, 903, 1989.
32. Robinson, T. E., Jurson, P. A., Bennett, J. A., and Bentgen, K. M., Persistent sensitization of dopamine neurotransmission in ventral striatum (nucleus accumbens) produced by prior experience with (+)-amphetamine: a microdialysis study in freely moving rats, *Brain Res.,* 462, 211, 1988.
33. Kalivas, P. W. and Duffy, P., Time course of extracellular dopamine and behavioral sensitization to cocaine. I. Dopamine axon terminals, *J. Neurosci.,* 13, 266, 1993.
34. Wolf, M. E., White, F. J., Nassar, R., Brooderson, R. J., and Khansa, M. R., Differential development of autoreceptor subsensitivity and enhanced dopamine release during amphetamine sensitization, *J. Pharmacol. Exp. Ther.,* 264, 249, 1993.
35. Hooks, M. S., Duffy, P., Striplin, C., and Kalivas, P. W., Behavioral and neurochemical sensitization following cocaine self administration, *Psychopharmacology,* 1994, in press.
36. Segal, D. S. and Kuczenski, R., In vivo microdialysis reveals a diminished amphetamine-induced dopamine response corresponding to behavioral sensitization produced by repeated amphetamine pretreatment, *Brain Res.,* 521, 330, 1992.
37. Segal, D. S. and Kuczenski, R., Repeated cocaine administration induces behavioral sensitization and corresponding decreased extracellular dopamine responses in caudate and accumbens, *Brain Res.,* 577, 351, 1992.
38. Kalivas, P. W. and Duffy, P., Time course of extracellular dopamine and behavioral sensitization to cocaine. II. Dopamine perikarya, *J. Neurosci.,* 13, 276, 1993.
39. Vezina, P. and Stewart, J., Amphetamine administered to the ventral tegmental area but not to the nucleus accumbens sensitizes rats to systemic morphine: lack of conditioned effects, *Brain Res.,* 516, 99, 1990.
40. Dougherty, G. G., Jr. and Ellinwood, E. H., Jr., Chronic d-amphetamine in nucleus accumbens: lack of tolerance or reverse tolerance of locomotor activity, *Life Sci.,* 28, 2295, 1981.
41. Kalivas, P. W. and Weber, B., Amphetamine injection into the A10 dopamine region sensitizes rats to peripheral amphetamine, and cocaine, *J. Pharmacol. Exp. Ther.,* 245, 1095, 1988.
42. Hitzemann, R., Wu, J., Hom, D., and Loh, H., Brain locations controlling the behavioral effects of chronic amphetamine intoxication, *Psychopharmacology,* 72, 93, 1980.

43. Hooks, M. S., Jones, G. H., Liem, B. J., and Justice, J. B., Jr., Sensitization and individual differences to intraperitoneal amphetamine, cocaine or caffeine following repeated intracranial amphetamine infusions, *Pharmacol. Biochem. Behav.*, 43, 815, 1992.
44. Vezina, P., Amphetamine injected into the ventral tegmental area sensitizes the nucleus accumbens dopaminergic response to systemic amphetamine: an *in vivo* microdialysis study in the rat, *Brain Res.*, 605, 332, 1993.
45. Perugini, M. and Vezina, P., Amphetamine administered to the ventral tegmental area sensitizes rats to the locomotor effects of nucleus accumbens amphetamine, *J. Pharmacol. Exp. Ther.*, 1994, in press.
46. Kalivas, P. W. and Duffy, P., Effect of acute and daily neurotensin and enkephalin treatments on extracellular dopamine in the nucleus accumbens, *J. Neurosci.*, 10, 2940, 1990.
47. Kalivas, P. W., Sensitization to repeated enkephalin administration into the ventral tegmental area of the rat. II. Involvement of the mesolimbic dopamine system, *J. Pharmacol. Exp. Ther.*, 225, 544, 1985.
48. Joyce, E. M. and Iversen, S. D., The effect of morphine applied locally to mesencephalic dopamine cell bodies on spontaneous motor activity in the rat, *Neurosci. Lett.*, 14, 207, 1979.
49. Vezina, P., Kalivas, P. W., and Stewart, J., Sensitization occurs to the locomotor effects of morphine and the specific mu opioid receptor agonist, DAGO, administered repeatedly to the VTA but not to the nucleus accumbens, *Brain Res.*, 417, 51, 1987.
50. Vezina, P. and Stewart, J., Conditioning and place-specific sensitization of increases in activity induced by morphine in the VTA, *Pharmacol. Biochem. Behav.*, 20, 925, 1984.
51. Kalivas, P. W. and Taylor, S., Effect of daily neurotensin injection into the ventral tegmental area, *Brain Res.*, 358, 70, 1985.
52. Elliot, P. J. and Nemeroff, C. B., Repeated neurotensin administration in the ventral tegmental area: effects on baseline and d-amphetamine-induced locomotor activity, *Neurosci. Lett.*, 68, 239, 1986.
53. Steketee, J. D. and Kalivas, P. W., Modulation of dopamine neurons by apamin, *J. Pharmacol. Exp. Ther.*, 254, 711, 1990.
54. Steketee, J. D. and Kalivas, P. W., Sensitization to psychostimulants and stress following injection of pertussis toxin into the A10 region, *J. Pharmacol. Exp. Ther.*, 259, 916, 1991.
55. Ben-Shahar, O. and Ettenberg, A., Repeated electrical stimulation of the VTA results in sensitization to the psychomotor stimulant effects of amphetamine, *Soc. Neurosci. Abst.*, 19, 334.9, 1993.
56. Kalivas, P. W., Neurotransmitter regulation of dopamine neurons in the ventral tegmental area, *Brain Res. Rev.*, 18, 75, 1993.
57. Steketee, J. D., Striplin, D. C., Murray, T. F., and Kalivas, P. W., Effect of pertussis toxin in the A10 region on dopamine transmission, *J. Neurochem.*, 58, 811, 1992.
58. Lacey, M. G., Mercuri, N. B., and North, R. A., Dopamine acts at D_2 receptors to increase potassium conductance in neurons of the rat substantia nigra, *J. Physiol. (Lond.)*, 392, 397, 1987.
59. Seutin, V., Verbank, P., Massotte, L., and Dresse, A., Acute amphetamine-induced subsensitivity of A10 dopamine autoreceptors in vitro, *Brain Res.*, 558, 141, 1991.
60. Antelman, S. M. and Chiodo, L. A., Dopamine autoreceptor subsensitivity: a mechanism common to the treatment of depression and the induction of amphetamine psychosis, *Biol. Psychiatr.*, 16, 717, 1981.
61. Henry, D. J., Greene, M. A., and White, F. J., Electrophysiological effects of cocaine in the mesoaccumbens dopamine system: repeated administration, *J. Pharmacol. Exp. Ther.*, 251, 833, 1989.
62. Ackerman, J. M. and White, F. J., A10 somatodendritic dopamine autoreceptor sensitivity following withdrawal from repeated cocaine treatment, *Neurosci. Lett.*, 117, 181, 1990.
63. White, F. J. and Wang, R. Y., Pharmacological characterization of dopamine autoreceptors in the rat ventral tegmental area: microiontophoretic studies, *J. Pharmacol. Exp. Ther.*, 231, 275, 1984.
64. Peris, J., Boyson, S. J., Cass, W. A., Curella, P., Dwoskin, L. P., Larson, G., Lin, L.-H., Yasuda, R. P., and Zahniser, N. R., Persistence of neurochemical changes in dopamine systems after repeated cocaine administration, *J. Pharmacol. Exp. Ther.*, 253, 38, 1990.
65. Nestler, E. J., Terwilliger, R. Z., Walker, J. R., Sevarino, K. A., and Duman, R. S., Chronic cocaine treatment decreases levels of the G protein subunits $G_{i\alpha}$ and $G_{o\alpha}$ in discrete regions of rat brain, *J. Neurochem.*, 55, 1079, 1990.
66. Striplin, C. D. and Kalivas, P. W., Correlation between behavioral sensitization to cocaine and G protein ADP-ribosylation in the ventral tegmental area, *Brain Res.*, 579, 181, 1992.

67. Shi, W. X. and Bunney, B. S., Role of intracellular cAMP and protein kinase A in actions of dopamine and neurotensin on midbrain dopamine neurons, *J. Neurosci.*, 12, 2433, 1992.
68. Kalivas, P. W., Blockade of neurotensin-induced motor activity by inhibition of protein kinase, *Psychopharmacology,* 1994, in press.
69. Stewart, J. and Vezina, P., Microinjections of SCH-23390 into the ventral tegmental area and substantia nigra pars reticulata attenuate the development of sensitizatin to the locomotor activating effects of systemic amphetamine, *Brain Res.*, 495, 401, 1989.
70. Fink, J. S. and Smith, G. P., Mesolimbocoritcal dopamine terminal fields are necessary for normal locomotor and investigatory exploration in rats, *Brain Res.*, 199, 359, 1980.
71. Llinas, R., Greenfield, S. A., and Jahnsen, H., Electrophysiology or pars compacta cells in the in vitro substantia nigra—a possible mechanism for dendritic release, *Brain Res.*, 294, 127, 1984.
72. Starr, M., Opposing roles of dopamine D_1 and D_2 receptors in nigral [^{3}H]aminobutyric acid release? *J. Neurochem.*, 49, 1042, 1987.
73. Timmerman, W., Zwaveling, J., and Westerink, B. H. C., Dopaminergic modulation of the GABA release in the substantia nigra reticulata, in *Monitoring Molecules in Neuroscience,* Rollema, H., Westerink, B. H. C., and Drijfhout, W. J., Eds., Krips Repro, Meppel, The Netherlands, 1991, 105.
74. Kalivas, P. W., Interactions between dopamine and excitatory amino acids in behavioral sensitization to psychostimulants, *Alc. Drug Dep.*, 1994, in press.
75. Stewart, J. and Druhan, J. P., The development of both conditioning and sensitization of the behavioral activating effects of amphetamine is blocked by the noncompetative NMDA receptor antagonist, MK-801, *Psychopharmacology,* 110, 125, 1993.
76. Wolf, M. E. and Khansa, M. R., Repeated administration of MK-801 produces sensitization to its own locomotor stimulant effects but blocks sensitization to amphetamine, *Brain Res.*, 562, 164, 1991.
77. Karler, R., Calder, L. D., Chaudhry, I. A., and Turkanis, S. A., Blockade of 'reverse tolerance' to cocaine and amphetamine by MK-801, *Life Sci.*, 45, 599, 1989.
78. Kalivas, P. W. and Alesdatter, J. E., Involvement of NMDA receptor stimulation in the VTA and amygdala in behavioral sensitization to cocaine, *J. Pharmacol. Exp. Ther.*, 267, 486, 1993.
79. Karler, R., Finnegan, K. T., and Calder, L. D., Blockade of behavioral sensitization to cocaine and amphetamine by inhibitors of protein synthesis, *Brain Res.*, 1994, in press.
80. Striplin, C. and Kalivas, P. W., Robustness of G protein changes in cocaine sensitization shown with immunoblotting, *Synapse,* 14, 10, 1993.
81. Terwilliger, R., Beitner-Johnson, D., Svarino, K. A., Crain, S. M., and Nestler, E. J., A general role for adaptations in G-proteins and cyclic AMP system in mediating the chronic actions of morphine and cocaine on neuronal function, *Brain Res.*, 548, 100, 1991.
82. Vrana, S. L., Vrana, K. E., Koves, T. R., Smith, J. E., and Dworkin, S. I., Chronic cocaine administration increases CNS tyrosine hydroxylase enzyme activity and mRNA levels and tryptophan hydroxylase enzyme activity levels, *J. Neurochem.*, 61, 2262, 1993.
83. Sorg, B. A., Chen, S.-Y., and Kalivas, P. W., Time course of tyrosine hydroxylase expression following behavioral sensitization to cocaine, *J. Pharmacol. Exp. Ther.*, 266, 424, 1993.
84. Beitner-Johnson, D., Guitart, X., and Nestler, E. J., Neurofilament proteins and the mesolimbic dopamine system: common regulation by chronic morphine and chronic cocaine in the rat ventral tegmental area, *J. Neurosci.*, 12, 2165, 1992.
85. Kolta, M. G., Shreve, P., De Souza, V., and Uretsky, N. J., Time course of the development of the enhanced behavioral and biochemical responses to amphetamine after pretreatment with amphetamine, *Neuropharmacology,* 24, 823, 1985.
86. Xia, Y., Goebel, D. J., Kapatos, G., and Bannon, M. J., Quantitation of rat dopamine transporter mRNA: effects of cocaine treatment and withdrawal, *J. Neurochem.*, 59, 1179, 1992.
87. Peris, J. and Zahniser, N. R., Persistent augmented dopamine release after acute cocaine requires dopamine receptor activation, *Pharmacol. Biochem. Behav.*, 32, 71, 1989.
88. Robinson, T. E. and Becker, J. B., Behavioral sensitization is accompanied by an enhancement in amphetamine-stimulated dopamine release from striatal tissue in vitro, *Eur. J. Pharmacol.*, 85, 253, 1982.
89. Castaneda, E., Becker, J. B., and Robinson, T. E., The long-term effects of repeated amphetamine treatment in vivo on amphetamine, KCl and electrical stimulation evoked striatal dopamine release in vitro, *Life Sci.*, 42, 2447, 1988.

90. Wilcox, R. A., Robinson, T. E., and Becker, J. B., Enduring enhancement in amphetamine-stimulated striatal dopamine release in vitro produced by prior exposure to amphetamine or stress *in vivo, Eur. J. Pharmacol.,* 124, 375, 1986.
91. Kuczenski, R., Biochemical actions of amphetamine and other stimulants, in *Stimulants: Neurochemical, Behavioral and Clinical Perspectives,* Creese, I., Ed., Raven Press, New York, 1983, 31.
92. Sulzer, D., Maidment, N. T., and Rayport, S., Amphetamine and other weak bases act to promote reverse transport of dopamine in ventral midbrain neurons, *J. Neurochem.,* 60, 527, 1993.
93. Kalivas, P. W. and Duffy, P., A comparison of axonal and somatodendritic dopamine release using *in vivo* dialysis, *J. Neurochem.,* 56, 961, 1991.
94. Westerink, B. H. C., Hofsteede, H. M., Damsma, G., and de Vries, J. R., The significance of extracellular calcium for the release of dopamine, acetylcholine and amino acids in conscious rats, evaluated by brain microdialysis, *N.-S. Arch. Pharmacol.,* 337, 373, 1988.
95. Warburton, E. C., Mitchell, S. N., and Joseph, M. H., Sensitization of dopamine release in the accumbens following amphetamine: calcium dependency accompanies altered effects on behavioral function, *Behav. Pharmacol.,* 4, 454, 1993.
96. Karler, R., Turkanis, S. A., Partlow, L. M., and Calder, L. D., Calcium channel blockers in behavioral sensitization, *Life Sci.,* 49, 165, 1991.
97. Sharpe, L. G., Pilotte, N. S., Mitchell, W. M., and DeSouza, E. B., Withdrawal of repeated cocaine decreases autoradiographic [^{3}H] mazindol-labelling of dopamine transporter in rat nucleus accumbens, *Eur. J. Pharmacol.,* 203, 141, 1991.
98. Farfel, G. M., Kleven, M. S., Woolverton, W. L., Seiden, L. S., and Perry, B. D., Effects of repeated injections of cocaine on catecholamine receptor binding sites, dopamine transporter binding sites and behavior in rhesus monkeys, *Brain Res.,* 578, 235, 1992.
99. Alburges, M. E., Narang, N., and Wamsley, J. K., Alterations in dopaminergic receptor system after chronic administration of cocaine, *Synapse,* 14, 314, 1993.
100. Yi, S.-J. and Johnson, K. M., Effects of acute and chronic administration of cocaine on striatal uptake, compartmentalization and release of [^{3}H]dopamine, *Neuropharmacology,* 29, 475, 1990.
101. Izenwasser, S., Werling, L. L., and Cox, B. M., Comparison of the effects of cocaine and other inhibitors of dopamine uptake in rat striatum, nucleus accumbens, olfactory tubercle, and medial prefrontal cortex, *Brain Res.,* 520, 303, 1990.
102. Kula, N. S. and Baldessarini, R. J., Lack of increase in dopamine transporter binding or function in rat brain tissue after treatment with blockers of neuronal uptake of dopamine, *Neuropharmacology,* 30, 89, 1991.
103. Reith, M. E. A. and Selmeci, G., Cocaine binding sites in mouse striatum, dopamine autoreceptors, and cocaine-induced locomotion, *Pharmacol. Biochem. Behav.,* 41, 227, 1992.
104. Benmansour, S., Tejani-Butt, S. M., Hauptmann, M., and Brunswick, D. J., Lack of effect of high-dose cocaine on monoamine uptake sites in rat brain as measured by quantitative autoradiography, *Psychopharmacology,* 106, 459, 1992.
105. Allard, P., Eriksson, K., Ross, S. B., and Marcusson, J O., Unaltered [3H]GBR-12935 binding after chronic treatment with dopamine active drugs, *Psychopharmacology,* 102, 291, 1990.
106. Parsons, L. H., Smith, A. D., and Justice, J. B., Jr., Basal extracellular dopamine is decreased in the rat nucleus accumbens during abstinence from chronic cocaine, *Synapse,* 9, 60, 1991.
107. Meiergerd, S. M., Hooks, M. S., and Schenk, J. O., The striatal transporter for dopamine in the rat may be kinetically up-regulated following three weeks of withdrawal from self-administration, *J. Neurochem.,* 1994, in press.
108. Meiergerd, S. M., McElvain, J. S., and Schenk, J. O., Effects of cocaine and repeated cocaine follwed by withdrawal: alterations of dopaminergic transporter turnover with no change in kinetics of substrate recognition, *Biochem. Pharmacol.,* 1994, in press.
109. Cass, W. A., Gerhardt, G. A., Gillespie, K., Curella, P., Mayfield, R. D., and Zahniser, N. R., Reduced clearance of exogenous dopamine in rat nucleus accumbens, but not in dorsal striatum following cocaine challenge in rats withdrawn from repeated cocaine, *J. Neurochem.,* 61, 273, 1993.
110. Cubeddu, L. X., Hoffmann, I. S., and Talmaciu, R. K., Is the release of dopamine from medial prefrontal cortex modulated by presynaptic receptors? Comparison with nigrostriatal and mesolimbic terminals, *Ann. N.Y. Acad. Sci.,* 452, 1990.
111. Dwoskin, L. P. and Zahniser, N. R., Robust modulation of [^{3}H]dopamine release from rat striatal slices by D-2 dopamine receptors, *J. Pharmacol. Exp. Ther.,* 239, 442, 1986.

112. Dwoskin, L. P., Peris, J., Yasuda, R. P., Philpott, K., and Zahniser, N. R., Repeated cocaine administration results in supersensitivity of striatal D-2 dopamine autoreceptors to pergolide, *Life Sci.,* 42, 255, 1988.
113. Yi, S.-J. and Johnson, K. M., Chronic cocaine treatment impairs the regulation of synaptosomal ^{3}H-DA release by D_2 autoreceptors, *Pharmacol. Biochem. Behav.,* 36, 457, 1990.
114. Fitzgerald, J. L. and Reid, J. J., Chronic cocaine treatment does not alter rat striatal D_2 autoreceptor sensitivity to pergolide, *Brain Res.,* 541, 327, 1991.
115. Sorensen, S. M., Jumphreys, T. M., and Palfreyman, M. G., Effect of acute and chronic MDL 73,147EF, a 5-HT3 receptor antagonist, on A9 and A10 dopamine neurons, *Eur. J. Pharmacol.,* 163, 115, 1989.
116. Wang, J. K. T., Presynaptic glutamate receptors modulate dopamine release from striatal synaptosomes, *J. Neurochem.,* 57, 819, 1991.
117. Galli, T., Godeheu, G., Artaud, F., Desce, J. M., Pittaluga, A., Barbeito, L., Glowinski, J., and Cheramy, A., Specific role of N-acetyl-aspartyl-glutamate in the *in vivo* regulation of dopamine release from dendrites and nerve terminals of nigrostriatal dopaminergic neurons in the cat, *Neuroscience,* 42, 19, 1991.
118. Imperato, A., Scrocco, M. G., Bacchi, S., and Angelucci, L., NMDA receptors and in vivo dopamine release in the nucleus accumbens and caudate, *Eur. J. Pharmacol.,* 187, 555, 1990.
119. Jones, S. M., Snell, L. D., and Johnson, K. M., Inhibition by phencyclidine of excitatory amino acid-stimulated release of neurotransmitter in the nucleus accumbens, *Neuropharmacology,* 26, 173, 1987.
120. Ritz, M. C., Cone, E. J., and Kuhar, M. J., Cocaine inhibition of ligand binding at dopamine, norepinephrine and serotonin transporters: a structure-activity study, *Life Sci.,* 46, 635, 1990.
121. Moghaddam, B., Stress preferentially increases extraneuronal levels of excitatory amino acids in the prefrontal cortex: comparison to hippocampus and basal ganglia, *J. Neurochem.,* 60, 1650, 1993.
122. Deutch, A. Y., Clark, W. A., and Roth, R. H., Prefrontal cortical dopamine depletion enhances the responsiveness of mesolimbic dopamine neurons to stress, *Brain Res.,* 521, 311, 1990.
123. Louilot, A., le Moal, M., and Simon, H., Opposite influences of dopaminergic pathways to the prefrontal cortex or the septum on the dopaminergic transmission in the nucleus accumbens. An *in vivo* voltammetric study, *Neuroscience,* 29, 45, 1989.
124. Sorg, B. A. and Kalivas, P. W., Effects of cocaine and footshock stress on extracellular dopamine levels in the medial prefrontal cortex, *Neuroscience,* 53, 695, 1993.
125. Sorg, B. A. and Kalivas, P. W., Effects of repeated cocaine on subsequent cocaine- and amphetamine-induced changes in extracellular dopamine levels in the medial prefrontal cortex, *Soc. Neurosci. Abst.,* 19, 755.3, 1993.
126. Hamamura, T. and Fibiger, H. C., Enhanced stress-induced dopamine release in the prefrontal cortex of amphetamine sensitized rats, *Eur. J. Pharmacol.,* 237, 65, 1993.
127. Henry, D. J. and White, F. J., Repeated cocaine administration causes persistent enhancement of D_1 dopamine receptor sensitivity within the rat nucleus accumbens, *J. Pharmacol. Exp. Ther.,* 258, 882, 1991.
128. Higashi, H., Inanaga, K., Nishi, S., and Uchimura, N., Enhancement of dopamine actions on rat nucleus accumbens neurones *in vitro* after methamphetamine pretreatment, *J. Physiol.,* 408, 587, 1989.
129. Beitner-Johnson, D., Guitart, X., and Nestler, E. J., Common intracellular actions of chronic morhine and cocaine in dopaminergic brain reward regions, in *The Neurobiology of Drug and Alcohol Addiction,* Kalivas, P.W. and Samson, H.H., Eds., New York, 1992, 70.
130. Mayfield, R. D., Larson, G., and Zahniser, N. R., Cocaine-induced behavioral sensitization and D_1 dopamine receptor function in rat nucleus accumbens and striatum, *Brain Res.,* 573, 331, 1992.
131. Kalivas, P. W., Churchill, L., and Klitenick, M. A., GABA and enkephalin projection from the nucleus accumbens and ventral pallidum to VTA, *Neuroscience,* 57, 1047, 1993.
132. Heimer, L. and Alheid, G. F., Piecing together the puzzle of basal forebrain anatomy, in *The Basal Forebrain: Anatomy to Function,* Napier, T.C., Kalivas, P. W. and Hanin, I., Eds., Plenum Press, New York, 1991, 1.
133. Mogenson, G. J., Brudzynski, S. M., Wu, M., Yang, C. R., and Yim, C. C. Y., From motivation to action: a review of dopaminergic regulation of limbic- nucleus accumbens-pedunculopontine nucleus circuitries involved in limbic-motor integration, in *Limbic Motor Circuits and Neuropsychiatry,* Kalivas, P. W. and Barnes, C. D., Eds., CRC Press, Boca Raton, 1993, 193.

134. Kalivas, P. W. and Barnes, C. D., *Limbic Motor Circuits in Neuropsychiatry*, CRC Press, Boca Raton, 1993.
135. Napier, T. C. and Chrobak, J. J., Evaluations of ventral pallidal dopamine receptor activiation in behaving rats, *Neuro Report,* 3, 609, 1992.
136. Maslowski, R. J. and Napier, T. C., Effects of D_1 and D_2 receptor agonists induce opposite changes in the firing rate of ventral pallidal neurons, *Eur. J. Pharmacol.,* 200, 103, 1991.
137. Bourdelais, A. J. and Kalivas, P. W., Modulation of extracellular GABA in the ventral pallidum using *in vivo* microdialysis, *J. Neurochem.,* 58, 2311, 1992.
138. Bourdelais, A. J., Physiological and pharmacological manipulations of the GABA-ergic pathway between the nucleus accumbens and the ventral pallidum, Washington State University, Ph.D. Thesis, 1991.

Chapter 7

Neurophysiological Alterations in the Mesocorticolimbic Dopamine System With Repeated Cocaine Administration

Francis J. White, Xiu-Ti Hu, Douglas J. Henry, and Xu-Feng Zhang

CONTENTS

I. INTRODUCTION

As the past decade witnessed a dramatic increase in cocaine abuse and related health problems, the scientific community became increasingly aware of the need to identify and characterize the physiological bases for the subjective and behavioral effects of cocaine. Such efforts concentrated on the powerful ability of cocaine to control behavior by inducing initial euphoria, as well as subsequent craving and dependence. Neuroscientists conducted a wide range of investigations to elucidate the mechanisms responsible for the effects of cocaine and how such mechanisms might be altered during repeated cocaine use and withdrawal. Our laboratory has participated in this effort by describing the neurophysiological effects of acute and chronic cocaine on the rat mesocorticolimbic dopamine (DA) system, which has long been associated with both the psychomotor stimulant and positive reinforcing effects of cocaine. Specifically, we have attempted to identify relevant alterations in the activity of neuronal populations at various levels of this system and the receptor mechanisms responsible for such alterations. This chapter reviews our investigations into the effects of repeated cocaine administration on the mesoaccumbens and mesocortical DA systems. The acute effects of cocaine on neuronal activity are altered by repeated treatment and the interested reader may wish to refer to other recent reviews describing the acute actions of cocaine,[1-3] as well as to original reports.[4,5] Throughout the following discussion, we will use the terms D_1 and D_2 receptors to refer to the broad subfamilies of cloned DA receptors, D_1 (D_1, D_5) and D_2 (D_2, D_3, D_4), because

1-8493-8311-0/95/$0.00+$.50

it is not possible to determine selectivity within these two subfamilies using current pharmacological tools.

II. DOPAMINE AND COCAINE SENSITIZATION

Repeated administration of cocaine to laboratory animals leads to a number of changes in the acute effects of the drug depending upon a number of factors, including dose, route, and frequency of administration, as well as the nature of the behavior and how it is measured. One of the best characterized effects of repeated administration of psychomotor stimulants is the development of reverse tolerance or "sensitization" to their ability to induce locomotion and repetitive stereotyped behaviors.[6,7] Sensitization has considerable relevance to psychoses[6,8] and drug addiction,[9,10] and many researchers have addressed the mechanisms underlying behavioral sensitization to psychomotor stimulants within the past few years. Our efforts have been directed at the elucidation of the electrophysiological correlates of cocaine sensitization as well as some aspects of cocaine withdrawal.

A. BEHAVIORAL SENSITIZATION

Before describing neurophysiological changes in the mesocorticolimbic DA system caused by repeated cocaine administration, we will first describe the cocaine treatment regimen used in these studies to induce sensitization. Unless noted otherwise, all of the studies have used a 14-day treatment regimen in which rats receive twice daily injections of cocaine HCl (10.0 mg/kg, i.p.) or vehicle (either deionized water or saline). Behavioral tests are conducted following a challenge injection of cocaine (10.0 mg/kg) using a 3-photobeam device placed around standard "shoe-box" shaped acrylic cages. Beam interruptions are detected by a personal computer using commercially available software which records both total beam interruptions and consecutive interruptions of adjacent beams (representing ambulation). For the experiments described here, separate groups of control and cocaine-treated rats were tested at each of 4 withdrawal times: 1 day (24 hr after the last treatment injection), 1 week (7 days), 1 month (28 to 30 days), or 2 months (60 to 64 days).

Repeated cocaine administration led to increased horizontal locomotor activity in response to cocaine challenge when rats were tested 1 day, 1 week, and 1 month after the last cocaine treatment injection, but not following a 2-month withdrawal (Figure 1). The extent of sensitization appeared to diminish as a function of withdrawal time. These findings are somewhat surprising in light of other reports in the literature which show more persistent sensitization following repeated cocaine administration[8,11] and suggest that sensitization is a long-lasting, if not permanent, effect.[6,7,10] Close scrutiny, however, reveals that demonstrations of more persistent sensitization utilized animals that were challenged repeatedly with cocaine and tested for sensitization at various withdrawal times. In contrast, our experiments used separate groups of rats that received only one challenge injection of cocaine. Given that cocaine sensitization can occur following only one exposure to cocaine,[12] and that repeated association of cocaine with the test environment increases the likelihood of conditioned ambulatory responding (i.e., context-dependent sensitization[13]), it is possible that repeated cocaine challenges at various withdrawal times were sufficient to maintain sensitization to the drug in these studies. Explicit tests of this possibility are needed. For the present purposes, the important point is that our treatment paradigm induced a relatively persistent, but not permanent, sensitization of locomotor activity in response to cocaine challenge.

B. DOPAMINE AUTORECEPTORS IN THE VENTRAL TEGMENTAL AREA (VTA)

Throughout the remainder of this chapter, we will describe our results from extracellular single cell recordings of neurons within the rat ventral tegmental area (VTA), nucleus

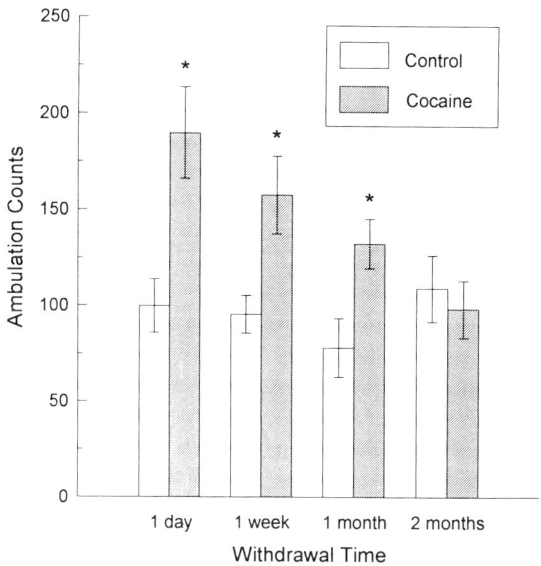

Figure 1. Sensitization to cocaine at various withdrawal times after repeated treatment with cocaine or deionized water (control). Data represent the mean ± SEM from 11 to 12 rats in each group. Separate groups of rats were tested at each time point. The behavioral test lasted for 90 min. Asterisks represent significantly ($p \leq 0.05$) greater ambulation in the cocaine group as compared to its appropriate control group by Student's *t*-test.

accumbens (NAc), and prefrontal cortex (PFC). All data were obtained from chloral hydrate-anesthetized rats using either single or multibarrel glass microelectrodes. Drugs were administered either intravenously (i.v.) through a catheterized tail vein or microiontophoretically through side barrels of the multibarrel pipettes. Details of our procedures are available elsewhere.[14,15]

Among the earliest hypotheses regarding possible mechanisms underlying behavioral sensitization was the idea that a subsensitivity of DA autoreceptors led to compensatory increases in DA activity and release following repeated stimulant administration.[16,17] In fact, repeated administration of cocaine readily induces subsensitivity of impulse-regulating DA autoreceptors within the VTA.[14] This can be observed with systemic administration of nonselective D_1/D_2 agonists such as apomorphine[18] or with agonists selective for the D_2 subfamily of DA receptors, such as quinpirole (Figure 2). In this respect, our findings differ from a recent study which reported that repeated amphetamine induced subsensitivity of substantia nigra DA neurons to apomorphine, but not quinpirole.[19] Whether this difference is related to nigral versus VTA DA neurons or to amphetamine versus cocaine treatments will need to be explored.

More direct proof of autoreceptor subsensitivity comes from experiments in which DA[14] or DA agonists were administered locally to VTA DA neurons using microiontophoresis. Repeated cocaine administration renders VTA somatodendritic DA autoreceptors subsensitive to the inhibitory effects of iontophoretic DA[14] and quinpirole (Figure 3). Therefore, despite the fact that cocaine only weakly enhances the stimulation of D_2 autoreceptors by DA,[4] such enhancement is sufficient to desensitize VTA DA neurons to DA and DA agonists after repeated cocaine administration.

The induction of autoreceptor subsensitivity appears to involve rather selective changes in DA receptors as opposed to general decreases in sensitivity to inhibitory transmitters. Thus, we found no change in the responsiveness of VTA DA cells to iontophoretically administered GABA.[14] This might seem surprising in view of other work indicating that

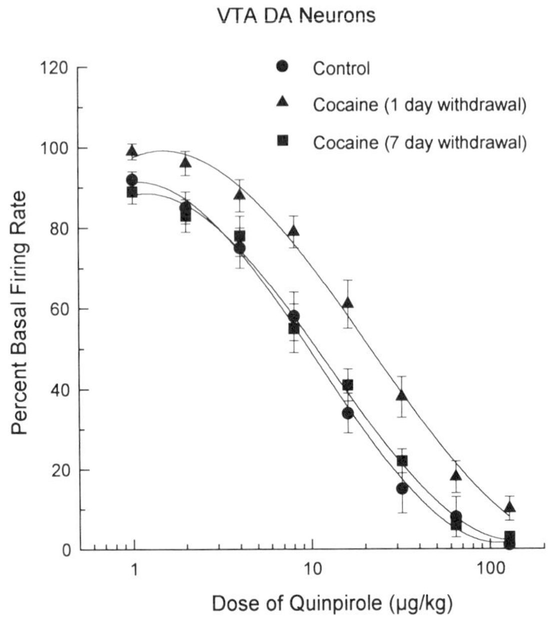

Figure 2. Subsensitivity of VTA DA neurons to intravenously injected quinpirole following a 1 day (24 to 36 hr), but not a 1 week (7 to 9 day) withdrawal from repeated cocaine administration. Quinpirole was injected through a tail vein using a cumulative dose strategy in which each injection doubled the previously administered dose. Injections were spaced 90 sec apart. Each point represents the mean ± SEM from 9 to 12 neurons in each group. Quinpirole-induced inhibition of firing was significantly ($p \leq 0.05$) reduced in the 1-day withdrawal group at each dose of quinpirole by Dunnett's test following repeated measures ANOVA.

repeated cocaine reduces the α-subunits of G_i and G_o proteins to which both D_2 and $GABA_B$ receptors are coupled in VTA DA neurons.[20,21] However, additional studies have demonstrated clearly that alterations in the levels of these G proteins occur only with high dose cocaine treatment regimens and persist for only a matter of hours following the last cocaine injection.[22]

Because somatodendritic autoreceptors are important in regulating the basal activity of VTA DA neurons, we next determined whether autoreceptor subsensitivity would result in an overall change in the basal activity of the VTA DA cell population. Thus, we conducted cell population analysis of the VTA following repeated cocaine treatment. This was achieved by defining a stereotaxic block of tissue contained within the VTA, passing a microelectrode 12 times through this area in a fixed pattern, counting the number of spontaneously active DA neurons encountered, and determining their basal firing rates. As shown in Table 1, at early time points after the cessation of repeated cocaine administration (24 to 72 hr), there is a significant increase in both the number of spontaneously active VTA DA neurons and their average basal firing rate.[14] It is known that a subpopulation of quiescent, hyperpolarized DA neurons exists in chloral hydrate-anesthetized rats.[23] In addition to increasing the basal firing rates of VTA DA neurons, repeated cocaine treatment apparently "recruits" a significant portion of these normally quiescent DA neurons.[14]

Because DA autoreceptor subsensitivity results in enhanced DA cell activity, it might appear to be a plausible substrate for behavioral sensitization to cocaine. Unlike the relatively persistent nature of sensitization, however, DA autoreceptor subsensitivity is a transient phenomenon which can be observed only for 1 to 4 days after discontinuation

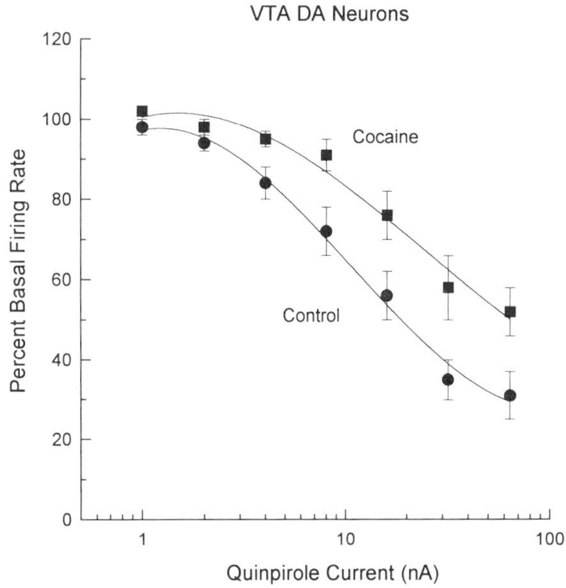

Figure 3. Subsensitivity of VTA DA neurons to iontophoretically administered quinpirole in cocaine-treated rats as compared to vehicle-treated control rats following 24 to 48 hr withdrawal. Each point represents the mean ± SEM from 9 neurons in each group. Quinpirole-induced inhibition of firing was significantly ($p \leq 0.05$) reduced at all currents above 2 nanoamperes (nA) by Dunnett's test following repeated measures ANOVA.

of cocaine injections.[18] Autoreceptor subsensitivity is present using quinpirole challenge following a 1-day (24 to 36 hr) withdrawal, but not after a 1-week (7- to 8-day) withdrawal period (Figure 2). These results argue against the direct involvement of desensitized somatodendritic impulse-regulating autoreceptors in the maintenance and expression of behavioral sensitization but, when considered along with other behavioral evidence, suggest their involvement in the initiation of sensitization processes (see additional discussion below).

It has been suggested previously that protracted withdrawal from repeated cocaine might produce DA autoreceptor supersensitivity, diminished DA cell activity, and various aspects of a cocaine abstinence syndrome.[24] Evidence for a supersensitivity of autoreceptors regulating the activity of nigral DA neurons has been found following 1 week of withdrawal from continuous cocaine administration.[25] In our studies using intermittent cocaine injections, however, VTA DA autoreceptor sensitivity returned to normal following longer withdrawal periods (7 to 8 days) with no evidence for supersensitivity at any time point tested.[18] Nevertheless, an analysis of the number of cells per microelectrode track revealed a decrease of spontaneously active DA neurons with no significant alteration in the basal firing rates of the remaining active neurons (Table 1).[26] This is consistent with other reports which indicate that similar periods of cocaine withdrawal reduce the basal level of extracellular DA[27,28] and DA transporter binding[29] in the NAc. Accordingly, it appears that VTA DA neurons are dysfunctional during cocaine withdrawal, which might well contribute to symptoms such as anergia, anhedonia, and cocaine craving observed during withdrawal.[24,30] Thus, determining the mechanism responsible for the reduced number of spontaneously active DA neurons could provide information which is essential to the successful pharmacological treatment of cocaine withdrawal symptoms.

Table 1. Effects of Repeated Cocaine Administration and Withdrawal on the Spontaneous Activity of VTA DA Neurons

Treatment Group	N[a]	Cells/Track	Rate (Spikes/sec)
Control, 1 day withdrawal	6	1.45 ± 0.09	4.19 ± 0.22
Cocaine, 1 day withdrawal	5	2.08 ± 0.14[b]	5.58 ± 0.16[b]
Control, 10-14 day withdrawal	10	1.39 ± 0.10	4.51 ± 0.22
Cocaine, 10-14 day withdrawal	8	0.81 ± 0.07[c]	4.57 ± 0.41

[a] Refers to the number of rats, 12 electrode tracks were made in each rat.
[b] Significantly greater than corresponding control, $p \leq 0.01$, Student's t-test.
[c] Significantly less than corresponding control, $p \leq 0.001$, Student's t-test.

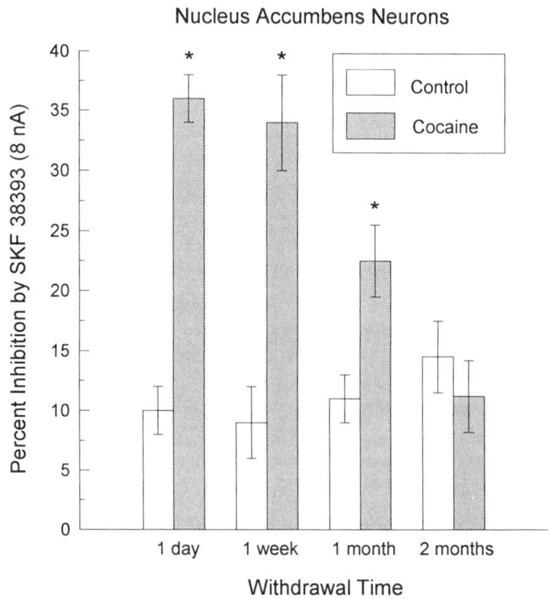

Figure 4. Supersensitivity of NAc neurons to iontophoretically administered SKF 38393 (8 nA current) at various withdrawal times after repeated cocaine administration. The percent inhibition of glutamate-induced activity produced by a 1 min iontophoretic application of SKF 38393 is plotted for groups of rats treated repeatedly with either cocaine or deionized water and withdrawn for different lengths of time. Each point represents the mean ± SEM from 10 to 15 neurons in each group. SKF 38393-induced inhibition of firing was significantly ($p \leq 0.01$) greater in the cocaine-treated group at all withdrawal times except 2 months by Dunnett's test following repeated measures ANOVA. Note the similarity to the behavioral results shown in Figure 1.

C. NUCLEUS ACCUMBENS DOPAMINE RECEPTORS

The studies described above indicate that repeated cocaine administration transiently desensitizes impulse-regulating DA autoreceptors within the VTA, decreasing autoinhibition and thereby increasing the activity of this cell population. We next focussed on the NAc to determine whether postsynaptic DA receptors also were desensitized following repeated cocaine administration. To our surprise, the results indicated that NAc neurons from cocaine-treated rats were supersensitive to the inhibitory effects of iontophoretic DA.[14]

Because NAc neurons are inhibited by iontophoretic administration of either D_1 or D_2 agonists,[31,32] the supersensitive response to DA following repeated cocaine treatment might be mediated by alterations in D_1 and/or D_2 receptors. In the next set of experiments,

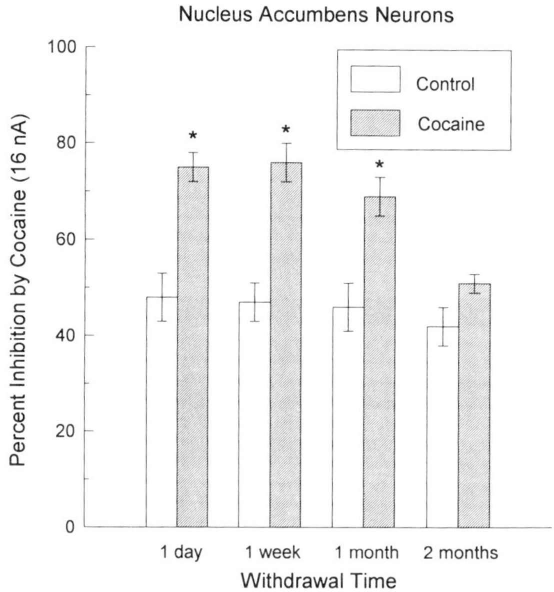

Figure 5. Supersensitivity of NAc neurons to iontophoretically administered cocaine (16 nA current) at various withdrawal times after repeated cocaine administration. The percent inhibition of glutamate-induced activity produced by a 1-min iontophoretic application of cocaine is plotted for groups of rats treated repeatedly with either cocaine or deionized water and withdrawn for different lengths of time. Each point represents the mean ± SEM from 8 to 12 neurons in each group. Cocaine-induced inhibition of firing was significantly ($p \leq 0.01$) greater in the cocaine-treated group at all withdrawal times except 2 months by Dunnett's test following repeated measures ANOVA. Note the similarity to the behavioral results shown in Figure 1.

selective agonists for D_1 (SKF 38393) and D_2 (quinpirole) receptors were applied iontophoretically onto NAc neurons. Repeated cocaine administration failed to alter the sensitivity of NAc neurons to quinpirole,[15] but produced a marked sensitization to the inhibitory efficacy of SKF 38393 (Figure 4), suggesting that the increased response of NAc neurons to DA was due to D_1 DA receptors. We have chosen to illustrate the results in this manner to point out the striking parallels between the time courses for behavioral sensitization and for D_1 sensitized responses of NAc neurons (compare Figures 1 and 4). Thus, D_1 receptor supersensitivity, like behavioral sensitization, was still evident after 1 week or 1 month, but not 2 months of withdrawal. These close temporal relationships strongly suggest that D_1 receptors play an important role in the expression of sensitized behavior. Such a role also is suggested by the marked increases in cocaine-induced inhibition of NAc neurons up to 1 month after the end of repeated cocaine injections (Figure 5 and compare with Figure 1). Other studies also provide evidence for D_1 receptor involvement in sensitized behaviors following repeated agonist treatments. Thus, repeated injections of the D_1 agonist SKF 38393 enhance the inhibitory responses of striatal neurons to iontophoretically applied D_1 agonists and also significantly sensitize D_1 agonist-induced grooming behavior.[32,33]

D. PREFRONTAL CORTEX NEURONS

Recently, we have extended our studies of repeated cocaine administration to include effects on DA-responsive neurons in layers V and VI of the medial PFC. Our preliminary findings indicate that unlike the NAc, PFC neurons develop tolerance to the inhibitory effects of DA when tested 24 to 48 hr after cessation of repeated cocaine treatment (Figure

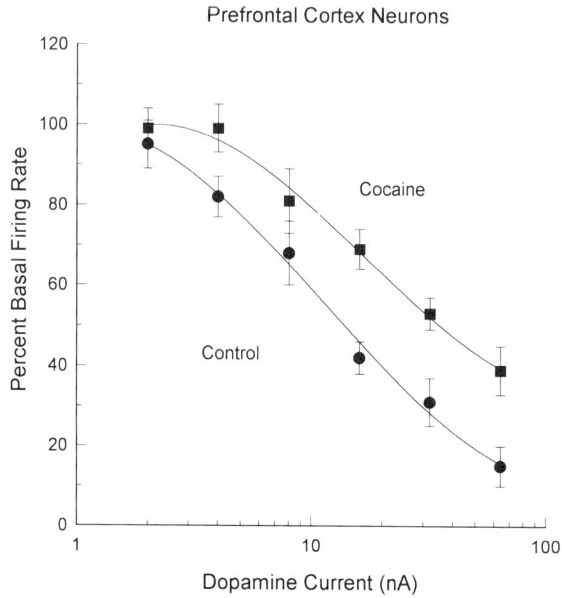

Figure 6. Subsensitivity of rat medial PFC neurons to iontophoretically administered DA following repeated cocaine administration. All neurons were recorded from layers V and VI and were DA-responsive. Experiments were conducted 24 to 48 hr after the last cocaine or deionized water (control) injection. Each point represents the mean ± SEM from 7 neurons in each group. DA-induced inhibition of firing was significantly ($p \leq 0.05$) reduced at all iontophoretic currents except 8 nA by Dunnett's test following repeated measures ANOVA.

6). We currently are investigating the time course for this effect as well as the pharmacological profile of the DA receptor involved. It is interesting to note that tolerance to the ability of cocaine to increase extracellular DA content within the PFC also appears to develop during repeated administration,[35] whereas an enhancement of this effect is usually observed within the NAc (discussed above). Thus, the mesocortical DA system is markedly different from the mesoaccumbens DA system with respect to its response to repeated cocaine administration. Given the known interactions between the cortical and subcortical DA systems, further elucidation of the effects of repeated cocaine on PFC neurons and its targets seems warranted (see further discussion below).

E. SUMMARY: SENSITIZATION AND DOPAMINE NEUROPHYSIOLOGY

Although our studies within the NAc clearly indicate a close temporal relationship between postsynaptic D_1 receptor sensitivity and the expression of cocaine sensitization, other work has revealed that the VTA plays a critical role in the induction of behavioral sensitization to psychomotor stimulants. Repeated administration of amphetamine directly into the VTA, but not the NAc, produces sensitized locomotor responses to systemic administration of psychomotor stimulants including cocaine.[35,36] For this reason, we have argued that VTA DA autoreceptor subsensitivity and the increases in basal DA transmission, although transient alterations, may be important "triggers" for more persistent alterations in DA terminal areas, including postsynaptic D_1 receptor supersensitivity. Furthermore, while repeated administration of the selective D_1 receptor agonist SKF 38393 can provide behavioral and electrophysiological evidence for D_1 receptor supersensitivity, this effect is quite transient.[33] Of course, neither acute nor repeated administration of SKF 38393 alters the activity of VTA DA neurons.[31] Additional evidence

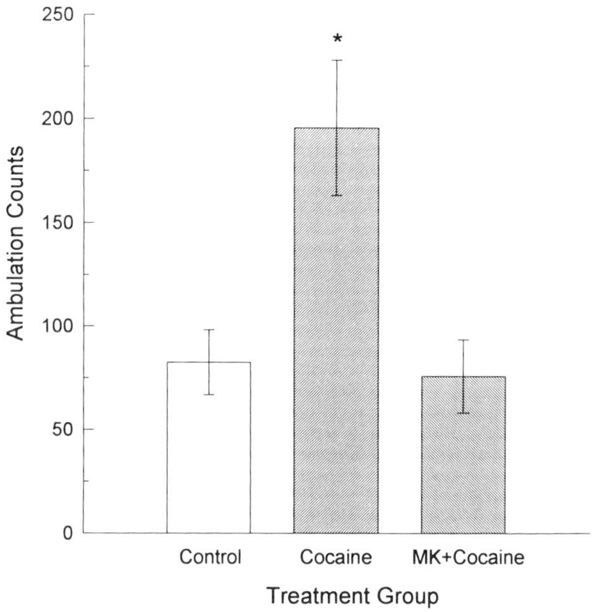

Figure 7. Prevention of cocaine sensitization by coadministration of MK-801. Cocaine-treated rats (n = 5) received five daily injections of 15.0 mg/kg cocaine preceded by a saline injection 30 min earlier, and control rats (n = 6) received two saline injections whereas MK+Cocaine rats (n = 6) received 0.1 mg/kg MK-801 30 min prior to cocaine (15.0 mg/kg). Following a 3-day withdrawal period, all rats received a challenge injection of cocaine (7.5 mg/kg) and were tested in activity chambers for 90 min. Each bar represents the mean ± SEM * – significant ($p < 0.01$) difference from control group by Dunnett's test following ANOVA.

suggests that persistent behavioral sensitization requires a sustained increase of DA neuronal activity, which need not result from autoreceptor subsensitivity. Thus, morphine sensitization is accompanied by increased basal activity of VTA DA cells, which occurs without any alteration of VTA autoreceptor sensitivity.[38]

NAc D_1 receptor supersensitivity is by no means the only alteration underlying the expression of cocaine sensitization. Indeed, sensitization is also accompanied by a persistent enhancement of the ability of cocaine to increase extracellular DA within the NAc as measured by microdialysis.[39,40] However, behavioral sensitization can occur during the early stages of cocaine withdrawal without simultaneous enhancement of extracellular DA concentration,[41,42] during the period when D_1 receptors are sensitized clearly. Thus, we believe that D_1 receptor supersensitivity may play an integral role in the expression of sensitized behavior. It also should be noted that adaptations of DA uptake[43] in nerve terminal DA autoreceptor function[40,44,45] and cocaine disposition within the brain and plasma[40,46] are also likely to play roles in the development and expression of sensitization. Moreover, it is now apparent that nondopaminergic systems, in particular the excitatory amino acid (EAA) system, may be altered in sensitized animals.

III. EXCITATORY AMINO ACIDS AND BEHAVIORAL SENSITIZATION

Although the overwhelming majority of studies of behavioral sensitization mechanisms have focused on alterations of dopaminergic systems, the recent finding that glutamate receptor antagonists, such as the noncompetitive *N*-methyl-D-aspartate (NMDA) receptor antagonist MK-801, can prevent the development of sensitization has led to speculation that EAA receptors are intricately involved in this form of neuronal plasticity. In 1989,

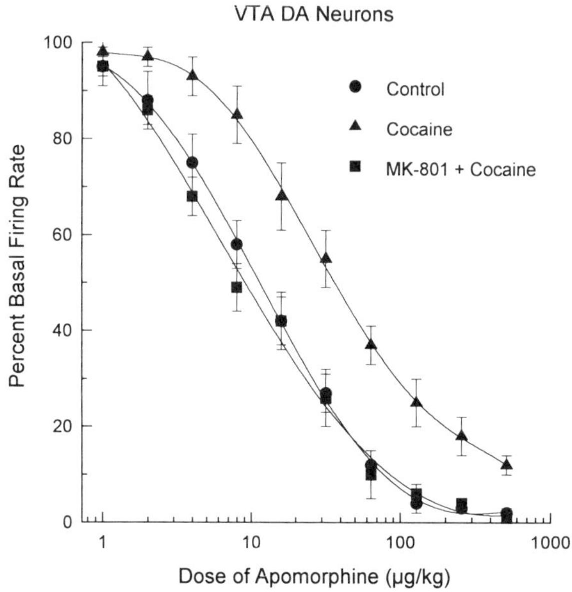

Figure 8. Coadministration of MK-801 prevents the induction of VTA DA autoreceptor subsensitivity during repeated cocaine administration. Treatment regimens are identical to those described in Figure 7 legend. This repeated cocaine treatment resulted in a substantial subsensitivity to apomorphine-induced inhibition of activity (n = 7 neurons), with a significant ($p \leq 0.01$) reduction observed at all doses of apomorphine greater than 2 mg/kg by Dunnett's test following repeated measures ANOVA. Coadministration of MK-801 (n = 6 neurons) completely prevented this effect, resulting in a dose-response curve nearly identical to that observed in control rats (n = 8 neurons).

Karler and associates provided the first evidence that coadministration of MK-801 with cocaine could block the development of sensitized stereotyped behavior in rats.[47] Subsequently, Wolf and Jeziorski[48] reported that MK-801 also prevented the development of cocaine sensitization measured as ambulatory locomotion following a 7-day withdrawal period. Recently, we have examined this effect following a 3-day withdrawal from a repeated cocaine regimen of five daily injections of 15.0 mg/kg (i.p.). As shown in Figure 7, coadministration of MK-801 (0.1 mg/kg) with cocaine completely prevents the development of sensitization. Based on this finding, we have begun a series of studies designed to examine the possible role of NMDA and other EAA receptors in neuronal alterations associated with behavioral sensitization. All of these studies have used the 5-day cocaine treatment regimen.

A. NMDA ANTAGONISTS PREVENT ELECTROPHYSIOLOGICAL CORRELATES OF COCAINE SENSITIZATION

We have reported previously that coadministration of MK-801 with *d*-amphetamine during repeated treatment prevents the development of both autoreceptor subsensitivity in the VTA and D_1 receptor supersensitivity in the NAc.[49] These findings now have been extended to cocaine sensitization. When tested 48 to 72 hr after the last of five daily cocaine injections, VTA DA neurons were significantly less sensitive to the inhibitory effect of apomorphine, and this effect was completely prevented by MK-801 coadministration (Figure 8). Daily administration of MK-801 alone failed to alter the

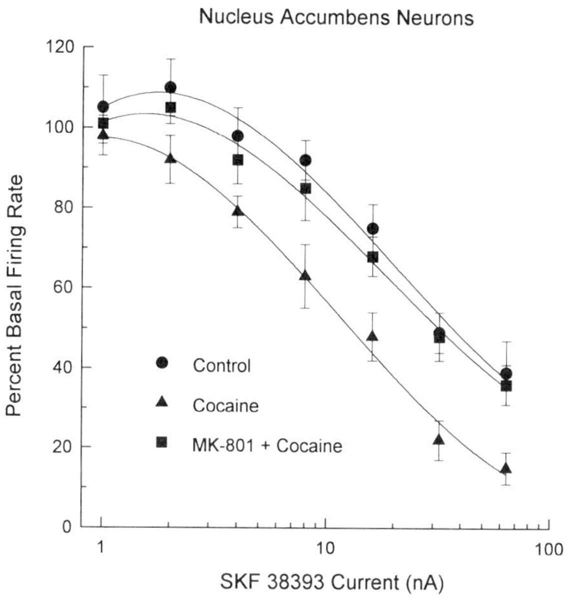

Figure 9. Coadministration of MK-801 prevents the development of D_1 receptor supersensitivity within the rat NAc during repeated cocaine administration. Treatment regimens are identical to those described in Figure 7 legend. This repeated cocaine treatment resulted in a substantial enhancement in the inhibitory effects of iontophoretically applied SKF 38393 (n = 6 neurons), with a significant ($p \leq 0.05$) increase observed at all currents greater than 2 nA by Dunnett's test following repeated measures ANOVA. Coadministration of MK-801 (n = 6 neurons) completely prevented this effect, resulting in a current-response curve nearly identical to that observed in control rats (n = 9 neurons).

sensitivity of VTA DA neurons to quinpirole.[49] Preliminary studies within the NAc suggest that coadministration of MK-801 also prevents the development of D_1 receptor supersensitivity typically observed after repeated cocaine administration. Thus, the 5-day regimen of cocaine injections markedly increased the sensitivity of NAc neurons to iontophoretically administered SKF 38393, an effect that was completely prevented by coadministration of MK-801 (Figure 9). Again, no effect of daily MK-801 injections was observed on NAc neurons.[49]

These findings lend further support to the notion that the induction of transient VTA DA autoreceptor subsensitivity and persistent postsynaptic NAc D_1 receptor supersensitivity may be related steps in a chain of events leading to behavioral sensitization. However, the prevention of such effects by NMDA receptor antagonists implies that these alterations of dopaminergic transmission are dependent upon NMDA receptor stimulation. Accordingly, it seems plausible that repeated administration of cocaine (and other psychomotor stimulants) would alter EAA neurotransmission. In an attempt to identify such alterations, we recently have studied the extent to which repeated cocaine administration might alter the sensitivity of the mesoaccumbens DA pathway to glutamate.

B. GLUTAMATE RESPONSES AFTER REPEATED COCAINE

We first sought to determine whether repeated administration of cocaine alters the responsiveness of neurons in the rat mesoaccumbens DA system to local administration of glutamate. We examined the sensitivities of both VTA DA and NAc neurons because

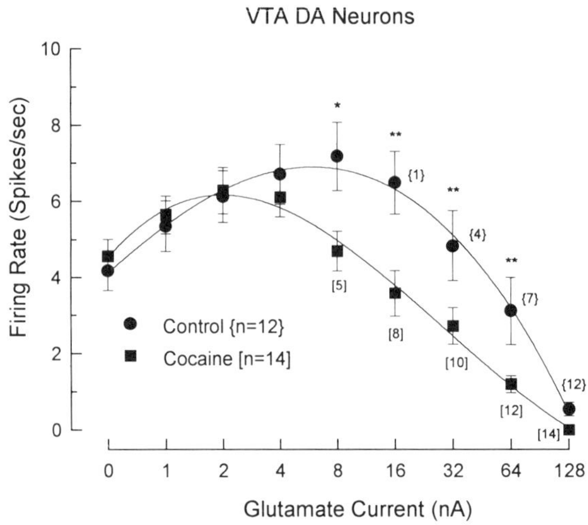

Figure 10. Repeated administration of cocaine increased the ability of iontophoretic glutamate to induce apparent depolarization block of VTA DA neurons. Note that there was no difference in the excitatory efficacy of glutamate at low iontophoretic currents (<8 nA), but at currents of 8 to 64 nA, the percentage of neurons entering depolarization block was significantly enhanced in the cocaine-treated rats by Dunnett's test following repeated measures ANOVA (* – $p \leq 0.05$, ** – $p \leq 0.01$). The numbers in brackets indicate the number of neurons that had entered a state of depolarization block at each current.

of their demonstrated roles in the initiation, development and expression of behavioral sensitization, and because they both receive prominent EAA projections from the various limbic and cortical structures.

1. Ventral Tegmental Dopamine Neurons

In control rats, iontophoretic administration of glutamate produced a current-dependent increase in the firing rate of VTA DA neurons (Figure 10). With continued administration of increasing currents, DA cells began to fire in bursts with decreasing spike amplitude. Eventually, the cells became overdepolarized and entered a state of apparent depolarization block.[50] To verify that the effect was due to depolarization block, we reversed glutamate-induced inactivation with the inhibitory amino acid GABA, which hyperpolarizes and inhibits the firing of midbrain DA cells under normal conditions.[50] A cell was considered to be in a state of depolarization block when the glutamate-induced firing rate fell below 50% of the basal rate during the 40-sec administration of a particular glutamate current, and continued to decline at higher currents. Indications of the percentage of neurons in a state of depolarization block at each current are provided in Figure 10.

Repeated cocaine treatment failed to alter the basal firing rate of VTA DA cells, which is somewhat surprising in view of our previous work (discussed above) indicating significant increases in basal firing rates following repeated administration of cocaine.[14] However, in that study, estimates of basal firing rates were derived from considerably larger samples since they were obtained during a cells/track analysis of the VTA. In addition, the previous study was conducted using a longer treatment regimen.

Repeated cocaine treatment also failed to alter the rate-enhancing effects of glutamate observed at low iontophoretic currents, but significantly decreased the currents at which VTA DA neurons entered a state of depolarization block (Figure 10). Taken together, these two findings suggest that the initial excitatory effect is mediated by mechanisms distinct from those responsible for the induction of depolarization block. Based upon *in vivo* intracellular recordings from midbrain DA neurons, Grace and Bunney[50] proposed that depolarization block is an extreme extension of the normal depolarization-induced transition from a slow, irregular mode of activity to one of burst firing, resulting from sustained elevations in intracellular Ca^{2+}. Because glutamate activates VTA DA neurons primarily by stimulating NMDA receptors[51,52] which increase Ca^{2+} influx, it seems reasonable to speculate that the enhanced induction of depolarization block in cocaine-treated rats may be due to NMDA receptors. This explanation also would fit well with the observation that Ca^{2+} antagonists have been shown to prevent the development of behavioral sensitization to amphetamine.[53] However, recent *in vitro* studies have demonstrated glutamate-induced depolarization block of VTA DA neurons in the absence of burst firing.[54] In addition, NMDA has been shown to sustain rhythms of DA cell burst firing without the induction of depolarization block, via gating of Na^+ and activation of an electrogenic Na^+/K^+ pump.[55] Thus, the precise mechanisms responsible for the ability of glutamate to induce depolarization block and for alterations in this effect following repeated cocaine administration remain a matter of speculation.

It is also possible that repeated cocaine administration alters the sensitivity of non-NMDA glutamate receptors on VTA DA neurons. Electrophysiological studies have demonstrated clearly that VTA DA neurons possess both NMDA and non-NMDA ionotropic receptors.[52,55-58] In fact, Wang and French[54] demonstrated that at low concentrations, glutamate excited VTA DA neurons via a preferential effect on NMDA receptors, with non-NMDA receptor involvement only at higher concentrations. Moreover, selective stimulation of both receptor classes led to apparent depolarization block. Since we observed no significant difference between the control and cocaine groups with respect to the rate-enhancing efficacy of low glutamate currents, it is conceivable that NMDA receptors were not altered and that differences in threshold for induction of depolarization block at high currents involved alterations of non-NMDA receptor function. Obviously, studies with selective glutamate receptor agonists and antagonists will be required to identify the contributions of various receptors and conductances to the observed effects.

Given that repeated cocaine administration leads to a reduced sensitivity of VTA DA autoreceptors and to enhanced basal activity of the DA neuron population, it is conceivable that the greater induction of depolarization block was due to increased excitability resulting from reduced tonic inhibition via D_2 autoreceptors. However, since coadministration of MK-801 prevented VTA autoreceptor subsensitivity, it is equally likely that an alteration of EAA transmission plays some role in the development of VTA DA autoreceptor subsensitivity, or that glutamate and DA receptor sensitivity on VTA DA neurons are interdependently modulated in some manner.

2. Nucleus Accumbens Neurons

Because most NAc neurons fire at very slow rates or were quiescent under our experimental conditions, we needed to apply a different strategy to study the effects of EAAs on NAc neuronal activity. First, quiescent NAc neurons were identified during electrode penetrations by leaking glutamate from one of the side barrels of our micropipette assemblies. Once a neuron was isolated, the current required to produce firing rates of 4 to 5 spikes/sec was determined, a technique which we typically employ in studying the pharmacological responses of NAc neurons.[15,34] Following a 4- to 5-min period with no glutamate administration, a current-response determination (1 to 128 nA) was made for each cell. For this analysis, glutamate was ejected using a 30-sec on/40-sec off pulse procedure in which each current was pulsed for at least two 30-sec periods.

The first indication that the responsiveness of NAc neurons to glutamate might be altered in cocaine-treated rats was that higher ejection currents often were needed to identify quiescent neurons. The results of subsequent tests indicated a clear subsensitivity

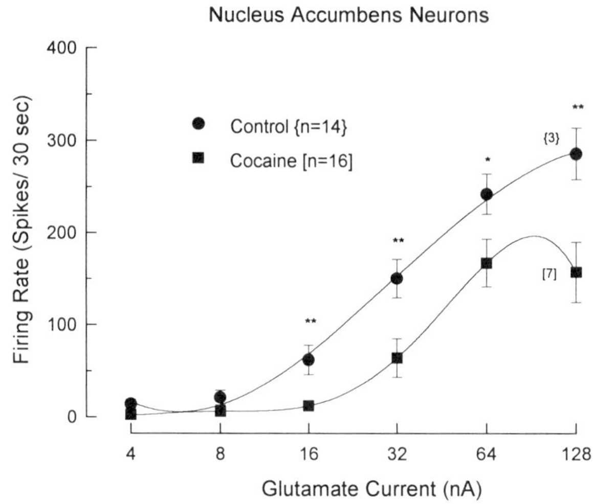

Figure 11. Repeated administration of cocaine significantly decreased the excitatory efficacy of iontophoretic glutamate on NAc neurons by Dunnett's test following repeated measures ANOVA (* – p <0.05, ** – p <0.01). Note that at the 128 nA current, more neurons in the cocaine-treated rats entered a state of depolarization block; however, this effect did not reach statistical significance by Fishers exact probability test. All cells tested with a current of 256 nA entered depolarization block.

of NAc neurons in the cocaine-treated rats compared to control animals. Thus, the mean (± S.E.M.) current required to achieve the targeted firing rate of 4 to 5 spikes/sec for the control group (34.8 ± 1.5 nA) was significantly less than that for the cocaine group (55.8 ± 3.5 nA).

Iontophoretic application of glutamate in control rats caused a current-dependent increase in the firing of NAc neurons, with little evidence of apparent depolarization block at currents below 256 nA. Only 3 of 14 neurons showed depolarization block at 128 nA, whereas every cell was inactivated at the 256 nA current. In cocaine-treated rats, there was a clear subsensitivity of NAc neurons to iontophoretic glutamate at currents of 16 to 128 nA (Figure 11). However, at 128 nA, there appeared to be an additional effect. Thus, 7 of the 16 NAc neurons tested in cocaine-treated rats entered a state of apparent depolarization block at the 128 nA glutamate current. Although this percentage was not statistically greater than that observed in the control group (3/14), it obviously contributed to the reduced excitation when firing rates were averaged (Figure 11). The other cells that were not inactivated were still excited at this current, contributing to the large variance. To determine whether the apparent subsensitivity of cocaine-treated NAc neurons to the 128 nA glutamate current was completely attributable to the greater percentage of cells that were in a state of depolarization block, we compared the remaining active cells in the two groups and found that cocaine-treated neurons were still subsensitive to the excitatory effects of glutamate. As in the VTA, iontophoretic administration of GABA reactivated NAc neurons during administration of constant glutamate currents, confirming that they were in a state of depolarization block.

We are currently investigating the EAA receptor subtypes that might be involved in the altered responsiveness of NAc neurons to glutamate following repeated cocaine administration. It is well known that activation of EAA-containing cortical, hippocampal, and amygdaloid afferents to the NAc results primarily in monosynaptic excitation.[59]

Pharmacological studies have found evidence for the existence of both NMDA and non-NMDA receptors mediating this effect.[60-65] Apparently, non-NMDA receptors are activated under normal levels of afferent activity whereas NMDA receptor involvement is most demonstrable under conditions of membrane depolarization when the Mg^{2+} block of the cation channel is relieved.[60,62] Accordingly, different glutamate receptors may be involved in the effects of low and high concentrations of extracellular glutamate produced by iontophoretic application, and repeated cocaine administration might alter more than one EAA receptor.

C. MECHANISMS OF ALTERED GLUTAMATE RESPONSES

The essential question which must now be addressed is how repeated cocaine administration induces alterations of EAA-mediated transmission within the mesoaccumbens pathway. Given that cocaine has not been shown to interact directly with glutamate receptors, multisynaptic circuitry may be implicated. With respect to the VTA, one interesting possibility is that receptor sensitivity is enhanced as a result of increased tonic activation, as has been demonstrated previously for NMDA receptors.[66] The EAA innervation of VTA DA neurons is derived, in large part, from the medial PFC.[67,68] Activation of this pathway causes burst firing of VTA DA neurons;[69] conversely, inactivation of the PFC reduces VTA DA neuron burst firing.[70] Within the PFC, DA inhibits the firing of pyramidal neurons in the DA-innervated deep layers,[71-73] including those neurons which project to the VTA.[74,75] As described above, repeated cocaine administration causes a subsensitivity of PFC neurons to DA-induced inhibition. Accordingly, it is possible that during repeated treatment with cocaine, DA-induced inhibition of PFC neurons projecting to the VTA diminishes, leading to greater activation of glutamate receptors on VTA DA neurons and the induction of receptor sensitization.

If we apply the same line of reasoning to the NAc, our results also could suggest altered excitatory input to this structure during repeated cocaine administration. Recall that the primary excitatory projections to the NAc originate in the PFC, hippocampus (ventral subiculum), and basolateral amygdala. As in the PFC, repeated administration of cocaine appears to produce tolerance to the inhibitory effects of DA on amygdaloid neurons.[76] If such an effect occurs on neurons projecting to the NAc, then the diminished inhibition might increase excitatory input to NAc neurons, perhaps resulting in reduced sensitivity of non-NMDA receptors, for which rapid desensitization is a well-documented phenomenon.[77,78]

It also is possible that the degree of the excitation caused by glutamate is not reduced by repeated cocaine treatment, but that such excitation (depolarization) is below the threshold for action potential generation because NAc neurons are hyperpolarized as a result of additional actions produced by repeated cocaine treatment. Although other investigators have reported that repeated methamphetamine treatment failed to alter the resting membrane potential of rat NAc neurons recorded in an *in vitro* slice preparation,[79] it is possible that enhanced hyperpolarization might exist *in vivo* if it was maintained by the tonic release of endogenous neurotransmitters. Obviously, DA would seem a likely candidate. Indeed, given that repeated cocaine administration leads to an increase of DA neuron basal activity and enhanced basal DA release in the NAc, there might well be increased hyperpolarization of NAc neurons mediated by supersensitive D_1 receptors. Additional studies are currently underway to determine whether the apparent decrease in sensitivity of NAc neurons to glutamate is a secondary result of increased D_1 receptor sensitivity and stimulation. However, the fact that D_1 receptor supersensitivity was prevented by NMDA antagonists suggests that events occurring at EAA receptors might precede, perhaps even cause, the changes in D_1 receptors.

D. FUNCTIONAL IMPLICATIONS OF ALTERED GLUTAMATE RESPONSES

The ultimate question to address is the extent to which alterations of EAA receptors within the mesoaccumbens DA system are related to cocaine sensitization and withdrawal. With respect to the VTA, it seems unlikely that increased sensitivity to the induction of depolarization block plays a role in behavioral sensitization observed in response to cocaine challenge, given that DA autoreceptor stimulation would provide an opposing hyperpolarization during the drug challenge. However, the enhanced likelihood of producing depolarization block might be related to symptoms observed during cocaine withdrawal. Recall that in cocaine-treated rats withdrawn for longer periods (10 to 14 days), there was a significant reduction in the number of spontaneously active VTA DA cells.[26] Additional studies are needed to determine whether this effect is due to depolarization block. It would certainly seem possible that upon heightened activation by cortical afferents, VTA DA neurons would be more likely to enter a state of depolarization block resulting in a diminished ability to transfer information to their postsynaptic targets.

Midbrain DA neurons respond to salient environmental stimuli with transient bursts of impulses which are thought to contribute to the setting of a motivational state by alerting other neuronal systems, in particular the NAc, that subserve reward-seeking and goal-directed activity.[80] If DA neurons were unable to provide this information due to depolarization block, then the animal would be less responsive to salient aspects of its environment. In this regard, it is noteworthy that during withdrawal from cocaine addiction, humans often experience reduced motivation, anergia, and anhedonia,[24,30] aspects of behavior thought to be modulated by VTA DA neurons.[9,10]

The NAc has been hypothesized to serve as an interface between limbic and motor systems modulating cognitive and motivational aspects of behavior as they are translated into specific actions.[81] Electrophysiological studies support the idea that the NAc and surrounding ventral striatal regions are involved in the "expectation of reward" and in execution of motor programs designed to obtain rewards.[82,83] Since the generation of anticipatory responses is likely to be derived from cortical and limbic glutamatergic inputs, tolerance to such afferent input during withdrawal from repeated cocaine administration might bias the organism against responding to salient information regarding reward expectancy and might thereby lead to a state of anergia and anhedonia. Upon subsequent cocaine challenge, the reduced responsiveness of the NAc to excitatory drive, coupled with the enhanced responsiveness to DA via D_1 DA receptors, would bias the NAc towards dopaminergic transmission which should enhance (i.e., sensitize) cocaine-induced cognitive and behavioral effects.

IV. SUMMARY AND CONCLUSIONS

It is becoming increasingly apparent that repeated cocaine administration leads to numerous neuroadaptations within mesocorticolimbic systems that are thought to be responsible for the subjective and behavioral effects of the drug. We have reviewed electrophysiological evidence for such alterations, with an emphasis on mechanisms of sensitization and withdrawal. Clearly, additional changes occur at other levels of neuronal function as reviewed elsewhere in this volume. Such changes may be responsible for, or result from, the neurophysiological changes which we have observed. As we have repeatedly argued, behavioral sensitization probably results from a sequence of alterations, some of which are necessary for the subsequent development of others.

At least two lines of evidence implicate the VTA as a site at which changes take place to induce behavioral sensitization. As discussed above, repeated administration of amphetamine or morphine into the VTA result in behavioral sensitization. Whether the same is true for cocaine has not yet been determined. Moreover, the mechanism underlying this effect remains a matter of speculation. In keeping with our line of thought, repeated injections of stimulants into the VTA might produce autoreceptor subsensitivity and enhance the activity of VTA DA neurons. Yet, the prevention of these effects and of behavioral sensitization by MK-801 argues for a primary role of NMDA receptors. Indeed, recent work by Kalivas and associates has demonstrated that intra-VTA injections of MK-801 prevent the short duration, context-dependent sensitization which occurs after a single cocaine injection (see Chapter 6). Whether the VTA is the site of preventive action of MK-801 during longer sensitization treatments is yet to be determined. Based upon the fact that lesions of the PFC prevent the development of amphetamine sensitization,[84] and our finding of altered glutamate responsiveness of VTA DA neurons in cocaine-sensitized rats, it seems apparent that the PFC-VTA excitatory projection is integral to the induction and development of sensitization.

A role for the NAc in the expression of sensitization is accepted widely. In studies complementary to those using intra-VTA and intra-NAc injections of amphetamine to test for the development of sensitization, other investigators have demonstrated that repeated systemic injections of amphetamine lead to sensitized locomotion when rats are challenged with intra-NAc amphetamine.[85,86] Our findings strongly implicate D_1 receptor supersensitivity in the expression of both cocaine and amphetamine[49] sensitization. How the changes in EAA receptor responsiveness within the NAc are also involved in sensitization remains to be determined. Apparently, EAA receptors within the NAc are not involved in the development of cocaine sensitization since intra-NAc administration of MK-801 failed to prevent cocaine sensitization.[87]

The precise roles of altered DA and EAA transmission in cocaine withdrawal also remain to be elucidated. Various studies, including those reviewed in this chapter, suggest that during the process of cocaine withdrawal, the mesoaccumbens DA system undergoes a time-dependent "down-regulation" with diminished basal activity of VTA DA neurons resulting in decreased release of DA within the NAc. Our latest results suggest that glutamatergic transmission might be enhanced during cocaine withdrawal, leading to a reduced responsiveness of EAA receptors on NAc neurons. Additional studies will be required to determine the cellular mechanisms responsible for such effects as well as their clinical relevance.

ACKNOWLEDGMENTS

This research was made possible by USPHS Grant DA 04093 and by Research Scientist Development Award DA 00207 (to FJW) from the National Institute on Drug Abuse. The scientific contributions of Drs. Janice M. Ackerman and Marina E. Wolf are gratefully acknowledged.

REFERENCES

1. White, F. J., Wachtel, S. R., Johansen, P. A., and Einhorn, L. E., Electrophysiological studies in the rat mesoaccumbens dopamine system: focus on dopamine receptor subtypes, interactions and the effects of cocaine, in *Neurophysiology of Dopaminergic Systems: Current Status and Future Perspectives,* Chiodo, L. A. and Freeman, A. S., Eds., Lakeshore Press, Inc., Detroit, 1987, 317.
2. White, F. J., Henry, D. J., , Hu, X.-T., Jeziorski, M., and Ackerman, J. M., Electrophysiological effects of cocaine in the mesoaccumbens dopamine system, in *Cocaine: Pharmacology, Physiology and Clinical Strategies,* Lakoski, J. M., Galloway, M. P., and White, F. J., Eds., CRC Press, Boca Raton, FL, 1992, 261.
3. White, F. J., The physiological basis of the rewarding effects of cocaine, *Behav. Pharmacol.,* 1, 303, 1990.
4. Einhorn, L. C., Johansen, P. A., and White, F. J., Electrophysiological effects of cocaine in the mesoaccumbens dopamine system: studies in the ventral tegmental area, *J. Neurosci.,* 8, 100, 1987.

5. White, F. J., Hu, X.-T., and Henry, D. J., Electrophysiological effects of cocaine in the rat nucleus accumbens: microiontophoretic studies, *J. Pharmacol. Exp. Ther.*, 266, 1075, 1993.
6. Robinson, T. E. and Becker, J. B., Enduring changes in brain and behavior produced by chronic amphetamine administration: a review and evaluation of animal models of amphetamine psychosis, *Brain Res. Rev.*, 11, 157, 1986.
7. Kalivas, P. W. and Stewart, J., Dopamine transmission in the initiation and expression of drug- and stress-induced sensitization of motor activity, *Brain Res. Rev.*, 16, 223, 1991.
8. Post, R. M., Weiss, S. R. B., and Pert, A., Sensitization and kindling effects of chronic cocaine administration, in *Cocaine: Pharmacology, Physiology and Clinical Strategies,* Lakoski, J. M., Galloway, M. P., and White, F. J., CRC Press, Boca Raton, FL, 1992, 115.
9. Wise, R. A. and Bozarth, M. A., A psychomotor stimulant theory of addiction, *Psychol. Rev.*, 94, 469, 1987.
10. Robinson, T. E. and Berridge, K. C., The neural basis of drug craving: an incentive-sensitization theory of addiction, *Brain Res. Rev.* 18, 247, 1993.
11. Shuster, L., Yu, G., and Bates, A., Sensitization of cocaine stimulation in mice, *Psychopharmacology,* 52, 185, 1977.
12. Weiss, S. R. B., Post, R. M., Pert, A., Woodward, R., and Murman, D., Context-dependent cocaine sensitization, *Pharmacol. Biochem. Behav.*, 34, 655, 1989.
13. Post, R. M. and Contel, N. R., Human and animal studies of cocaine: implications for development of behavioral pathology, in *Stimulants: Neurochemical, Behavioral and Clinical Perspectives,* Creese, I., Ed., Raven, New York, 1983, 169.
14. Henry, D. J., Greene, M. A., and White, F. J., Electrophysiological effects of cocaine in the mesoaccumbens dopamine system: repeated administration, *J. Pharmacol. Exp. Ther.*, 251, 833, 1989.
15. Henry, D. J. and White, F. J., Repeated cocaine administration causes a persistent enhancement of D1 dopamine receptor supersensitivity within the rat nucleus accumbens, *J. Pharmacol. Exp. Ther.*, 258, 882, 1991.
16. Muller, P. and Seeman, P., Presynaptic subsensitivity as a possible basis for sensitization by long-term dopamine mimetics, *Eur. J. Pharmacol.*, 55, 149, 1979.
17. Antelman, S. M. and Chiodo, L. A., Dopamine autoreceptor subsensitivity: a mechanism common to the treatment of depression and the induction of amphetamine psychosis, *Biol. Psychiatry,* 16, 717, 1981.
18. Ackerman, J. M. and White, F. J., A10 somatodendritic dopamine autoreceptor sensitivity following withdrawal from repeated cocaine treatment, *Neurosci. Lett.*, 117, 181, 1990.
19. Pitts, D. K., Freeman, A. S., Kelland, M. D., and Chiodo, L. A., Repeated amphetamine: reduced dopamine neuronal responsiveness to apomorphine but not quinpirole, *Eur. J. Pharmacol.*, 162, 167, 1989.
20. Nestler, E. J., Terwilliger, R. Z., Walker, J. R., Sevarino, K. A., and Duman, R. A., Chronic cocaine treatment decreases levels of the G proteins subunits $G_{i\alpha}$ and $G_{o\alpha}$ in discrete regions of rat brain, *J. Neurochem.*, 55, 1079, 1990.
21. Striplin, C. D. and Kalivas, P. W., Correlation between behavioral sensitization to cocaine and G protein ADP-ribosylation in the ventral tegmental area, *Brain Res.*, 579, 181, 1992.
22. Striplin, C. D. and Kalivas, P. W., Robustness of G protein changes in cocaine sensitization shown with immunoblotting, *Synapse,* 14, 10, 1993.
23. White, F. J. and Wang, R. Y., Comparison of the effects of chronic haloperidol treatment on A10 dopamine neurons in the rat, *Life Sci.*, 32, 983, 1983.
24. Gawin, F.H. and Ellinwood, E. H., Jr., Cocaine and other stimulants: actions, abuse and treatment, *N. Engl. J. Med.*, 318, 1173, 1988.
25. Zhang, H., Lee, T. H., and Ellinwood, E. H., Jr., The progressive changes of neuronal activities of the nigral dopaminergic neurons upon withdrawal from continuous infusion of cocaine, *Brain Res.*, 594, 315, 1992.
26. Ackerman, J. M. and White, F. J., Decreased activity of rat A10 dopamine neurons following withdrawal from repeated cocaine, *Eur. J. Pharmacol.*, 218, 171, 1992.
27. Parsons, L. H., Smith, A. D., and Justice, J. B., Jr., Basal extracellular dopamine is decreased in the rat nucleus accumbens during abstinence from chronic cocaine, *Synapse,* 9, 60, 1991.
28. Weiss, F., Markou, A., Lorang, M. T., and Koob, G. F., Basal extracellular dopamine levels in the nucleus accumbens are decreased during withdrawal after unlimited-access self-administration, *Brain Res.*, 593, 314, 1992.

29. Pilotte, N. A., Sharpe, L. G., and Kuhar, M. J., Withdrawal of repeated intravenous infusion of cocaine persistently reduces binding to dopamine transporters in the nucleus accumbens of Lewis rats, *J. Pharmacol. Exp. Ther.*, 269, 963, 1994.
30. Gawin, F. H., Cocaine addiction: psychology and neurophysiology, *Science,* 251, 1580, 1991.
31. White, F. J. and Wang, R. Y., Electrophysiological evidence for the existence of both D-1 and D-2 dopamine receptors in the rat nucleus accumbens, *J. Neurosci.*, 6, 274, 1986.
32. Wachtel, S. R., Hu, X.-T., Galloway, M. P., and White, F. J., D1 dopamine receptor stimulation enables the postsynaptic, but not autoreceptor, effects of D2 dopamine agonists in nigrostriatal and mesoaccumbens dopamine systems, *Synapse,* 4, 327, 1989.
33. White, F. J., Hu, X.-T., and Brooderson, R. J., Repeated stimulation of D_1 dopamine receptors enhances the effects of DA agonists, *Eur. J. Pharmacol.*, 191, 497, 1990.
34. Hu, X.-T., Brooderson, R. J., and White, F. J., Repeated stimulation of D_1 dopamine receptors causes time-dependent alterations in the sensitivity of both D_1 and D_2 dopamine receptors within the rat striatum, *Neuroscience,* 50, 137, 1992.
35. Sorg, B. A. and Kalivas, P. W., Effects of cocaine and foot-shock stress on extracellular dopamine levels in the medial prefrontal cortex, *Neuroscience,* 53, 695, 1993.
36. Kalivas, P. W. and Weber, B., Amphetamine injection into the ventral mesencephalon sensitizes rats to peripheral amphetamine and cocaine, *J. Pharmacol. Exp. Ther.*, 245, 1095, 1988.
37. Vezina, P. and Stewart, J., Amphetamine administered to the ventral tegmental area but not to the nucleus accumbens sensitizes rats to systemic morphine: lack of conditioned effects, *Brain Res.*, 516, 99, 1990.
38. Jeziorski, M. and White, F. J., Dopamine receptor antagonists prevent expression, but not development, of morphine sensitization, *Eur. J. Pharmacol.*, in press.
39. Kalivas, P. W. and Duffy, P., The effect of acute and daily cocaine treatment on extracellular dopamine in the nucleus accumbens, *Synapse,* 5, 48, 1990.
40. Pettit, H. O., Pan, H.-T., Parsons, L. H., and Justice, J. B., Jr., Extracellular concentrations of cocaine and dopamine are enhanced during chronic cocaine administration, *J. Neurochem.*, 55, 798, 1990.
41. Kalivas, P. W. and Duffy, P., Time course of extracellular dopamine and behavioral sensitization to cocaine. I. Dopamine axon terminals, *J. Neurosci.*, 13, 266, 1993.
42. Segal, D. S. and Kuczenski, R., Repeated cocaine administration induces behavioral sensitization and corresponding decreased extracellular dopamine responses in caudate and accumbens, *Brain Res.*, 577, 351, 1992.
43. Izenwasser, S. and Cox, B. M., Daily cocaine treatment produces a persistent reduction of [^{3}H]dopamine uptake in vitro in rat nucleus accumbens but not in striatum, *Brain Res.*, 531, 338, 1990.
44. Brock, J. W., Ng, J. P., and Justice, J. B., Jr., Effect of chronic cocaine on dopamine synthesis in the nucleus accumbens as determined by microdialysis perfusion with NSD-1015, *Neurosci. Lett.*, 117, 234, 1990.
45. Yi, S.-J. and Johnson, K. M., Chronic cocaine treatment impairs the regulation of synaptosomal ^{3}H-DA release by D_2 autoreceptors, *Pharmacol. Biochem. Behav.*, 36, 457, 1990.
46. Cass, W. A. and Zahniser, N. R., Cocaine levels in striatum and nucleus accumbens: augmentation following challenge injection in rats withdrawn from repeated cocaine administration, *Neurosci. Lett.*, 152, 177, 1993.
47. Karler, R., Calder, L. D., Chaudhry, I. A., and Turkanis, S. A., Blockade of 'reverse tolerance' to cocaine and amphetamine by MK-801, *Life Sci.*, 45, 599, 1989.
48. Wolf, M. E. and Jeziorski, M., Coadministration of MK-801 with amphetamine, cocaine or morphine prevents rather than transiently masks the development of behavioral sensitization, *Brain Res.*, 613, 291, 1993.
49. Wolf, M. E., White, F. J., and Hu, X.-T., MK-801 prevents alterations in the mesoaccumbens dopamine system associated with behavioral sensitization to amphetamine, *J. Neurosci.*, 14, 1735, 1994.
50. Grace, A. A. and Bunney, B. S., Induction of depolarization block in midbrain dopamine neurons by repeated administration of haloperidol: analysis using *in vivo* intracellular recording, *J. Pharmacol. Exp. Ther.*, 238, 1092, 1986.
51. Mereu, G., Costa, E., Armstrong, D. M., and Vicini, S., Glutamate receptor subtypes mediate excitatory synaptic currents of dopamine neurons in midbrain slices, *J. Neurosci.*, 11, 1359, 1991.

52. Wang, T. and French, E. D., L-glutamate excitation of A10 dopamine neurons is preferentially mediated by activation of NMDA receptors: extra- and intracellular electrophysiological studies in brain slices, *Brain Res.,* 627, 299, 1993.
53. Karler, R., Calder, L. D., and Turkanis, S. A., Calcium channel blockers and behavioral sensitization, *Life Sci.,* 49, 165, 1991.
54. Wang, T. and French, E. D., Electrophysiological evidence for the existence of NMDA and non-NMDA receptors on rat ventral tegmental dopamine neurons, *Synapse,* 13, 270, 1993.
55. Johnson, S. W., Seutin, V., and North, R. A., Burst firing in dopamine neurons induced by *N*-methyl-D-aspartate: role of electrogenic sodium pump, *Science,* 258, 665, 1992.
56. Seutin, V., Verbanck, P., Massotte, L., and Dresse, A., Evidence for the presence of *N*-methyl-D-aspartate receptors in the ventral tegmental area of the rat: an electrophysiological *in vitro* study, *Brain Res.,* 514, 147, 1990.
57. Mercuri, N. B., Stratta, F., Calabresi, P., and Bernardi, G., A voltage-clamp analysis of NMDA-induced responses on dopaminergic neurons of the rat substantia nigra zona compacta and ventral tegmental area, *Brain Res.,* 593, 51, 1992.
58. Suaud-Chagny, M. F., Chergui, K., Chouvet, G., and Gonon, F., Relationship between dopamine release in the rat nucleus accumbens and the discharge activity of dopaminergic neurons during local in vivo application of amino acids in the ventral tegmental area, *Neuroscience,* 49, 63, 1992.
59. Mogenson, G. J. and Yim, C. C., Neuromodulatory functions of the mesolimbic dopamine system: electrophysiological and behavioural studies, in T*he Mesolimbic Dopamine System: From Motivation to Action,* Willner, P. and Scheel-Kruger, J., Eds., J. Wiley, Chichester, 1991, 106.
60. Horne, A. L., Woodruff, G. N., and Kemp, J. A., Synaptic potentials mediated by excitatory amino acid receptors in the nucleus accumbens of the rat, in vitro, *Neuropharmacology,* 29, 917, 1990.
61. Pennartz, C. M. A., Boeijinga, P. H., and Lopes da Silva, F. H., Locally evoked potentials in slices of the rat nucleus accumbens: NMDA and non-NMDA receptor mediated components and modulation by GABA, *Brain Res.,* 529, 30, 1990.
62. Pennartz, C. M. A., Beoijinga, P. H., Kitai, S. T., and Lopes da Silva, F. H., Contribution of NMDA receptors to postsynaptic potentials and paired-pulse facilitation in identified neurons of the rat nucleus accumbens in vitro, *Exp. Brain Res.,* 86, 190, 1991.
63. Pennartz, C. M. A. and Kitai, S. T., Hippocampal inputs to identified neurons in an in vitro slice preparation of the rat nucleus accumbens: evidence for feed-forward inhibition, *J. Neurosci.,* 11, 2838, 1991.
64. Uchimura, N., Higashi, H., and Nishi, S., Membrane properties and synaptic responses of the guinea pig nucleus accumbens neurons in vitro, *J. Neurophysiol.,* 61, 769, 1989.
65. Uchimura, N. and North, R. A., Baclofen and adenosine inhibit synaptic potentials mediated by γ-aminobutyric acid and glutamate release in rat nucleus accumbens, *J. Pharmacol Exp. Ther.,* 258, 663, 1991.
66. Sah, P., Hestrin, S., and Nicoll, R. A., Tonic activation of NMDA receptors by ambient glutamate enhances excitability of neurons, *Science,* 246, 815, 1989.
67. Christie, M. J., Bridge, S., James, L. B., and Beart, P. M., Excitotoxin lesions suggest an aspartatergic projection from rat medial prefrontal cortex to ventral tegmental area, *Brain Res.,* 333, 169, 1985.
68. Sesack, S. R. and Pickel, V. M., Prefrontal cortical terminals synapse on dopaminergic neurons in the ventral tegmental area and on unlabeled neuronal targets of catecholaminergic terminals in the nucleus accumbens septi, *J. Comp. Neurol.,* 320, 145, 1992.
69. Gariano, R. F. and Groves, P. M., Burst firing induced in midbrain dopamine neurons by stimulation of the medial prefrontal and anterior cingulate cortices, *Brain Res.,* 462, 194, 1988.
70. Svensson, T.H. and Tung, C.-S., Local cooling of pre-frontal cortex induces pacemaker-like firing of dopamine neurons in rat ventral tegmental area in vivo, *Acta Physiol. Scand.,* 136, 135, 1989.
71. Bunney, B. S. and Aghajanian, G. K., Dopamine and norepinephrine innervated cells in the rat prefrontal cortex: pharmacological differentiation using microiontophoretic techniques, *Life Sci.,* 19, 1783, 1976.
72. Ferron, A., Thierry, A. M., Le Douarin. C., and Glowinski. J., Inhibitory influence of the mesocortical dopaminergic system on spontaneous activity or excitatory response induced from the thalamic mediodorsal nucleus in the rat medial prefrontal cortex, *Brain Res.,* 302, 257, 1984.
73. Sesack, S. R. and Bunney, B. S., Pharmacological characterization of the receptor mediating electrophysiological responses to dopamine in the rat medial prefrontal cortex: a microiontophoretic study, *J. Pharmacol. Exp. Ther.,* 248, 1323, 1989.

74. Godbout, R., Mantz, J., Pirot, S., Glowinski, J., and Thierry, A. M., Inhibitory influence of the mesocortical dopaminergic neurons on their target cells: electrophysiological and pharmacological characterization, *J. Pharmacol. Exp. Ther.*, 258, 728, 1991.
75. Pirot, S., Godbout, R., Mantz, J., Tassin, J.-P., Glowinski, J., and Thierry, A.-M., Inhibitory effects of ventral tegmental area stimulation on the activity of prefrontal cortical neurons: evidence for the involvement of both dopaminergic and GABAergic components, *Neuroscience,* 49, 857, 1992.
76. Callahan, P. M., Mah, E. J., and Cunningham, K. A., Electrophysiological sensitivity of amygdala neurons to dopamine and serotonin following repeated cocaine administration, *Soc. Neurosci. Abstr., 19*, 1855, 1993.
77. Mayer, M.L. and Westbrook, G.L., The physiology of excitatory amino acids in the vertebrate central nervous system, *Prog. Neurobiol.,* 8, 197, 1987.
78. Trussel, L. O. and Fischbach, G. D., Glutamate receptor desensitization and its role in synaptic transmission, *Neuron,* 3, 209, 1989.
79. Higashi, H., Inanaga, K., Nishi, S., and Uchimura, N., Enhancement of dopamine actions on rat nucleus accumbens neurones *in vitro* after methamphetamine pre-treatment, *J. Physiol.,* 408, 587, 1989.
80. Schultz, W. and Romo, R., Dopamine neurons of monkey midbrain: contingencies of responses to stimuli eliciting immediate behavioral reactions, *J. Neurophysiol.,* 63, 607, 1990.
81. Mogenson, G. J., Limbic-motor integration, in *Progress in Psychobiology and Physiological Psychology,* Epstein, A.N., Ed., Academic Press, New York, 1987, 117.
82. Schultz, W., Apicella, P., Scarnati, E., and Ljungberg, T., Neuronal activity in monkey ventral striatum related to expectation of reward, *J. Neurosci.,* 12, 4595, 1992.
83. Chang, J.-Y, Sawyer, S. F., Lee, R.-S., and Woodward, D. J., Electrophysiological and pharmacological evidence for the role of the nucleus accumbens in cocaine self-administration in freely moving rats, *J. Neurosci.,* 14, 1224, 1994.
84. Wolf, M. E., Dahlin, S. L., Hu, X.-T., Xue, C.-J., Effects of lesions of prefrontal cortex, amygdala, or fornix on behavioral sensitization to amphetamine: comparison with N-methyl-D-aspartate antagonists, *Neuroscience,* in press.
85. Paulson, P. E. and Robinson, T. E., Sensitization to systemic amphetamine produces an enhanced locomotor response to a subsequent intra-accumbens amphetamine challenge in rats, *Psychopharmacology,* 104, 140, 1991.
86. Kolta, M. G., Shreve, P., and Uretsky, N. J., Effect of pretreatment with amphetamine on the interaction between amphetamine and dopamine neurons in the nucleus accumbens, *Neuropharmacology,* 28, 9, 1989.
87. Kalivas, P. W. and Alesdatter, J. E., Involvement of *N*-methyl-D-aspartate receptor stimulation in the ventral tegmental area and amygdala in behavioral sensitization to cocaine, *J. Pharmacol. Exp. Ther.,* 267, 486, 1993.

Chapter 8

Modulation of Serotonin Function by Acute and Chronic Cocaine: Neurophysiological Analyses

Kathryn A. Cunningham

CONTENTS

I. INTRODUCTION

Potentiation of dopamine (DA) neurotransmission, particularly within mesolimbic pathways which originate in the ventral tegmental area (VTA) and terminate in the nucleus accumbens (NAc), is a pivotal mediator of the behavioral effects of cocaine.[1,2] Cocaine binds to the DA transporter[3] and inhibits DA reuptake from the synapse,[4] prolonging the stimulation of DA receptors. The potentiation of mesolimbic DA transmission appears to mediate the locomotor stimulatory,[5-8] rewarding,[9-11] and discriminative effects of cocaine.[12-19]

These same pathways also play an important role in the alterations associated with intermittent, repeated injections of cocaine (1 to 2/day for several days) which are well documented to result in sensitization of locomotor activity and stereotypy.[20,21] *In vivo* and *in vitro* analyses have demonstrated enhancements in stimulated DA release in mesolimbic circuits after repeated, intermittent cocaine exposure,[22-25] a change which might be driven partially by altered D_2 autoreceptor control of DA neurons.[26] Single unit recordings of DA VTA neurons support the hypothesis that somatodendritic autoreceptors become subsensitive following chronic cocaine exposure; this change would contribute to the subsequent enhancement of stimulated DA release seen in terminal fields.[23] In concert with this, the postsynaptic responsiveness of D_1 receptors in the NAc is enhanced.[27] Thus, during early withdrawal, the subsensitivity of VTA DA neurons[26] coupled with the

1-8493-8311-0/95/$0.00+$.50

supersensitivity of D_1 receptors in the NAc[27,28] would be expected to produce enhanced DA neurotransmission as shown to occur *in vivo,*[24,25] a situation which would promote the development of sensitization.

Behavioral analyses suggest that, since intra-VTA amphetamine does not elicit hyperactivity, but multiple intra-VTA amphetamine infusions sensitize rats to systemic amphetamine and cocaine,[29] the VTA may play an initiating role in behavioral sensitization. On the other hand, intra-accumbens amphetamine results in hyperlocomotion, but sensitization does not develop upon repeated NAc amphetamine injections.[29] Thus, both DA somata and terminals (presynaptic) as well as DA receptors (postsynaptic) are sites for the indirect DA agonist actions of cocaine, and the VTA and NAc may play differential roles in sensitization.

However, cocaine is not a selective DA reuptake inhibitor, and does not share an identical behavioral profile with such compounds, particularly with regard to abuse liability.[30,31] The differentiation between cocaine and the selective DA reuptake inhibitors could be related to other neural actions. Cocaine is a local anesthetic.[32] It also binds to the transporters[3] for serotonin (5-hydroxytryptamine; 5-HT) and norepinephrine (NE)[4] to block reuptake with potency and efficacy equal to its effects on DA reuptake. In addition, cocaine has affinity for the 5-HT_3 receptor,[33] M_1 and M_2 muscarinic receptors,[34] and σ-receptors.[35] Thus, the actions of cocaine on each of these non-DA systems probably contributes to its overall behavioral profile, and alterations in the normal function of these systems after chronic administration of cocaine may be relevant behaviorally.

In addition to the well-documented involvement of the NAc in the *in vivo* effects of cocaine (discussed above), other limbic brain regions (e.g., the amygdala) have been implicated in this regard. Intermittent subconvulsive doses of cocaine elicit limbic responses very similar to the kindling seen with progressive electrical stimulation of the limbic forebrain, which has been especially well studied in the amygdaloid nuclei.[36-38] The relationship between these phenomena (sensitization and limbic kindling) and their underlying mechanisms are unclear; however, neither the local anesthetic nor monoaminergic stimulant properties of cocaine appear to account totally for its effects on limbic excitability.[37,39] The fact that repeated amphetamine exposure results in sensitization but fails to induce limbic kindling[37] supports a dissociation between stimulant-induced sensitization and electrical kindling.[40] However, there is not a consensus[41] and recent evidence suggests that stimulation of the amygdala enhances the development of sensitization to amphetamine.[42] Thus, these studies illustrate that other mesolimbic sites are targets for the actions of cocaine and that the amygdala itself may be particularly important.

The amygdala, a primary component of the limbic system, is a collection of nuclei and tracts densely innervated by monoamine neurons[43] and thought to be involved in such psychological processes as attention, mood, and emotion as well as the responsiveness to environmental stimuli.[44,45] A modulatory role has been proposed for the amygdala in several behavioral effects of psychostimulants. For example, disruption of DA function following 6-hydroxydopamine (6-OHDA) lesion of the amygdala facilitated self-administration of a low dose of amphetamine,[46] while ibotenic acid lesion of the substantia innominata (part of the "extended amygdala"[47]) decreased cocaine self-administration.[48] Microinjection of the DA D_1 receptor antagonist SCH 23390 into the amygdala blocked the discriminative stimulus,[49] reinforcing,[50,51] and hypermotive effects of cocaine.[51] When 5-HT systems of the amygdala were disrupted by 5,7-dihydroxytryptamine (5,7-DHT) lesion, the breakpoint on a progressive ratio schedule of cocaine self-administration increased, suggesting an enhanced motivation to acquire cocaine.[52] Thus, both DA and 5-HT systems of the amygdala affect cocaine-induced behavior, and further investigation of these effects in the amygdala is important for a full understanding of the actions of this abused drug.

The goal of the present chapter is to review current evidence with regard to the electrophysiological significance of the interaction of cocaine with central 5-HT systems, in particular raphe-amygdala pathways, and to provide a focus for future research on the role of mesolimbic 5-HT-DA interactions in the *in vivo* effects of acute and chronic exposure to cocaine.

II. CELLULAR EFFECTS OF ACUTE COCAINE ON SEROTONIN SYSTEMS

A. SEROTONIN SYSTEMS

1. Anatomy of Serotonin Neurons and Projections

The dorsal (DR) and median (MR) raphe nuclei of the midbrain contain the majority of 5-HT neurons which ascend to innervate the mesencephalon and most forebrain regions, including mesolimbic DA circuits.[43,53,54] Of interest in the actions of cocaine, the VTA, NAc, and amygdala receive 5-HT afferents from DR and MR,[53,55] and 5-HT terminals are known to make synaptic contacts on somatodendritic processes of both VTA DA and non-DA neurons.[56] There is also a reciprocal innervation of the raphe from other 5-HT and non-5-HT raphe neurons,[57] DA and non-DA VTA neurons,[58] and limbic nuclei.[55,59,60] Thus, the anatomy of the 5-HT system supports the hypothesis that 5-HT has a wide influence on mesolimbic brain function; 5-HT-DA interactions could take place at the level of 5-HT raphe neurons and/or DA VTA neurons and at the level of the mesolimbic monoamine terminals (e.g., NAc and amygdala). Cocaine is likely to alter the balance of the individual actions, and any interplay between these actions, at multiple sites in the brain.

2. Serotonin Receptors and Localization

There is a large diversity of 5-HT receptors, and two subfamilies have been identified. Most of the currently identified 5-HT receptors are G-protein-coupled and include the 5-HT_1 (5-HT_{1A}, 5-HT_{1B}, and 5-HT_{1D}), 5-HT_2 (5-HT_{2A}, 5-HT_{2C}), 5-HT_4, 5-HT_5, 5-HT_6, and 5-HT_7 receptors; the 5-HT_3 receptor is a ligand-gated ion channel (for recent reviews, see Branchek[61] and Martin and Humphrey[62]).

The raphe nuclei contain abundant 5-HT_{1A} receptors[63,64] which function as impulse-modulating somatodendritic autoreceptors[65] to decrease firing rates as a consequence of increased K^+ conductance and membrane hyperpolarization.[66] There is also a high density of 5-HT_{1B} receptors in the raphe.[64] The response of raphe neurons to 5-HT_{1B} stimulation is somewhat region-specific; MR 5-HT neurons seem to be more responsive to the excitatory effects of systemic 5-HT_{1B} agonists *in vivo* than those of the DR.[67,68] On the other hand, the raphe has few 5-HT_{2A} receptors, and no 5-HT_{2C} or 5-HT_3 receptors have been visualized there.[33,64]

Serotonin receptors also are localized differentially in the VTA and limbic regions. For example, a majority of the 5-HT_1 receptors in the VTA are of the 5-HT_{1B} subtype, while 5-HT_{1A} receptor numbers are low or below detection limits.[64,69] The 5-HT_{1B} receptor has been shown to control 5-HT terminal release[70] and appears to function in the VTA to control γ-aminobutyric acid (GABA) release[71]; similarly, 5-HT_{1B} receptors control GABA and glutamate release in a physiologically relevant manner in locus coeruleus.[72] Heterogenous distributions of 5-HT receptor subtypes are found in limbic regions. For example, unlike the nuclei which contain 5-HT and DA cell bodies, the NAc has appreciable levels of 5-HT_{2A}, 5-HT_{2C}, and 5-HT_3 receptors,[33,73] but few 5-HT_{1A} receptors.[63] The amygdala similarly shows significant numbers of 5-HT_{1B}, 5-HT_{2A}, 5-HT_{2C}, and 5-HT_3; unlike the NAc, the amygdala also shows a high concentration of 5-HT_{1A} receptors.[73-76] Thus, with regard to the behavioral consequences of its indirect actions on 5-HT receptors, cocaine

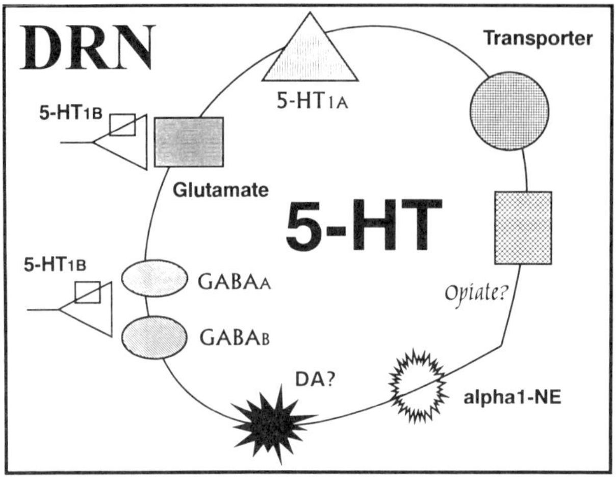

Figure 1. Raphe 5-HT neuron. Schematic illustration of receptors and putative innervation of raphe 5-HT neurons which might contribute to the action of cocaine on this population of neurons. See text for explanation of the influence of each receptor on the function of 5-HT neurons.

might be expected to enhance the normal actions of 5-HT at a potentially large number of 5-HT receptors found in raphe, VTA, and their limbic targets.

B. ACUTE EFFECTS OF COCAINE ON SEROTONIN NEURONS

1. Basic Cellular Physiology of Serotonin Neurons

Serotonin neurons of the raphe are spontaneously active *in vivo* with regular, slow firing rates (1 to 2 spikes/sec) which are dependent upon intrinsic pacemaker potentials[77] and the action of NE at an excitatory α_1-adrenoceptor[78] to suppress resting K^+ currents and transient outward A-currents.[78] A number of modulatory forces impinge on raphe 5-HT neurons (Figure 1) including the G-protein-linked somatodendritic 5-HT_{1A} autoreceptor, activation of which increases K^+ conductance and suppresses spontaneous neural activity.[66,79] Agonists for the 5-HT_{1A} receptor and drugs which increase synaptic accumulation of 5-HT, such as 5-HT releasers and reuptake inhibitors, result in activation of this receptor and a cessation of activity.

In addition to the 5-HT_{1A} autoreceptor, a phosphoinositide-linked α_1-NE receptor mediates other modulatory effects of NE, although the impact of these receptors on 5-HT raphe activity *in vivo* is relatively uncharacterized.[80] Both N-methyl-D-aspartate (NMDA) and non-NMDA synaptic potentials have also been identified in DR 5-HT neurons *in vitro*.[81] Although the direct influence of excitatory amino acid receptors has not been studied *in vivo,* glutamate has a stimulatory action on 5-HT raphe neurons as intra-DR glutamate injection increases both 5-HT and DA release in the NAc.[82] Since 5-HT_{1B} receptors recently have been shown to inhibit glutamate- (and GABA-) mediated synaptic potentials in the locus coeruleus,[72] 5-HT_{1B} receptors (which are abundant in raphe) also could modulate glutamate or GABA responses of 5-HT neurons in the DR.

A monosynaptic GABA pathway has been proposed to mediate habenula-evoked inhibition of 5-HT neurons; a polysynaptic pathway in which an excitatory (e.g., Substance P or excitatory amino acid) neuron synapses onto a GABA interneuron to inhibit 5-HT neuronal activity also has been suggested.[83,84] In fact, both $GABA_A$ and $GABA_B$ receptors are found in abundance in the raphe,[85] and GABA agonists and antagonists modulate raphe 5-HT output.[81,84] Dopamine terminals and D_1 and D_2 receptors have been found in the raphe[86-88] and, although an early study showed that intra-raphe DA microinfusion significantly altered EEG,[89] no further information exists concerning the modulatory action of DA on 5-HT neurons. Terminals and receptors for opiates[92,93] and tachykinins[90,91] also have been reported, and 5-HT neurons could respond to 5-HT_3 stimulation,[94] although no 5-HT_3 receptors have been visualized in the raphe.[33]

2. Role of Serotonin Reuptake Inhibition

Single unit extracellular recordings in anesthetized rats have proven a useful model to study the effects of cocaine in intact animals. Such studies have demonstrated that the intravenous administration of (–)-cocaine, but not the local anesthetic procaine, potently and reversibly depresses the activity of 5-HT neurons in the DR of the rat.[95-97] Interestingly, while cocaine is relatively equipotent as an inhibitor of DA, 5-HT and NE uptake[4] as well as ligand binding to DA, 5-HT, and NE transporter sites,[3] cocaine is a more efficacious and potent inhibitor of the impulse activity of 5-HT neurons[95-97] than that of either DA[98] or NE neurons.[99] Iontophoretic application of cocaine depresses the activity of 5-HT neurons with prolonged application resulting in a progressive increase in the magnitude of the drug-induced depression; cocaine also potentiates the inhibition induced by iontophoretic 5-HT.[95]

Pharmacological analyses *in vivo* suggest that, as a consequence of cocaine-induced reuptake inhibition, the rise in synaptic 5-HT stimulates impulse-modulating 5-HT_{1A} autoreceptors on 5-HT DR cell bodies to dampen cellular activity. Administration of the nonselective 5-HT_{1A} antagonist spiperone, which has been shown to block 5-HT_{1A}-mediated inhibitory responses of 5-HT DR neurons,[100] reversed the cocaine-induced suppression of firing.[96] Following depletion of 5-HT with the synthesis inhibitor p-chlorophenylalanine (PCPA), the sensitivity to cocaine-induced depression was reduced significantly, suggesting that the effects of cocaine are dependent in part on endogenous 5-HT stores.[96]

In a raphe slice preparation, cocaine prolonged the 5-HT-mediated inhibitory postsynaptic potentials (IPSPs) at concentrations (EC_{50} = 1.7 μM) which have little or no effect on the membrane potential of 5-HT neurons.[101] At higher concentrations, cocaine caused a hyperpolarization (EC_{50} = 4.2 μM) and decreased the amplitude of the 5-HT-mediated IPSP (EC_{50} = 5.7 μM);[101] these actions of cocaine appear to be mediated by effects upon 5-HT reuptake. Thus, autoregulatory feedback mechanisms recruited by increased synaptic concentrations of 5-HT and subsequent stimulation of impulse-modulating 5-HT_{1A} autoreceptors on 5-HT cell bodies are important mediators of cocaine-induced inhibitory actions on these neurons.[95,96,101]

3. Role of Mesohabenular Feedback

Several observations prompted further investigation of regulation by afferent projections to DR 5-HT neurons which might be integral to the rate-suppressive effects of cocaine. First, although nonselective 5-HT_{1A} antagonists (e.g., propranolol or spiperone) were found to *reverse* cocaine-induced inhibition,[96] *pretreatment* with the selective 5-HT_{1A} antagonist WAY 100478 [(+)-WAY 100135] only partially blocked the effects of cocaine at doses which completely blocked an equipotent dose of the selective 5-HT_{1A} agonist 8-OHDPAT.[102] Second, while iontophoretic cocaine inhibited the spontaneous activity of 5-HT neurons, rarely was *complete* suppression observed (at low currents). Third, 5-HT depletion with PCPA significantly attenuated cocaine-induced depression; however, this diminution was surmountable at higher doses of cocaine. Although this residual effect of cocaine may be related to a failure to completely eliminate raphe 5-HT content with PCPA, cocaine might retain its ability to depress 5-HT activity if afferent pathways to the DR (unaffected by PCPA) play a role in the rate-suppressant effects of cocaine. If this is the case, such afferent input would be required to elicit complete suppression of activity as occurs following systemic, but not iontophoretic, cocaine, and a 5-HT_{1A} antagonist might not completely prevent its actions.

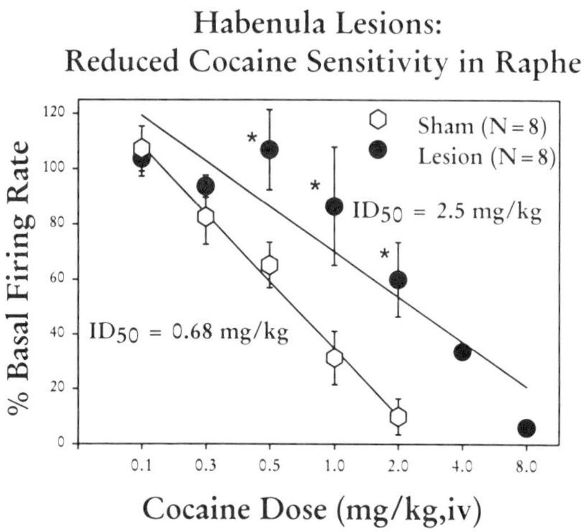

Figure 2. Habenula lesions: dose-response curves for inhibition of 5-HT DR neurons by cocaine. Rats were anesthetized with urethane and bilateral radiofrequency lesions of habenula were made 1 to 4 hr prior to single unit recordings of 5-HT DR neurons. Once a 5-HT neuron was identified and a stable basal rate established, cocaine was administered intravenously at 2-min intervals such that each dose doubled the previously administered dose. Open and closed circles represent the mean percentage of basal firing rate (± SEM) observed at each dose of cocaine in sham or habenula-lesioned rats, respectively. The dose-effect curve for cocaine was shifted to the right in lesioned rats (*–$p \leq 0.05$). [Adapted from Paris and Cunningham].[110]

Prominent afferent innervation of the DR as well as mesencephalic catecholamine nuclei arises in the habenula.[103] Ascending efferent projections to limbic regions do not originate in the habenular complex, although many prominent afferents of the habenula arises in forebrain, including the amygdala, NAc, and striatum as well as in monoamine-containing midbrain regions.[103,104] The habenula is thought to be part of the circuit by which DA systems control the activity of 5-HT raphe systems.[105] In fact, the influence of the (lateral) habenula on DR glucose utilization appears to be mediated via the D_1 receptor.[106] Stimulation of the lateral habenula is known to inhibit the spontaneous activity of 5-HT[107] and DA neurons.[108] As such, the indirect DA agonist actions of cocaine might trigger these habenular feedback circuits, causing an increase in habenular activity and a consequent decrease in DR 5-HT firing, although cocaine applied microiontophoretically is associated with an inhibition of the activity of some habenular neurons.[109]

As a first step in testing the dependence of the effects of cocaine on long-loop feedback to raphe, radiofrequency lesions of the habenular nuclei were made 1 to 4 hr ("short-term") or 7 days ("long-term") prior to recordings of 5-HT DR neurons.[110] Control animals underwent identical surgical procedures without current delivery. In rats with short-term (but not long-term) lesions, the cocaine-induced depression was significantly attenuated (Figure 2). A population sampling of DR neurons indicated that the number of spontaneously active cells was decreased significantly following short-term (but not long-term) habenular lesions; when cells were distinguished based upon firing rates, the number of slow-firing (<1.2 Hz), presumably 5-HT cells, decreased in lesion versus control rats (Figure 3).[110] Thus, disruption of habenular function resulted in a change in the basal characteristics of 5-HT neurons and reduced the responsiveness of these neurons to inhibition by cocaine.[110] Interestingly, the brain eventually compensated for the effects of habenular lesions as raphe responsiveness to cocaine recovered by 7 days.[110]

The prominent actions of cocaine on mesolimbic DA function might feedback via the habenula to partially mediate the suppressive actions of cocaine on raphe 5-HT cells. While this hypothesis remains to be tested systematically, the present data elucidate both the process of afferent regulation of 5-HT neurons as well as the neural effects of cocaine. Cocaine is a highly efficacious reinforcer, and the combined effects of cocaine on both

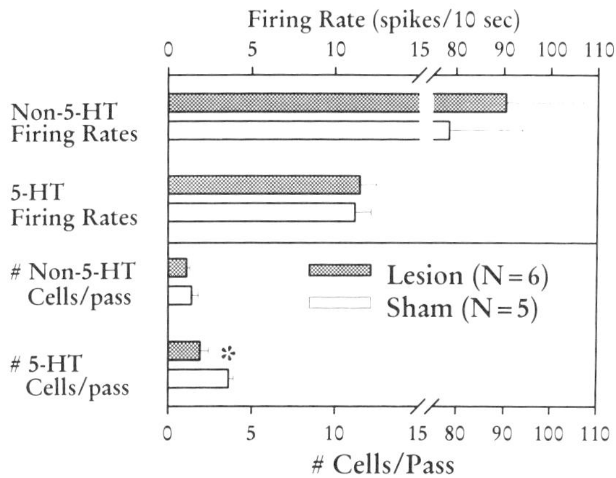

Figure 3. Habenula lesions: population studies of 5-HT DR neurons. In sham or habenula-lesioned rats, population cell counts were conducted as described in text. The number of 5-HT neurons encountered in lesioned rats was significantly reduced (*– $p \leq 0.05$). [Adapted from Paris and Cunningham, 1994].[110]

DA and 5-HT systems could account for the fact that selective DA or 5-HT uptake inhibitors do not share total similarity with cocaine.

C. ACUTE EFFECTS OF COCAINE ON AMYGDALA NEURONS

1. Basic Cellular Physiology of Amygdala Neurons

The amygdaloid nucleus is actually a collection of nuclei, subnuclei, and their numerous intra-amygdaloid connections. Studies of the neuronal morphology and physiology of the amygdala have identified several classes of neurons, most notably studied in the basolateral amygdala (BLA). The majority of cells (Class I) are large, typically pyramidal or stellate in shape with spine-laden dendrites (see McDonald[111] for review); the remaining neurons are small with spine-sparse dendrites and appear to be interneurons (Class II) or neurogliaform cells (Class III). Class I neurons appear to be projection neurons and are the major cell population of the amygdala in which glutamate can be localized.[112] Class II and Class III neurons are immunoreactive for GABA and acetylcholine as well as a number of neuropeptides (e.g., somatostatin, vasoactive intestinal peptide, and cholecystokinin).[111,113,114] Although the two major cell types do not differ in resting membrane potential, Class I and II neurons can be differentiated by other electrophysiological characteristics *in vitro,* in particular distinctions in input resistance, duration of action potentials, and firing patterns.[115]

In addition to these *in vitro* characteristics, the activity of amygdala neurons has been studied *in vivo* in intact rats and cats. Both BLA and central (CEA) nuclei of the amygdala exhibit spontaneous activity when recorded in anesthetized,[116,117] transected,[117-119] or paralyzed preparations.[120-122] The firing rates of amygdala cells vary widely between 0 and 60 spikes/sec, although lower firing rates predominate both in awake and anesthetized animals.[123,124] Amygdala neurons respond with excitation, inhibition, and mixed inhibition/excitation in response to electrical stimulation of afferent projection neurons as well as a variety of sensory stimuli.[124-126]

Serotonergic innervation of the amygdala arises from both dorsal and median raphe cell groups.[43,53,54] The terminals of 5-HT neurons in the amygdala are widespread and quite dense, particularly in basal and central nuclei.[43] Stimulation of the DR results in an inhibition of spontaneous activity in the BLA[118] while iontophoresis of 5-HT[117,118] as well as systemic administration of the 5-HT precursor 5-hydroxytryptophan depress the spontaneous activity of amygdala neurons.[36,127] In contrast, while similar dense inputs to the amygdala arise in midbrain catecholamine neurons,[43] administration of the catecholamine precursor *l*-3,4-dihydroxyphenylalanine (*l*-DOPA) accelerates cell firing—an effect which is prevented by blockade of amino acid decarboxylase.[127] Since *l*-DOPA is the precursor

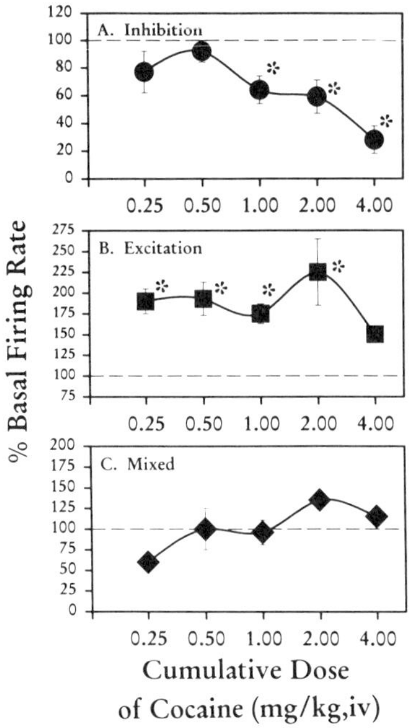

Figure 4. Effects of intravenous cocaine on the activity of amygdala neurons. After stable baseline firing was established, intravenous cocaine was administered in a cumulative fashion. Symbols represent the mean percentage of basal firing rate (± SEM) observed at each dose of cocaine. A. Inhibition: Cells which responded to cocaine with > 20% suppression of firing were located either in the BLA (n = 9) or CEA (n = 6). B. Excitation: Cells which responded to cocaine with > 20% increase in firing rates were observed either in the BLA (n = 4) or CEA (n = 5). C. Mixed: Cells which responded to cocaine with > 20% decrease followed by a > 20% increase in firing rate were found either in the BLA (n = 4), CEA (n = 1), or medial nuclei (n = 1). Asterisks denote significant ($p \leq 0.05$) difference from basal firing rates.

for both DA and NE, its excitatory effects on amygdala activity cannot be attributed specifically to either one of these neurotransmitters; however, microiontophoretic DA is known to inhibit single cell activity in the BLA and CEA.[128,129] Eidelberg and Woodbury[36] suggest that their data provide evidence for a dual antagonistic synaptic input upon amygdala neurons with 5-HT and a catecholamine (NE?) serving as an inhibitory and excitatory neurotransmitter, respectively. Since amygdala neurons exhibit a high degree of convergence of different inputs upon the same unit,[124] these data raise the possibility that the effects of cocaine could be more pronounced, or absent, in those neurons receiving convergent monoaminergic input and provide a means by which the effects of cocaine on monoamine systems can be studied.

2. Effects of Systemic and Iontophoretic Cocaine

Initial studies determined the response of amygdala neurons to intravenous cocaine injections.[130] Single-barrel glass micropipettes and conventional extracellular recording procedures were used to record from single neurons in the amygdala of anesthetized rats. Of a total of 32 neurons recorded, 18 were located in the BLA, 13 were in the CEA, and 1 was in the medial amygdaloid nucleus. The spontaneous activity of these neurons was slow and irregular while triphasic (+/–/+) waveforms were most typical, as has been shown previously.[124] Firing rates (# spikes/10 sec interval) averaged 10.7 ± 4.9 and 15.3 ± 8.7 in the BLA and CEA, respectively. Intravenous administration of cocaine had significant effects on the neural activity of neurons located in amygdala (Figure 4). An inhibition of cell firing was the predominant effect (47%; Figure 4A), although a significant percentage of cells was excited by cocaine (28%; Figure 4B), and mixed inhibitory/excitatory responses were also observed (19%; Figure 4C); only a small fraction of recorded neurons were nonresponsive to cocaine administration (6%).[130] Thus, systemic cocaine administration can activate or dampen cellular activity of amygdala neurons.

These findings with cocaine can be contrasted to those obtained with amphetamine in similar preparations. Amphetamine has been reported to depress 75% of neurons recorded in the amygdala; the onset of this response was slow and the duration of the drug-induced inhibition was significantly longer than in other sites investigated.[120] While excitation was reported to occur in only a small subpopulation of amygdala neurons,[120] Wepsic and

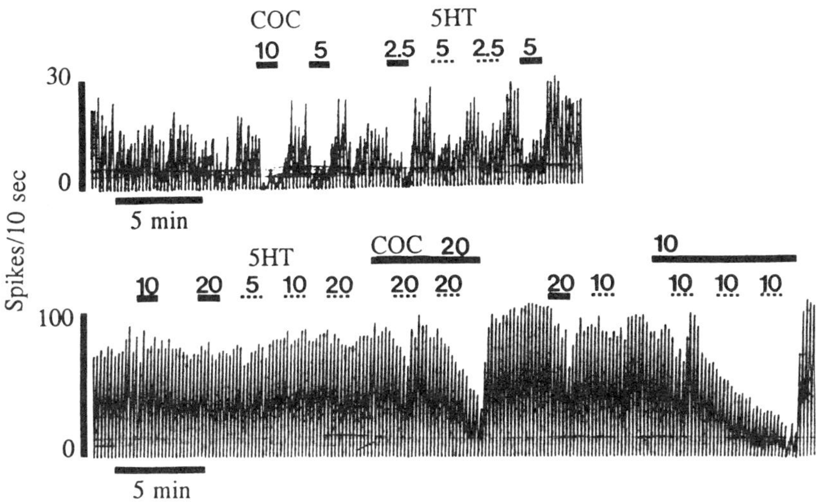

Figure 5. Effects of microiontophoretic cocaine and 5-HT on the activity of amygdala neurons. Top. Microiontophoretic application of cocaine (COC; 0.01 M; thick bars) inhibited the activity of this BLA neuron with an onset and offset similar to that observed with 5-HT (5-HT; 0.04 M; dots). Bottom. In another cell, neither cocaine nor 5-HT significantly affected cell firing at currents up to 20 nA. However, there is a dramatic synergistic effect of co-iontophoresis of cocaine and 5-HT. This cell was located on the border of the BLA and CEA. In these histograms, the duration of drug ejection is indicated by the bar above the record; numbers refer to iontophoretic current in nanoamperes (nA) used for drug ejection.

Austin[131] found that systemic amphetamine excited the majority of cells recorded in a paralyzed cat preparation. Differences in species, sex, amphetamine dose, and other variables could account for these divergent results. Interestingly, Wepsic and Austin[131] did find that amphetamine decreased the amplitude of responses recorded in the amygdala upon septal stimulation, an observation which is consonant with the predominantly inhibitory effects seen in studies conducted by Rebec and colleagues.[120,121]

Because cocaine acts at the transporter on cell bodies and terminals of monoamine neurons, the diverse response to systemic cocaine probably is related to both suppressed neural activity of monoamine soma and reuptake inhibition at monoamine terminals in amygdala. To determine the extent to which local actions of cocaine in the amygdala account for its systemic effects and to assess the normal actions of 5-HT in the amygdala as well as the interactions of cocaine with 5-HT, microiontophoretic techniques have been employed in conjunction with single unit extracellular recordings. Multi-barrel glass micropipettes filled with cocaine (0.01 M, pH 4), 5-HT (0.04 M, pH 4) and *l*-glutamate (0.01 M, pH 8) were lowered into amygdala. A low retaining current of –10 (+ 10 nA for glutamate) was applied to each drug barrel and the duration of iontophoretic drug ejection was controlled to provide a uniform 60 sec interval between drug applications. For these studies, microiontophoretic glutamate was occasionally used to activate amygdala cells so that both spontaneously active and quiescent neurons could be studied. Because ejection currents greater than 64 nA for cocaine have been found to exert apparent local anesthetic effects (reduced amplitude and widening of the duration of the waveform),[95,132] only currents between 2.5 and 40 nA were used.

Of the 24 neurons recorded, 13 were located in BLA, 8 in CEA, 2 in the medial dorsal, and 1 in the lateral amygdaloid nuclei. The majority of both spontaneously active and glutamate-driven cells were inhibited by iontophoretic cocaine (66%) and 5-HT (95%). Average maximal inhibition (50 to 70%) occurred with ejection currents between 10 to 40 nA. Figure 5 (top) illustrates a BLA neuron which was particularly sensitive to

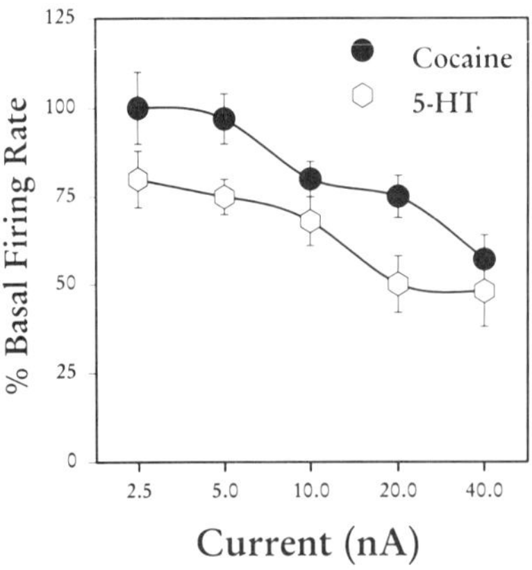

Figure 6. Current-dependent effects of iontophoretic cocaine and 5-HT on neuronal activity of amygdala neurons. Filled and open symbols represent the mean percentage of basal firing rate (± SEM) observed during iontophoresis of a given current of cocaine or 5-HT, respectively. Data were collected from a total of 24 cells.

cocaine; almost complete suppression was observed at 10 nA in this cell. Serotonin also elicited a current-dependent inhibition of firing in this neuron. In contrast, the amygdala cell found on the border between the BLA and CEA (Figure 5, bottom) was relatively non-responsive to cocaine and 5-HT. However, co-iontophoresis of cocaine and 5-HT resulted in a dramatic and synergistic suppression of cell firing. Both iontophoretic cocaine and 5-HT inhibit amygdala activity in a current-dependent manner (Figure 6).

The responsivity of amygdala neurons to microiontophoretic cocaine probably depends upon the degree of convergence of diverse monoamine inputs upon the same unit as well as the relative potency and efficacy of cocaine to inhibit reuptake in this region. Microiontophoretic 5-HT inhibits the activity of amygdala neurons as shown previously[36,117-119,127,129] and replicated here, and thus 5-HT is likely to contribute to the inhibitory actions of microiontophoretic cocaine. Dopamine is also inhibitory in this nucleus,[36,128,129] and low currents of cocaine, which are not inhibitory, potentiate the responsiveness of amygdala cells to DA.[129] The coejection of DA and 5-HT also results in an additive or synergistic inhibitory effect (data not shown) supporting the hypothesis that the ability of iontophoretic cocaine to inhibit amygdala neurons is dependent on the simultaneous inhibition of reuptake for both 5-HT and DA. Such a phenomenon also was observed previously for cocaine effects in the NAc.[132,133] In support of this, a study of cocaine applied to a slice preparation of NAc indicated that both the DA D_2 receptor-stimulated depolarization and the DA D_1 receptor- and 5-HT (5-HT_2) receptor-induced hyperpolarizations were enhanced by cocaine application, with 5-HT responses increased more than DA responses.[134]

The ultimate impact of cocaine on limbic neurons is complex, in part because its actions are overlaid upon the normal functions of, and interactions between, 5-HT and DA. An abundance of both 5-HT (5-HT_{1A}, 5-HT_{1B}, 5-HT_{2A}, 5-HT_{2C}, and 5-HT_3) receptors[73-76] and DA (D_1 and D_2) receptors are localized within the amygdala.[135] The role of particular 5-HT (as well as DA) receptors in the predominantly inhibitory actions of these transmitters in the amygdala is largely undefined at present, although such studies are currently underway.[136]

Only suppression of firing of amygdala neurons was observed in response to microiontophoretic cocaine; this stands in contrast to the mixed effects of intravenous cocaine injection on amygdala neurons. The neural explanation for excitatory and mixed responses observed following systemic cocaine administration must include a consideration of the consequences of cocaine actions on neuronal pathways which impinge on amygdala and take into account such factors as the relative impact of cessation of monoamine neuronal activity on the function of amygdala neurons.[69,124,137] Alternatively, some amygdala neurons might respond to stimulation of resident 5-HT and/or DA receptors with excitation; although no excitatory responses to either monoamine have been reported to occur in the amygdala to date, further microiontophoretic studies of cocaine and monoamine interactions in the amygdala as well as studies of the influence of stimulation of monoamine cell bodies are needed to clarify this hypothesis.

III. CELLULAR EFFECTS OF CHRONIC COCAINE ON SEROTONIN SYSTEMS

A. BEHAVIORAL SENSITIZATION TO COCAINE

Intermittent, repeated administration of cocaine results in the enhancement of behaviors including horizontal and vertical motor activity and the development of stereotyped behavioral patterns, including sniffing and head/limb movements (for reviews, see Post and Contel,[20] Robinson and Becker,[138] and Kalivas et al.[139]). To study potential modifications in 5-HT neuronal function associated with cocaine sensitization, male Sprague-Dawley rats (250 to 350 g) were injected twice daily (at 9:00 a.m. and 4:00 p.m.) intraperitoneally (i.p.) with either cocaine (15 mg/kg) or saline vehicle (0.9% NaCl) for 7 days. In some experiments, rats were replaced into home cages and behaviors were observed and rated using the Kilbey-Ellinwood scale for stimulant-induced stereotypic behaviors[6] 15 min after the first and last injections on days 1 and 7 of treatment.[6,69,140] The experimenter was unaware of the treatment during injections, observational ratings and subsequent electrophysiological recordings. For a representative group of rats (n = 6) injected with saline on day 1, a median score of 4 (range: 2 to 4) was obtained following the first injection; this median did not differ significantly from the median score (range: 4 to 5) of rats injected with a first injection of cocaine (n = 6; Figure 7, inset). Furthermore, the median scores of saline-treated rats did not differ between the first (day 1) and last injection (day 7). However, a significant ($p \leq 0.05$) increase in the median behavioral score for cocaine-treated rats was observed between the first and last injection (8; range: 7 to 8; Figure 7, inset). This enhanced behavioral response was characterized by hyperactivity as well as repetitive head and forelimb movements, and indicated that sensitization had developed.[6,69,140]

That this regimen of cocaine administration produced sensitization of cocaine-induced locomotor stimulation is shown most clearly in Figure 7. After a 30-min habituation period in the activity chambers on day 1 and day 7 of treatment, rats were injected with saline or cocaine and horizontal activity was measured during a 60-min period. The time course of activity observed following an injection of saline on day 7 did not differ from that on day 1 (Figure 7). However, while cocaine-injected rats exhibited a modest but significant increase in activity on day 1 (compared to saline injected animals), horizontal activity doubled on day 7 in response to repeated treatment with the same dose of cocaine (Figure 7). Thus, sensitization can be effectively determined using either observational stereotypy scales or locomotor activity monitors. All cocaine-treated animals utilized in the following studies exhibited behaviors characteristic of sensitization as measured by one means or the other.

B. SENSITIZATION AND CELLULAR PHYSIOLOGY OF SEROTONIN NEURONS

1. Population Characteristics of Serotonin Neurons

Sensitization has been linked to aberrations of DA function, but 5-HT malfunctions could also contribute to this behavioral phenomenon. The highest concentration of 5-HT transporters are found in the raphe,[141,142] and cocaine has a dramatic inhibitory effect on these 5-HT raphe neurons.[95,96,101,143] To determine whether the 5-HT system exhibits electrophysiological deficits during withdrawal from a sensitizing regimen of cocaine administration, population sampling of 5-HT cells was conducted 24 hr after the last treatment injection. For these studies, a single barrel electrode was passed several times

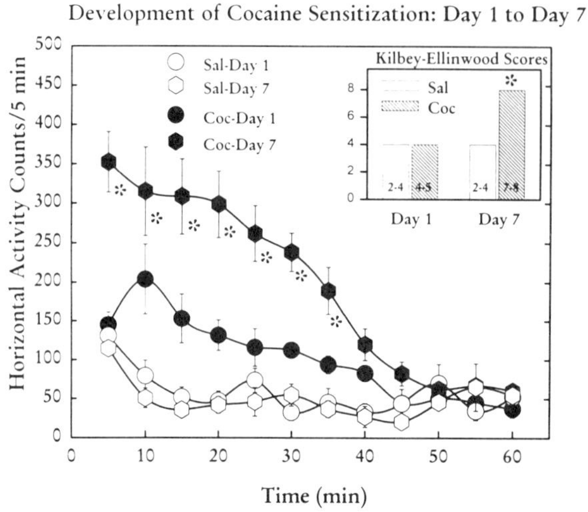

Figure 7. Development of behavioral sensitization to cocaine. Horizontal activity was measured on the first (day 1) and last (day 7) days of treatment with either saline or cocaine (15 mg/kg, i.p., BID for 7 days) in plexiglass enclosures equipped with horizontal photobeams (n = 6/group). The ordinate represents mean activity counts (± SEM) per 5-min bin over the course of a 60-min observation. Open and closed symbols refer to saline- and cocaine-treated rats, respectively; day 1 and day 7 counts are represented as circles and hexagons, respectively. Asterisks denote significant ($p \leq 0.05$) difference between day 7 and day 1 for cocaine-treated rats. Inset: Behavioral sensitization scored on the Kilbey-Ellinwood scale[6] on days 1 and 7 are represented as median scores. Numbers in bars represent the range of scores recorded for each treatment group on days 1 and 7. The asterisk indicates stereotypy scores that differed significantly ($p \leq 0.05$) between days 1 and 7. The animals were assigned a score from 1 to 9 based upon the following behavioral characteristics: *1* — asleep, lying down, eyes closed; *2* — inactive, lying down, eyes open; *3* — in place activities, normal grooming, chewing cage litter, eating, or drinking; *4* — normal, alert, active, moving about cage, sniffing, rearing; *5* — hyperactive with rapid changes in position; *6* — slow-patterned, repetitive exploration of the cage at normal levels of activity; *7* — fast-patterned, repetitive exploration of the cage with hyperactivity; *8* — restricted, remaining in same place in cage with fast, repetitive head and/or forelimb movement (includes licking, chewing, gnawing stereotypies); and *9* — dyskinetic-reactive with backing up, jumping, seizures, abnormally-maintained postures.

through the DR, and 5-HT neurons were readily identified by their characteristic action potentials (1 to 2 ms, +/– spikes) and regular, slow firing rates (0.2 to 2.5 Hz). In contrast, *non*-5-HT neurons exhibit short duration spikes (<1 ms, +/– waveform), irregular rhythms, and fast firing rates (>2.5 Hz).[144]

Characteristic 5-HT cells were counted and their firing rates were determined 24 hr after the last treatment injection. At this early withdrawal interval, the number of 5-HT DR neurons and their firing rates were significantly decreased (Figure 8). The decreased number of 5-HT neurons encountered in the DR coupled with their reduction in firing rate suggests either that some 5-HT cells became quiescent or begin to exhibit erratic, slowed or increased firing rates during cocaine treatment, or that there were changes in waveform which resulted in a failure to classify such neurons as characteristic 5-HT cells. It is difficult to speculate on this point at the present time; however, one way to test this hypothesis would be to sample populations of both 5-HT and *non*-5-HT neurons concurrently in treated animals. An increase in the number of *non*-5-HT cells found and/or changes in their firing rates would support this possibility.

The slowing of 5-HT neuronal activity during withdrawal from chronic cocaine raises the possibility that basal somatodendritic and/or terminal release of 5-HT also might be

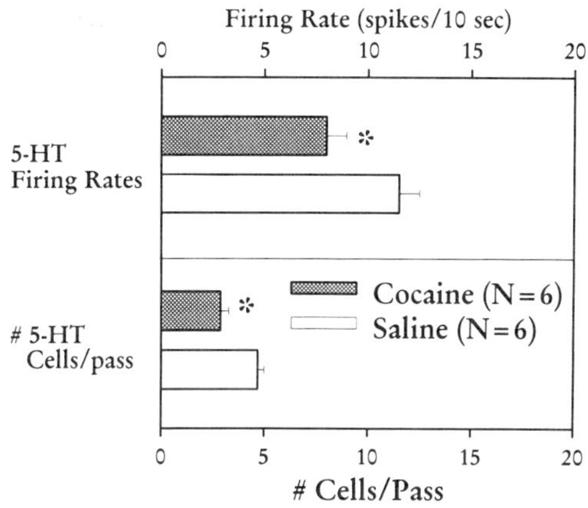

Figure 8. Cocaine sensitization: population studies of 5-HT DR Neurons. Rats were treated with saline or cocaine (15 mg/kg, i.p., BID for 7 days) and prepared for single unit recording 24 hr after the last injection. Population cell counts were conducted as described in text. The number of 5-HT neurons and their firing rates were significantly ($p \leq 0.05$) reduced in the cocaine-sensitized rats as denoted by asterisks.

reduced during withdrawal from chronic cocaine. In support of this hypothesis, Parsons and Justice[145] have recently measured 5-HT content in dialysate from DR, VTA, and NAc and found that basal extracellular 5-HT tended to be reduced in or following cocaine sensitization. A reduction of basal 5-HT levels in the NAc also have been observed during withdrawal from a repeated regimen of cocaine self-administration.[146] There is also behavioral evidence to suggest that the pharmacological sensitivity of 5-HT_1 and 5-HT_2 receptors is increased in cocaine-treated rats during withdrawal.[147,148] Thus, electrophysiological, neurochemical, and behavioral evidence suggest that some withdrawal-related behavioral and psychiatric changes could be due to reduced 5-HT function and resultant modifications in the sensitivity of 5-HT receptors.

2. Pharmacological Sensitivity of Serotonin Neurons

To test the hypothesis that the pharmacological sensitivity of DR 5-HT neurons might be altered during withdrawal, recordings of 5-HT neurons were again made at 24 hr withdrawal following chronic cocaine. An enhancement in the electrophysiological responsivity of 5-HT neurons to intravenous administration of cocaine, the selective 5-HT reuptake inhibitor fluoxetine and the selective 5-HT_{1A} agonist 8-OHDPAT was observed.[69] The dose-response curves for each of these compounds were shifted to the left such that the dose of drug required to inhibit cell firing to 50% of basal activity was reduced significantly (Figure 9). Although increased brain levels of cocaine are probably at least partially responsible for enhanced sensitivity of DR neurons to cocaine challenge in sensitized rats,[24] there is currently no evidence to suggest that repeated cocaine exposure can result in increased brain levels of fluoxetine or 8-OHDPAT, although this remains a possibility. Nevertheless, the enhanced inhibitory response to cocaine, fluoxetine and 8-OHDPAT suggests that supersensitivity of somatodendritic autoregulatory mechanisms which control impulse flow in 5-HT neurons has developed as a consequence of repeated cocaine exposure.[69]

Because these compounds were administered intravenously, the actual site(s) of action for cocaine to produce the shift of 5-HT neuronal sensitivity could be raphe 5-HT cells themselves or afferent circuits which impinge on the raphe. Cocaine sensitization is also associated with an increased density of 5-HT transporters in the DR and several cortical areas (e.g., the medial prefrontal cortex), as measured by [^{3}H]-imipramine[69] and [^{3}H]-paroxetine autoradiography (Paris, Goeders, and Cunningham, unpublished observations). Thus, the enhanced behavioral and electrophysiological sensitivity of 5-HT systems which accompany sensitization might be attributed to changes in 5-HT reuptake processes in the raphe. Such changes occur in the absence of overt cocaine-induced

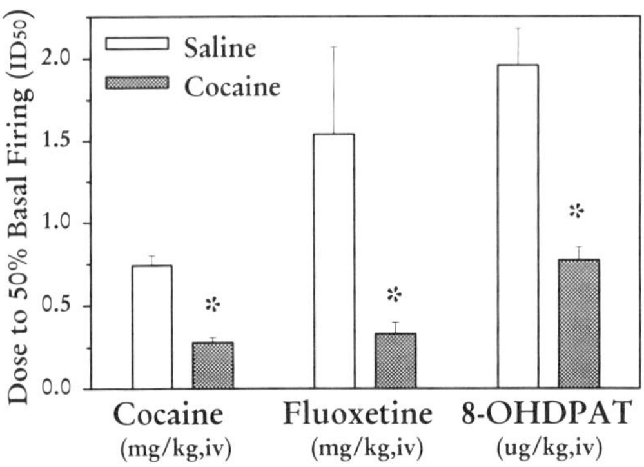

Figure 9. Cocaine sensitization: enhanced pharmacological sensitivity in raphe. Rats were treated with saline or cocaine (15 mg/kg, i.p., BID for 7 days) and prepared for single unit recording 24 hr after the last injection. After establishment of a stable baseline, drugs were administered intravenously in a cumulative fashion. The dose to inhibit neural activity of 5-HT cells by 50% was determined (ID_{50}) for each cell. The ordinate represents the mean ID_{50} (± SEM) for saline- or cocaine-treated rats administered either cocaine, fluoxetine, or 8-OHDPAT. Asterisks indicate ID_{50} values which differ between treatment groups ($p \leq 0.05$). [Adapted from Cunningham et al, 1992].[69]

neurotoxicity of 5-HT systems, as measured *ex vivo* by neurochemical[149] or immunohistochemical assays.[140] Furthermore, 5-HT transporters reportedly decrease in the striatum after 36-hr withdrawal from repeated cocaine administration,[150] suggesting that withdrawal intervals are an important variable to consider. This increased 5-HT autoregulation could contribute to the *decrease* in 5-HT neurotransmission which appears to have developed in sensitized rats.[69,145,146]

On the other hand, the density of [^{3}H]-8-OHDPAT-labeled 5-HT_{1A} receptors was unchanged in raphe nuclei following chronic cocaine treatment,[69] although the sensitivity to 8-OHDPAT of 5-HT_{1A} autoreceptors on DR 5-HT neurons was enhanced in cocaine-treated compared to saline-treated rats. Either cocaine-induced alterations in cellular processes distal to 5-HT_{1A} receptor activation, such as G-protein-linked second messenger function[151-153] or disruption of the functional interaction between 5-HT transporters and 5-HT_{1A} autoreceptors[154,155] could contribute to the enhanced inhibitory response to 8-OHDPAT. Interference with long-loop feedback is also an unexplored possibility and might be important given that a significant decrease in 5-HT_{1A} receptors in the central medial amygdala has been observed.[69]

An apparent modification of 5-HT autoregulation of raphe neurons develops in response to chronic cocaine administration. Cocaine also interferes with NE reuptake, and modifications of NE regulation of 5-HT neurons could also occur as a consequence of chronic cocaine treatment. The response to NE was enhanced in NE-containing cells of the locus coeruleus,[156] although subsensitivity *in vivo* had previously been reported.[157] Thus, numerous influences on 5-HT neurons have the potential to modulate 5-HT output to limbic regions, and cocaine may potentially interfere with several of these. For example, chronic cocaine exposure also has been associated with increased substance P levels in the DR[90] and reduced ^{3}H-naloxone binding to opiate receptors in the DR,[93] thus, other neuromodulators of 5-HT raphe function also must be considered.

C. SENSITIZATION AND CELLULAR PHYSIOLOGY OF AMYGDALA NEURONS

Both 5-HT and DA systems of the amygdala affect aspects of the behavioral response to stimulants. Amygdala susceptibility to electrical and cocaine-induced "kindling" and the similarities between kindling and sensitization suggest that this limbic region might be particularly vulnerable to repeated cocaine exposure. Stimulation of the amygdala enhanced the development of sensitization to amphetamine,[42] and amygdala lesions appear

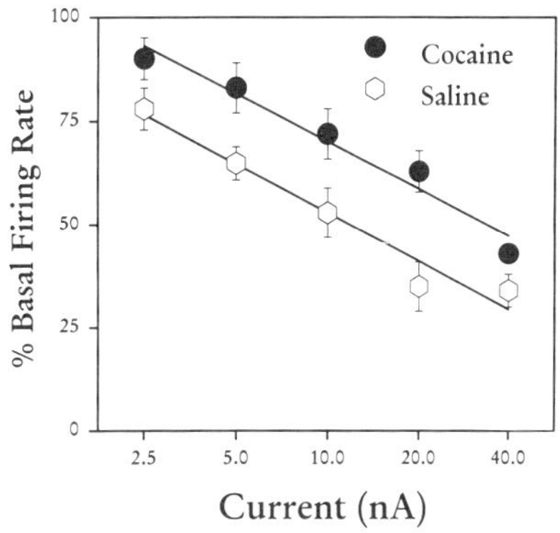

Figure 10. Cocaine sensitization: 5-HT sensitivity in the amygdala. Rats were treated with saline or cocaine (15 mg/kg, i.p., BID for 7 days) and prepared for single unit recording 24 hr after the last injection. The sensitivity to iontophoretic 5-HT was established in glutamate-activated cells in the BLA. Open and closed symbols represent the mean percentage of basal firing rate (± SEM) observed at each 5-HT current in saline- (n = 11 cells) or cocaine-treated rats (n = 9 cells). There was a significant effect of treatment on the sensitivity of amygdala neurons to 5-HT [$F(1,18) = 7.22$, $p = 0.015$].

to block cocaine sensitization.[158,159] Because of the potential importance of the amygdala in cocaine sensitization, this limbic area was chosen for assessment of neuronal sensitivity to 5-HT following chronic cocaine administration. In this study, behaviorally sensitized rats were anesthetized and prepared for single unit recordings of the amygdala using multi-barrel pipettes containing 5-HT. Microiontophoretic 5-HT again elicited current-related inhibitory effects on amygdala neural firing; after 24 hr withdrawal from chronic cocaine treatment, a shift to the right in the 5-HT current-response relationship was observed for these neurons (Figure 10). The 5-HT receptors which mediate the inhibitory response to 5-HT in the amygdala are not yet thoroughly characterized, however, this cocaine-induced desensitization to 5-HT is interesting in light of decreased 5-HT_{1A} receptors observed in amygdala under the same conditions.[69] The diminished functional response to 5-HT may be a consequence of reductions in the number or affinity of the 5-HT_{1A} receptor. Of course, there are a number of other 5-HT receptors which probably contribute to the inhibitory actions of 5-HT in this region, and alterations in 5-HT_{1B}, 5-HT_{2A}, 5-HT_{2C}, and 5-HT_3 receptors which are found in the amygdala could account for the observed changes in response to 5-HT following chronic cocaine exposure.

The electrophysiological effects of chronic cocaine on amygdala neuronal sensitivity to 5-HT can be contrasted with cocaine-induced changes observed in other limbic nuclei. During withdrawal, a desensitization to 5-HT also was observed in the medial prefrontal cortex,[160] another limbic forebrain region implicated in the behavioral effects of cocaine.[161,162] On the other hand, *supersensitivity* to 5-HT was observed in the NAc.[163] These three limbic regions are all innervated by midbrain 5-HT neurons, although some distinctions in the relative density of innervation as well as the origin (dorsal vs. median raphe) and type of innervating fibers (fine versus beaded) exist. Clearly, not all limbic regions respond identically when exposed to chronic cocaine. Although there are probably a number of viable explanations, the relative concentrations of 5-HT (and DA) transporters and any potential differential contributions of terminal autoreceptor control of release/synthesis in these regions may contribute to the differential responses to chronic cocaine exposure.

IV. CONCLUSIONS

Currently available data demonstrate that both acute and chronic exposure to cocaine significantly alter the electrophysiological function of 5-HT systems. Under normal conditions, raphe 5-HT neurons fire in a pacemaker-like fashion; 5-HT release in DR and limbic terminals is a function of this neural activity balanced by the somatodendritic and terminal autoregulatory mechanisms. 5-Serotonin transporters and 5-HT_{1A} receptors found on 5-HT neurons and a host of other neurotransmitter receptors in the raphe may potentially influence raphe 5-HT neuronal function. Acute intravenous cocaine injection increases extracellular 5-HT in cell body areas and completely suppresses the activity of 5-HT neurons due to recruitment of short-loop autoregulatory (5-HT_{1A}) feedback mechanisms and mesohabenular, long-loop feedback circuits.

Serotonin transporters and numerous 5-HT receptor subtypes are localized in the amygdaloid nuclei and these neurons respond to intravenous cocaine with either inhibitory, excitatory, or mixed excitatory responses; potentiation of the normally inhibitory actions of 5-HT in these cells contributes to the systemic effects of cocaine.

After chronic cocaine exposure, reductions of the firing rate and number of 5-HT cells, enhancement of autoregulatory mechanisms in DR 5-HT neurons and the development of subsensitive 5-HT receptors in the amygdala suggest that altered 5-HT function contributes to the development of behavioral sensitization and might play a role in the aberrant behavioral and psychological sequelae observed in cocaine abusers during early abstinence.

Single unit electrophysiological studies of neural activity *in vivo* have identified significant alterations in 5-HT function following cocaine exposure. Although the present research represents a focus on raphe-amygdala circuits, there is also evidence to support modifications of 5-HT function in other mesolimbic neurons (e.g., the NAc and medial prefrontal cortex). As our knowledge about the neural consequences of cocaine administration increases, a clearer perspective will be gained with regard to the integration of these 5-HT changes with those that are observed in the DA system and how these alterations combine to produce both the acute and chronic behavioral effects associated with cocaine.

ACKNOWLEDGMENTS

The research and preparation of this manuscript was supported by USPHS Awards DA 05708 and DA 06511, the John Sealy Memorial Endowment Fund and the National Alliance for Research on Schizophrenia and Affective Disorders. The author would like to thank Mary Thomas for her careful critique of this manuscript, and Patrick M. Callahan, Nicholas E. Goeders, Eva J. Mah, Joseph M. Paris, and Rolf T. Windh for their expert contributions to this research.

REFERENCES

1. Swanson, L. W., The projections of the ventral tegmental area and adjacent regions: A combined fluorescent retrograde tracer and immunofluorescence study in the rat, *Brain Res. Bull.*, 9, 321, 1982.
2. Fallon, J. H. and Moore, R. Y., Catecholamine innervation of basal forebrain. IV. Topography of the dopamine projection to the basal forebrain and striatum, *J. Comp. Neurol.*, 180, 545, 1978.
3. Ritz, M. C., Lamb, R. J., Goldberg, S. R., and Kuhar, M. J., Cocaine receptors on dopamine transporters are related to self-administration of cocaine, *Science,* 237, 1219, 1987.
4. Koe, B. K., Molecular geometry of inhibitors of the uptake of catecholamines and serotonin in synaptosomal preparations of rat brain, *J. Pharmacol. Exp. Ther.*, 199, 649, 1976.
5. Scheel-Kruger, J., Braestrup, C., Nielson, M., Golembiowska, K., and Mogilnicka, E., Cocaine: discussion on the role of dopamine in the biochemical mechanism of action, in *Cocaine and Other Stimulants,* Ellinwood, E. H. and Kilbey, M. M., Eds., Plenum, New York, 1976, 373-407.
6. Kilbey, M. M. and Ellinwood, E. H., Chronic administration of stimulant drugs: response modification, in *Cocaine and Other Stimulants,* Ellinwood, E.H. and Kilbey, M.M., Eds., Plenum, New York, 1976, 409-429.
7. Delfs, J. M., Schreiber, L., and Kelley, A. E., Microinjection of cocaine into the nucleus accumbens elicits locomotor activation in the rat, *J. Neurosci.*, 10, 303, 1990.
8. Kelly, P. H. and Iversen, S. D., Selective 6-OHDA-induced destruction of mesolimbic dopamine neurons: abolition of psychostimulant-induced locomotor activity in rats, *Eur. J. Pharmacol.*, 40, 45, 1976.

9. de Wit, H. and Wise, R. A., Blockade of cocaine reinforcement in rats with the dopamine receptor blocker pimozide, but not with the noradrenergic blockers phentolamine or phenoxybenzamine, *Can. J. Psychol.,* 31, 195, 1977.
10. Roberts, D. C. S., Koob, G. F., Klonoff, P., and Fibiger, H. C., Extinction and recovery of cocaine self-administration following 6-hydroxydopamine lesions of the nucleus accumbens, *Pharmacol. Biochem. Behav.,* 12, 781, 1980.
11. Pettit, H. O., Ettenberg, A., Bloom, F. E., and Koob, G. F., Destruction of dopamine in the nucleus accumbens selectively attenuates cocaine but not heroin self-administration in rats, *Psychopharmacology,* 84, 167, 1984.
12. Colpaert, F. C., Niemegeers, C. J. E., and Janssen, P. A. J., Discriminative stimulus properties of cocaine: Neuropharmacological characteristics as derived from stimulus generalization experiments, *Pharmacol. Biochem. Behav.,* 10, 535, 1979.
13. Callahan, P. M., Appel, J. B., and Cunningham, K. A., Dopamine D_1 and D_2 mediation of the discriminative stimulus properties of *d*-amphetamine and cocaine, *Psychopharmacology,* 103, 50, 1991.
14. McKenna, M. L. and Ho, B. T., The role of dopamine in the discriminative stimulus properties of cocaine, *Neuropharmacology,* 19, 297, 1980.
15. Witkin, J. M., Nichols, D. E., Terry, P., and Katz, J. L., Behavioral effects of selective dopaminergic compounds in rats discriminating cocaine injections, *J. Pharmacol. Exp. Ther.,* 257, 706, 1991.
16. Wood, D. M. and Emmett-Oglesby, M. W., Mediation in the nucleus accumbens of the discriminative stimulus produced by cocaine, *Pharmacol. Biochem. Behav.,* 33, 453, 1989.
17. Callahan, P. M., De La Garza, R., II, and Cunningham, K. A., Discriminative stimulus properties of cocaine: modification by dopamine D_1 receptors in nucleus accumbens, *Psychopharmacology,* 115, 110, 1994.
18. Dworkin, S. I. and Smith, J. E., Neurobehavioral pharmacology of cocaine, in *Mechanisms of Cocaine Abuse and Toxicity,* National Institute on Drug Abuse Monograph 88, Clouet, D., Asghar, K., and Brown, R., Eds., U.S. Government Printing Office, Washington, D.C., 1988, 185-198.
19. Dworkin, S. I. and Bimle, C., 6-Hydroxydopamine lesions of the nucleus accumbens attenuate the discriminative stimulus effects of *d*-amphetamine, *Drug Dev. Res.,* 16, 435, 1989.
20. Post, R. M. and Contel, N. R., Human and animal studies of cocaine: implications for development of behavioral pathology, in *Stimulants: Neurochemical, Behavioral and Clinical Perspectives,* Creese, I., Ed., Raven Press, New York, 1983, 169-203.
21. Segal, D. S. and Kuczenski, R., Individual differences in responsiveness to single and repeated amphetamine administration: behavioral characteristics and neurochemical correlates, *J. Pharmacol. Exp. Ther.,* 242, 917, 1987.
22. King, G. R., Kuhn, C., and Ellinwood, E. H., Jr., Dopamine efflux during withdrawal from continuous or intermittent cocaine, *Psychopharmacology,* 111, 179, 1993.
23. Kalivas, P. W. and Duffy, P., Effects of daily cocaine and morphine treatment on somatodendritic and terminal field dopamine release, *J. Neurochem.,* 50, 1498, 1988.
24. Pettit, H. O., Pan, H.-T., Parsons, L. H., and Justice, J. B., Jr., Extracellular concentrations of cocaine and dopamine are enhanced during chronic cocaine administration, *J. Neurochem.,* 55, 798, 1990.
25. Kalivas, P. W. and Duffy, P., Effect of acute and daily cocaine treatment on extracellular dopamine in the nucleus accumbens, *Synapse,* 5, 48, 1990.
26. Ackerman, J. M. and White, F. J., A10 somatodendritic dopamine autoreceptor sensitivity following withdrawal from repeated cocaine treatment, *Neurosci. Lett.,* 117, 181, 1990.
27. Henry, D. J. and White, F. J., Repeated cocaine administration causes persistent enhancement of D_1 dopamine receptor sensitivity within the rat nucleus accumbens, *J. Pharmacol. Exp. Ther.,* 258, 882, 1991.
28. Henry, D. J., Greene, M. A., and White, F. J., Electrophysiological effects of cocaine in the mesoaccumbens dopamine system: repeated administration, *J. Pharmacol. Exp. Ther.,* 251, 833, 1989.
29. Kalivas, P. W. and Weber, B., Amphetamine injection into the ventral mesencephalon sensitizes rats to peripheral amphetamine and cocaine, *Pharmacol. Biochem. Behav.,* 245, 1095, 1988.
30. Lamb, R. J. and Griffiths, R. R., Self-administration in baboons and the discriminative stimulus effects in rats of bupropion, nomifensine, diclofensine and imipramine, *Psychopharmacology,* 102, 183, 1990.

31. Bergman, J., Madras, B. K., Johnson, S. E., and Spealman, R. D., Effects of cocaine and related drugs in nonhuman primates. III. Self-administration by squirrel monkeys, *J. Pharmacol. Exp. Ther.*, 251, 150, 1989.
32. Ritchie, J. M. and Greengard, P., On the mode of action of local anesthetics, *Annu. Rev. Pharmacol.*, 6, 405, 1966.
33. Kilpatrick, G. J., Jones, B. J., and Tyers, M. B., Binding of the 5-HT_3 ligand GR65630, to rat area postrema vagus nerve and the brain of several species, *Eur. J. Pharmacol.*, 159, 157, 1989.
34. Sharkey, J., Ritz, M. C., Schenden, J. A., Hanson, R. C., and Kuhar, M. J., Cocaine inhibits muscarinic cholinergic receptors in heart and brain, *J. Pharmacol. Exp. Ther.*, 246, 1048, 1988.
35. Sharkey, J., Glen, K. A., Wolfe, S., and Kuhar, M. J., Cocaine binding at sigma receptors, *Eur. J. Pharmacol.*, 149, 171, 1988.
36. Eidelberg, E. and Woodbury, C. M., Electrical activity in the amygdala and its modification by drugs: possible nature of synaptic transmitters, a review, in *The Neurobiology of the Amygdala Complex,* Eleftheriou, B. E., Ed., Plenum Press, New York, 1972, 609-622.
37. Lesse, H. and Harper, R. K., Frequency related, bidirectional limbic reponses to cocaine: comparisons with amphetamine and lidocaine, *Brain Res.*, 335, 21, 1985.
38. Post, R. M., Cocaine psychoses: a continuum model, *Am. J. Psychiatry,* 132, 225, 1975.
39. Stripling, J. S. and Hendricks, C., Facilitation of kindling by convulsions induced by cocaine or lidocaine but not pentylenetetrazol, *Pharmacol. Biochem. Behav.*, 15, 793, 1981.
40. Rackham, A. and Wise, R. A., Independence of cocaine sensitization and amygdaloid kindling in the rat, *Physiol. Behav.*, 22, 631, 1979.
41. Kirkby, R. D. and Kokkinidis, L., Evidence for a relationship between amphetamine sensitization and electrical kindling of the amygdala, *Exp. Neurol.*, 97, 270, 1987.
42. Gelowitz, D. L. and Kokkinidis, L., The effects of amygdaloid stimulation on amphetamine-elicited locomotor sensitization, *Brain Res. Bull.*, 32, 561, 1993.
43. Fallon, J. H., Histochemical characterization of dopaminergic, noradrenergic and serotonergic projections to the amygdala, *INSERM Symp.*, 20, 175, 1981.
44. Gloor, P., Inputs and outputs of the amygdala: what the amygdala is trying to tell the rest of the brain, in *Limbic Mechanisms: The Continuing Evolution of the Limbic System Concept,* Livingston, K. E. and Hornykiewicz, O., Eds., Plenum Press, New York, 1978, 189-209.
45. Kaada, B. R., Stimulation and regional ablation of the amygdaloid complex with reference to functional representations, in *The Neurobiology of the Amygdala Complex,* Eleftheriou, B. E., Ed., Plenum Press, New York, 1972, 205-281.
46. Deminiere, J.-M., Taghzouti, K., Tassin, J. P., Moal, M. L., and Simon, H., Increased sensitivity to amphetamine and facilitation of amphetamine self administration after 6-hydroxydopamine lesions of the amygdala, *Psychopharmacology,* 94, 232, 1988.
47. Alheid, G. F. and Heimer, L., New perspectives in basal forebrain organization of special relevance for neuropsychiatric disorders: the striatopallidal, amygdaloid, and corticopetal components of substantia innominata, *Neuroscience,* 27, 1, 1988.
48. Robledo, P. and Koob, G. F., Two discrete nucleus accumbens projection areas differentially mediate cocaine self-administration in the rat, *Behav. Brain Res.*, 55, 159, 1993.
49. Callahan, P. M., Bryan, S. K., and Cunningham, K. A., Microinjection of the D_1 antagonist SCH 23390 into the amygdala blocks the discriminative stimulus properties of cocaine, *Pharmacol. Biochem. Behav.*, in press, 1994.
50. Caine, S. B., Heinrichs, S. C., Williams, C. E., Coffin, V. L., and Koob, G. F., D-1 receptor antagonist injected into the accumbens shell or central amygdala increases cocaine self-administration, *Soc. Neurosci. Abst.*, 1856, 1993, 1993.
51. McGregor, A. and Roberts, D. C. S., Dopaminergic antagonism within the nucleus accumbens or the amygdala produces differential effects on intravenous cocaine self-administration under fixed and progressive ratio schedules of reinforcement, *Brain Res.*, 624, 245, 1993.
52. Loh, E. A. and Roberts, D. C. S., Break points on a progressive ratio schedule reinforced by intravenous cocaine increase following depletion of forebrain serotonin, *Psychopharmacology,* 101, 262, 1990.
53. Vertes, R. P., A PHA-L analysis of ascending projections of the dorsal raphe nucleus in the rat, *J. Comp. Neurol.*, 313, 643, 1991.
54. Moore, R. Y., Halaris, A. E., and Jones, B. E., Serotonin neurons of the midbrain raphe: ascending projections, *J. Comp. Neurol.*, 180, 417, 1978.

55. Simon, H., Moal, M., and Calas, A., Efferents and afferents of the ventral tegmental-A10 region studied after local injection of 3H-leucine and horseradish peroxidase, *Brain Res.*, 178, 17, 1979.
56. Herve, D., Pickel, V. M., Joh, T. H., and Beaudet, A., Serotonin axon terminals in the ventral tegmental area of rat: fine structure and synaptic input to dopaminergic neurons, *Brain Res.*, 435, 71, 1987.
57. Kapadia, S. E., de Lanerolle, N. C., and LaMotte, C. C., Immunocytochemical and electron microscopic study of serotonin neuronal organization in the dorsal raphe nucleus of the monkey, *Neuroscience*, 15, 729, 1985.
58. Kalen, P., Skagerberg, G., and Lindvall, O., Projections from the ventral tegmental area and mesencephalic raphe to the dorsal raphe nucleus in the rat, *Exp. Brain Res.*, 73, 69, 1988.
59. Arnsten, A. F. T. and Goldman-Rakic, P. S., Selective prefrontal cortex projections to the region of the locus coeruleus and raphe nuclei in the rhesus monkey, *Brain Res.*, 306, 9, 1984.
60. Van Bockstaele, E. J., Biswas, A., and Pickel, V. M., Topography of serotonin neurons in the dorsal raphe nucleus that send axon collaterals to the rat prefrontal cortex and nucleus accumbens, *Brain Res.*, 624, 188, 1993.
61. Branchek, T., Receptor diversity: more serotonin receptors, *Curr. Biol.*, 3, 315, 1993.
62. Martin, G. R. and Humphrey, P. P. A., Receptors for 5-hydroxytryptamine: current perspectives on classification and nomenclature, *Neuropharmacology*, 33, 261, 1994.
63. Chalmers, D. T. and Watson, S. J., Comparative anatomical distribution of 5-HT_{1A} receptor mRNA and 5-HT_{1A} binding in rat brain—a combined in situ hybridisation/in vitro receptor autoradiographic study, *Brain Res.*, 561, 51, 1991.
64. Pazos, A. and Palacios, J. M., Quantitative autoradiographic mapping of serotonin receptors in the rat brain. I. Serotonin-1 receptors, *Brain Res.*, 346, 205, 1985.
65. Sprouse, J. S. and Aghajanian, G. K., (–)-Propranolol blocks the inhibition of serotonergic dorsal raphe cell firing by 5-HT_{1A} selective agonists, *Eur. J. Pharmacol.*, 128, 295, 1986.
66. Williams, J. T., Colmers, W. F., and Pan, Z. Z., Voltage-and ligand-activated inwardly rectifying currents in dorsal raphe neurons in vitro, *J. Neurosci.*, 8, 3499, 1988.
67. Sinton, C. M. and Fallon, S. L., Electrophysiological evidence for a functional differentiation between subtypes of the 5-HT_1 receptor, *Eur. J. Pharmacol.*, 157, 173, 1988.
68. Sprouse, J. S. and Aghajanian, G. K., Electrophysiological responses of serotoninergic dorsal raphe neurons to 5-HT_{1A} and 5-HT_{1B} agonists, *Synapse*, 1, 3, 1987.
69. Cunningham, K. A., Paris, J. M., and Goeders, N. E., Chronic cocaine enhances serotonin autoregulation and serotonin uptake binding, *Synapse*, 11, 112, 1992.
70. Wilkinson, L. O. and Middlemiss, D. N., The autoregulation of 5-HT release from nerve terminals in vitro, *Adv. Biosci.*, 85, 49, 1992.
71. Johnson, S. W., Mercuri, N. B., and North, R. A., 5-$Hydroxytryptamine_{1B}$ receptors block the $GABA_B$ synaptic potential in rat dopamine neurons, *J. Neurosci.*, 12, 2000, 1992.
72. Bobker, D. H. and Williams, J. T., Serotonin agonists inhibit synaptic potentials in the rat locus ceruleus in vitro via 5-$hydroxytryptamine_{1A}$ and 5-$hydroxytryptamine_{1B}$ receptors, *J. Pharmacol. Exp. Ther.*, 250, 37, 1989.
73. Pazos, A., Cortes, R., and Palacios, J. M., Quantitative autoradiographic mapping of serotonin receptors in rat brain. II. Serotonin-2 receptors, *Brain Res.*, 346, 231, 1985.
74. Sijbesma, H., Schipper, J., Cornelissen, J. C. H. M., and de Kloet, E. R., Species differences in the distribution of central 5-HT_1 binding sites: a comparative autoradiographic study between rat and guinea pig, *Brain Res.*, 555, 295, 1991.
75. Rocha, B., Di Scala, G., Rigo, M., Hoyer, D., and Sandner, G., Effect of 5,7-dihydroxytryptamine lesion on mianserin-induced conditioned place aversion and on 5-$hydroxytryptamine_{1c}$ receptors in the rat brain, *Neuroscience*, 56, 687, 1993.
76. Kilpatrick, G. J., Jones, B. J., and Tyers, M. B., Identification and distribution of 5-HT_3 receptors in rat brain using radioligand binding, *Nature*, 330, 746, 1987.
77. Burlhis, T. M. and Aghajanian, G. K., Pacemaker potentials of serotonergic dorsal raphe neurons: contribution of a low-threshold Ca^{2+} conductance, *Synapse*, 1, 582, 1987.
78. Aghajanian, G. K., Modulation of a transient outward current in serotonergic neurones by $alpha_1$-adrenoceptors, *Nature*, 315, 501, 1985.
79. Innis, R. B. and Aghajanian, G. K., Pertussis toxin blocks 5-HT_{1A} and $GABA_B$ receptor-mediated inhibition of serotonergic neurons, *Eur. J. Pharmacol.*, 143, 195, 1987.

80. Freedman, J. E. and Aghajanian, G. K., Role of phosphoinositide metabolites in the prolongation of afterhyperpolarizations by alpha$_1$-adrenoceptors in rat dorsal raphe neurons, *J. Neurosci.*, 7, 3897, 1987.
81. Pan, Z. Z. and Williams, J. T., GABA and glutamate mediated synaptic potentials in rat dorsal raphe neurons in vitro, *J. Neurophysiology,* 61, 719, 1989.
82. Yoshimoto, K. and McBride, W. J., Regulation of nucleus accumbens dopamine release by the dorsal raphe nucleus in the rat, *Neurochem. Res.*, 17, 401, 1992.
83. Kalen, P., Strecker, R. E., Rosengren, E., and Bjorklund, A., Regulation of striatal serotonin release by the lateral habenula dorsal raphe pathway in the rat as demonstrated by in vivo microdialysis: role of excitatory amino acids and GABA, *Brain Res.*, 492, 187, 1989.
84. Nishikawa, T. and Scatton, B., Evidence for a GABAergic inhibitory influence on serotonergic neurons originating from the dorsal raphe, *Brain Res.*, 279, 325, 1983.
85. Bowery, N. G., Hudson, A. L., and Price, G. W., $GABA_A$ and $GABA_B$ receptor site distribution in the rat central nervous system, *Neuroscience,* 20, 365, 1987.
86. Dabadie, H., Mons, N., and Geffard, M., Simultaneous detection of tryptamine and dopamine in rat substantia nigra and raphe nuclei using specific antibodies, *Brain Res.*, 512, 138, 1990.
87. Unguez, G. A. and Schneider, J. S., Dopaminergic dorsal raphe neurons in cats and monkeys are sensitive to the toxic effects of MPTP, *Neurosci. Lett.*, 94, 218, 1988.
88. Mansour, A., Meador-Woodruff, J. H., Bunzow, J. R., Civelli, O., Akil, H., and Watson, S. J., Localization of dopamine D_2 receptor mRNA and D_1 and D_2 receptor binding in the rat brain and pituitary: an *in situ* hybridization-receptor autoradiographic analysis, *J. Neurosci.*, 10, 2587, 1990.
89. Key, B. J. and Krzywoskinski, L., Electrocortical changes induced by the perfusion of noradrenaline, acetycholine and their antagonists directly into the dorsal raphe nucleus of the cat, *Br. J. Pharmacol.*, 61, 297, 1977.
90. Pradhan, S., Hanson, G., and Lovenberg, W., Inverse relation of substance P-like immunoreactivity in DRN to serotonin levels in pons-medulla following administration of cocaine and 5-hydroxytryptophan, *Biochem. Pharmacol.*, 30, 1071, 1981.
91. Helke, C. J., Krause, J. E., Mantyh, P. W., Couture, R. L., and Bannon, M. J., Diversity in mammalian tachykinin peptidergic neurons: multiple peptides, receptors, and regulatory mechanisms, *FASEB J.*, 4, 1606, 1990.
92. Wang, Q. -P., Nakai, Y., and Shioda, S., Enkephalinergic synaptic axon terminals on serotoninergic neurons in the dorsal raphe nucleus of the rat: An electron microscopic study by the double immunostaining method, *Neurosci. Lett.*, 124, 129, 1991.
93. Hammer, R. P., Jr., Cocaine alters opiate receptor binding in critical brain reward regions, *Synapse*, 29, 55, 1989.
94. Costall, B., Jones, B. J., Kelly, M. E., Naylor, R. J., Onaivi, E. S., and Tyers, M. B., Sites of action of ondansetron to inhibit withdrawal from drugs of abuse, *Pharmacol. Biochem. Behav.*, 36, 97, 1990.
95. Cunningham, K. A. and Lakoski, J. M., Electrophysiological effects of cocaine and procaine on dorsal raphe serotonin neurons, *Eur. J. Pharmacol.*, 148, 457, 1988.
96. Cunningham, K. A. and Lakoski, J. M., The interaction of cocaine with serotonin dorsal raphe neurons: single unit extracellular recording studies, *Neuropsychopharmacology,* 3, 41, 1990.
97. Pitts, D. K. and Marwah, J., Cocaine modulation of central monoaminergic neurotransmission, *Pharmacol. Biochem. Behav.*, 26, 453, 1987.
98. Einhorn, L. C., Johansen, P. A., and White, F. J., Electrophysiological effects of cocaine in the mesoaccumbens dopamine system: studies in the ventral tegmental area, *J. Neurosci.*, 8, 100, 1988.
99. Pitts, D. K. and Marwah, J., Electrophysiological actions of cocaine on noradrenergic neurons in rat locus coeruleus, *J. Pharmacol. Exp. Ther.*, 240, 345, 1987.
100. Lum, J. T. and Piercey, M. F., Electrophysiological evidence that spiperone is an antagonist of 5-HT_{1A} receptors in the dorsal raphe nucleus, *Eur. J. Pharmacol.*, 149, 9, 1988.
101. Pan, Z. Z. and Williams, J. T., Differential actions of cocaine and amphetamine on dorsal raphe neurons in vitro, *J. Pharmacol. Exp. Ther.*, 251, 56, 1989.
102. Windh, R. T. and Cunningham, K. A., Cocaine inhibits dorsal raphe 5-HT activity through 5-HT_{1A}-dependent and -independent mechanisms, *Soc. Neurosci. Abstr.*, 23, 1850, 1993.
103. Herkenham, M. and Nauta, W. J. H., Efferent connections of the habenular nuclei in the rat, *J. Comp. Neurol.*, 187, 19, 1979.
104. Cragg, B. G., The connections of the habenula in the rabbit, *Exp. Neurol.*, 3, 388, 1961.

105. McCulloch, J., Savaki, H. E., and Sokoloff, L., Influence of dopaminergic systems on the lateral habenular nucleus of the rat, *Brain Res.*, 194, 117, 1980.
106. Ramm, P., Beninger, R. J., and Frost, B. J., Functional activity in the lateral habenular and dorsal raphe nuclei following administration of several dopamine receptor antagonist, *Can. J. Physiol. Pharmacol.*, 62, 1530, 1984.
107. Stern, W. C., Johnson, A., Bronzino, J. D., and Morgane, P. J., Neuropharmacology of the afferent projections from the lateral habenula and substantia nigra to the anterior raphe in the rat, *Neuropharmacology*, 20, 979, 1981.
108. Christoph, G. R., Leonizo, R. J., and Wilcox, K. S., Stimulation of the lateral habenula inhibits dopamine containing neurons in the substantia nigra and ventral tegmental area of the rat, *J. Neurosci.*, 6, 613, 1986.
109. Dougherty, P. M., Qiao, J.-T., Wiggins, R. C., and Dafny, N., Microiontophoresis of cocaine, desipramine, sulpride, methysergide, and naloxone in habenula and parafasciculus, *Eur. Neurol.*, 108, 241, 1990.
110. Paris, J. M. and Cunningham, K. A., Habenula lesions decrease the responsiveness of dorsal raphe serotonin neurons to cocaine, *Pharmacol. Biochem. Behav.*, 49, 555, 1994.
111. McDonald, A.J., Cell types and intrinsic connections of the amygdala, in *The Amygdala: Neurobiological Aspects of Emotion, Memory, and Mental Dysfunction*, Aggleton, J. P., Ed., Wiley-Liss, New York, 1992, 67-96.
112. McDonald, A. J., Beitz, A. J., Larson, A. A., Kuriyama, R., Sellitto, C., and Madl, J. E., Co-localization of glutamate and tublin in putative excitatory neurons of the hippocampus and amygdala: an immunohistochemical study using monoclonal antibodies, *Neuroscience*, 30, 405, 1989.
113. McDonald, A. J., Coexistence of somatostain with neuropeptide Y, but not with cholecystokinin or vasoactive intestinal plrptide, in neurons of the rat amygdala, *Brain Res.*, 500, 37, 1989.
114. McDonald, A. J. and Pearson, J. C., Coexistence of GABA and peptide immunoreactivity in non-pyramidal neurons of the basolateral amygdala, *Neurosci. Lett.*, 100, 53, 1989.
115. Rainnie, D. G., Asprodini, E. K., and Shinnick-Gallagher, P., Intracellular recordings from morphologically identified neurons of the basolateral amygdala, *J. Neurophysiol.*, 69, 1350, 1993.
116. Straughan, D. W. and Legge, K. F., The pharmacology of amygdaloid neurons, *J. Pharm. Pharmacol.*, 17, 675, 1965.
117. Wang, R. Y. and Aghajanian, G. K., Enhanced sensitivity of amygdaloid neurons to serotonin and norepinephrine after chonic antidepressant treatment, *Commun. Psychopharmacol.*, 4, 83, 1980.
118. Wang, R. Y. and Aghajanian, G. K., Inhibition of neurons in the amygdala by dorsal raphe stimulation: mediation through a direct serotonergic pathway, *Brain Res.*, 120, 85, 1977.
119. Wang, R. Y., de Montigny, C., Gold, B. I., Roth, R. H., and Aghajanian, G. K., Denervation supersensitivity to serotonin in rat forebrain: single cell studies, *Brain Res.*, 178, 479, 1979.
120. Bashore, T. R., Rebec, G. V., and Groves, P. M., Alterations of spontaneous activity in the caudate-putmen, nucleus accumbens and amygdaloid complex of rats produced by d amphetamine, *Pharmacol. Biochem. Behav.*, 8, 467, 1978.
121. Rebec, G. V. and Lee, E. H., Tolerance to amphetamine induced inhibition of neuronal activity in the central amygdaloid nucleus, *Pharmacol. Biochem. Behav.*, 19, 219, 1983.
122. Rebec, G. V., Alloway, K. D., and Bashore, T. R., Differential actions of classical and atypical antipsychotic drugs on spontaneous neuronal activity in the amygdaloid complex, *Pharmacol. Biochem. Behav.*, 14, 49, 1981.
123. Bordi, F., LeDoux, J., Clugnet, M. C., and Pavlides, C., Single-unit activity in the lateral nucleus of the amygdala and overlying areas of the striatum in freely behaving rats: rates, discharge patterns, and responses to acoustic stimuli, *Behav. Neurosci.*, 107, 757, 1993.
124. Le Gal La Salle, G., and Ben-Ari, Y., Unit activity in the amygdaloid complex: a review, in *The Amygdaloid Complex*, Ben-Ari, Y., Eds., Elsevier/North-Holland Biomedical Press, Amsterdam, 1981, 221-237.
125. Saphier, D., Mor, G., and Feldman, S., Neurogenic stimuli alter preoptic area and amygdala unit activity: central effects of olfactory projections on paraventricular nucleus units, *Exp. Neurol.*, 100, 71, 1988.
126. Maeda, H., Morimoto, H., and Yanagimoto, K., Response characteristics of amygdaloid neurons provoked by emotionally significant environmental stimuli in cats, with special reference to response durations, *Can. J. Physiol. Pharmacol.*, 71, 374, 1993.

127. Eidelberg, E., Goldstein, G. P., and Deza, L., Evidence for serotonin as a possible inhibitory transmitter in some limbic structures, *Exp. Brain Res.*, 4, 73, 1967.
128. Ben-Ari, Y. and Kelly, J. S., Dopamine evoked inhibition of single cells of the feline putamine and basolateral amygdala, *J. Physiol. (Lond.)*, 256, 1, 1976.
129. Callahan, P. M. and Cunningham, K. A., Electrophysiological effects of cocaine in amygdala nuclei: relationship to dopamine and serotonin actions, *Soc. Neurosci. Abstr.*, 21, 679, 1991.
130. Cunningham, K. A., Callahan, P. M., and Lekan, H. L., Cocaine: electrophysiological characterization of effects in amygdala nuclei, *Soc. Neurosci. Abstr.*, 15, 1098, 1989.
131. Wepsic, J. G. and Austin, G. M., The neurophysiological effects of amphetamine upon the cat amygdala, in *The Neurobiology of the Amygdala Complex*, Eleftheriou, B. E., Ed., Plenum Press, New York, 1972, 623-640.
132. White, F. J., Wachtel, S. R., Johansen, P. A., and Einhorn, L. C., Electrophysiological studies in the rat mesoaccumbens dopamine system: focus on dopamine receptor subtypes, interactions, and the effects of cocaine, in *Neurophysiology of Dopaminergic Systems: Current Status and Clinical Perspectives*, Chiodo, L. A. and Freeman, A. S., Eds., Lakeshore Publishing, Detroit, MI, 1987, 317-365.
133. White, F. J., Hu, X.-T., and Henry, D. J., Electrophysiological effects of cocaine in the rat nucleus accumbens: microiontophoretic studies, *J. Pharmacol. Exp. Ther.*, 266, 1075, 1993.
134. Uchimura, N. and North, R. A., Actions of cocaine on rat nucleus accumbens neurones in vitro, *Br. J. Pharmacol*, 99, 736, 1990.
135. Scibilia, R. J., Lachowicz, J. E., and Kilts, C. D., Topographic nonoverlapping distribution of D_1 and D_2 dopamine receptors in the amygdaloid nuclear complex of the rat brain, *Synapse*, 11, 146, 1992.
136. Mah, E. J. and Cunningham, K. A., Electrophysiological studies of the serotonergic (5-HT) modulation of glutamate-evoked activity in amygdala neurons, *Soc. Neurosci. Abst.*, 23, 1554, 1993.
137. Pitts, D. K. and Marwah, J., Reciprocal pre- and postsynaptic actions of cocaine at a central noradrenergic synapse, *Exp. Neurol.*, 98, 518, 1987.
138. Robinson, T. E. and Becker, J. B., Enduring changes in brain and behavior produced by chronic amphetamine administration: a review and evaluation of animal models of amphetamine psychosis, *Brain Res. Rev.*, 11, 157, 1986.
139. Kalivas, P. W., Sorg, B. A., and Hooks, M. S., The pharmacology and neural circuitry of sensitization to psychostimulants, *Behav. Pharmacol.*, 4, 315, 1993.
140. Paris, J. M., Callahan, P. M., Lee, J. M., and Cunningham, K. A., Behavioral sensitization to cocaine is not associated with changes in serotonin (5-HT) fiber immunoreactivity in rat forebrain, *Brain Res. Bull.*, 27, 843, 1991.
141. Fuxe, K., Calza, L., Benfenati, F., Zini, I., and Agnati, L. F., Quantitative autoradiographic localization of [^{3}]imipramine binding sites in the brain of the rat: relationship to ascending 5-hydroxytryptamine neuron systems, *Proc. Natl. Acad. Sci. USA*, 80, 3836, 1983.
142. Gobbi, M., Cervo, L., Taddei, C., and Mennini, T., Autoradiographic localization of [^{3}H]paroxetine specific binding in the rat brain, *Neurochem. Int.*, 16, 247, 1990.
143. Pitts, D. K. and Marwah, J., Electrophysiological effects of cocaine on central monoaminergic neurons, *Eur. J. Pharmacol.*, 131, 95, 1986.
144. Aghajanian, G. K., Wang, R. Y., and Baraban, J., Serotonergic and non-serotonergic neurons of the dorsal raphe: Reciprocal changes in firing induced by peripheral nerve stimulation, *Brain Res.*, 153, 169, 1978.
145. Parsons, L. H. and Justice, J. B., Jr., Serotonin and dopamine sensitization in the nucleus accumbens, ventral tegmental area, and dorsal raphe nucleus following repeated cocaine administration, *J. Neurochem.*, 61, 1611, 1993.
146. Parsons, L. H., Koob, G. F., and Weiss, F., Withdrawal from extended-access cocaine self-administration leads to decreased extracellular serotonin levels in the nucleus accumbens of the rat, in *Problems of Drug Dependence, 1994; Proceedings of the 55th Annual Scientific Meeting*, National Institute on Drug Abuse Monograph, Harris, L. S., Ed., U.S. Government Printing Office, Washington, D.C., in press, 1995.
147. Darmani, N. A., Martin, B. R., and Glennon, R. A., Repeated administration of low doses of cocaine enhances the sensitivity of 5-HT_2 receptor function, *Pharmacol. Biochem. Behav.*, 41, 519, 1992.
148. Paris, J. M., Callahan, P. M., Lee, J. M., and Cunningham, K. A., Cocaine: behavioral sensitization and serotonin immunoreactivity in the rat forebrain, *Soc. Neurosci. Abst.*, 16, 1990.

149. Kleven, M. S., Woolverton, W. L., and Seiden, L. S., Lack of long-term monoamine depletions following repeated or continuous exposure to cocaine, *Brain Res. Bull.,* 21, 233, 1988.
150. Tung, F., Bonadonna, A. M., Rittenhouse, P. A., Bakkum, E., Van De Kar, L. D., and Battaglia, G., Cocaine effects on serotonin and dopamine recognition sites in rat brain: *in vitro* and *in vivo* studies, *Soc. Neurosci. Abst.,* 16, 1990.
151. Nestler, E. J., Terwilliger, R. Z., Walker, J. R., Sevarino, K. A., and Duman, R. S., Chronic cocaine treatment decreases levels of the G protein subunits G_{ia} and G_{oa} in discrete regions of rat brain, *J. Neurochem.,* 55, 1079, 1990.
152. Terwilliger, R. Z., Beitner-Johnson, D. B., Sevarino, K. A., Crain, S. M., and Nestler, E. J., A general role for adaptations in G-proteins and the cyclic AMP system in mediating the chronic actions of morphine and cocaine on neuronal function, *Brain Res.,* 548, 100, 1991.
153. Steketee, J. D., Striplin, C. D., Murray, T. F., and Kalivas, P. W., Possible role for G-proteins in behavioral sensitization to cocaine, *Brain Res.,* 545, 287, 1991.
154. Limberger, N., Starke, K., and Singer, E. A., Serotonin uptake blockers influence serotonin autoreceptors by increasing the biophase concentration of serotonin and not through a "molecular link", *Naunyn-Schmiedeberg's Arch. Pharmacol.,* 342, 363, 1990.
155. Raiteri, M., Bonanno, G., Marchi, M., and Maura, G., Is there a functional linkage between neurotransmitter uptake mechanisms and presynaptic receptors? *J. Pharmacol. Exp. Ther.,* 231, 671, 1984.
156. Harris, G. C. and Williams, J. T., Sensitization of locus ceruleus neurons during withdrawal from chronic stimulants and antidepressants, *J. Pharmacol. Exp. Ther.,* 261, 476, 1992.
157. Pitts, D. K. and Marwah, J., Chronic cocaine reduces adrenoceptor elicited mydriasis and inhibition of locus coeruleus neurons, *Eur. J. Pharmacol.,* 160, 201, 1989.
158. Post, R. M., Weiss, S. R. B., Pert, A., and Uhde, T. W., Chronic cocaine administration: sensitization and kindling effects, in *Cocaine: Clinical and Biobehavioral Aspects,* Fisher, S., Raskin, A., and Uhlenhuth, E. H., Eds., Oxford University Press, New York, 1987, 109-173.
159. Post, R. M., Lockfeld, A., Squillace, K. M., and Contel, N. R., Drug-environment interaction: context dependency of cocaine-induced behavioral sensitization, *Life Sci.,* 28, 755, 1981.
160. Wang, R. Y. and Zhang, J. Y., Repeated cocaine administration induces desensitization of the response of medial prefrontal cortical (mPFC) cells to cocaine, dopamine (DA) and serotonin (5-HT), *Soc. Neurosci. Abstr.,* 19, 1856, 1994.
161. Goeders, N. E. and Smith, J. E., Cortical dopaminergic involvement in cocaine reinforcement, *Science,* 221, 773, 1983.
162. Goeders, N. E. and Smith, J. E., Intracranial cocaine self-administration into the medial prefrontal cortex increases dopamine turnover in the nucleus accumbens, *J. Pharmacol. Exp. Ther.,* 265, 592, 1993.
163. White, F. J., Henry, D.J., Hu, X.-T., Jeziorski, M., and Ackerman, J. M., Electrophysiological effects of cocaine in the mesoaccumbens dopamine system, In *Cocaine: Pharmacology, Physiology and Clinical Strategies*, Lakoski, J. M., Galloway, M. P., and White, F. J., Eds., CRC Press, Boca Raton, FL, 1991.

Chapter 9

Cocaine Interactions With the Endogenous Opioid System

Ellen M. Unterwald

CONTENTS

I. INTRODUCTION

Although cocaine has a variety of pharmacological actions, its reinforcing effects are believed to be mediated by its ability to inhibit the uptake of dopamine.[1] Cocaine binds directly to the dopamine transporter and prevents dopamine reuptake into presynaptic dopaminergic neurons, thereby increasing the concentration of dopamine within the synapse.[2-4] The mesolimbic dopaminergic system is thought to play a primary role in mediating the reinforcing properties of many, if not all, drugs of abuse including both cocaine and opiates.[5,6] The actions of cocaine on dopaminergic projections to the nucleus accumbens appear necessary for cocaine reward.[7] Other brain regions such as the medial prefrontal cortex,[8] ventral pallidum,[9] and olfactory tubercle[10] also have been implicated in cocaine reinforcement.

The relationship between opioid and dopaminergic systems has received much attention due to their collective role in the regulation of motivated and emotional behaviors and in the control of locomotor activity. The interaction of these two systems is supported by anatomical, behavioral, and neurochemical evidence. Moderate to high levels of endogenous opioid peptides derived from proenkephalin and prodynorphin are found in brain regions containing dopaminergic neurons.[11,12] Likewise, these regions contain significant densities of opioid receptors.[13,14] Using neurological lesioning techniques, some of the opioid receptors in the caudatoputamen have been localized to presynaptic dopaminergic terminals.[15-17] In other regions such as the nucleus accumbens, opioid receptors appear to reside on elements immediately postsynaptic to dopamine terminals.[18] Therefore, opioid receptors probably are co-localized with pre- or postsynaptic dopamine receptors and/or dopamine transporter sites, providing a rather direct mechanism for regulation of opioid systems by dopaminergic activity.

Behavioral data suggest a functional interaction between cocaine and the endogenous opioid system. In animal models of drug reinforcement, dopamine receptor antagonists attenuate cocaine-induced reward supporting the role of dopamine in cocaine reinforcement.[19] In a similar manner, opioid receptor antagonists also can block the rewarding effects of cocaine,[20-22] suggesting an interaction between central opioid and

1-8493-8311-0/95/$0.00+$.50

dopaminergic systems in cocaine reinforcement. Cocaine self-administration in rats is attenuated by pretreatment with the opioid antagonist, naltrexone.[23] In monkeys, self-administration of cocaine is reduced by co-administration of the mixed agonist-antagonist opioids, nalbuphine and butorphanol, although the attenuation of cocaine self-administration in this case might be mediated by the agonist properties of these agents.[24] Cocaine-induced hyperactivity also is blocked by co-administration of the opioid antagonist, naloxone.[21] Taken together, these data suggest that the endogenous opioid system could be involved in mediating some of the effects of cocaine.

The schedule by which cocaine is administered appears to be a critical factor in determining the effects of cocaine on both neurochemistry and behavior.[25] For example, rats injected once daily with cocaine for 3 days exhibit sensitization to cocaine's inhibition of [^{3}H]dopamine uptake by cocaine,[26] whereas rats receiving a constant infusion of cocaine for 7 days via osmotic minipumps show tolerance to the inhibition of [^{3}H]dopamine uptake by cocaine.[27] The effect of cocaine on locomotor activity is also dependent on dosing schedule. Animals treated with a continuous infusion of cocaine show tolerance to the locomotor-activating effects of a challenge dose of cocaine, whereas animals treated with the same total daily dose for the same number of days but administered by intermittent injections show sensitization.[28] Similar findings concerning intermittent vs. steady-state effects of cocaine have been reported from neuroendocrine studies.[29,30] Therefore, differences in the schedule of cocaine administration might underlie some of the inconsistencies in the literature regarding the effects of cocaine on a variety of systems.

This chapter will review current research on the regulation of cellular and molecular components of the endogenous opioid system as a consequence of exposure to cocaine, following a brief overview of opioid receptors and opioid peptides.

A. THE ENDOGENOUS OPIOID SYSTEM

Opioids produce their characteristic effects by interacting with specific, high-affinity receptors. At least three types of opioid receptors have been defined, namely the μ-, δ, and κ-receptors, which differ in their distribution and affinity for various opioid ligands. In addition, subtypes of each of these receptors have been postulated based on pharmacological data.[31-34] Morphine has its highest affinity for the μ-receptor, which is thought to mediate the effects of this and other related opioid alkaloids including analgesia, respiratory depression, and euphoria. The μ-, δ-, and κ-receptors are coupled to adenylyl cyclase via pertussis toxin-sensitive G proteins. Activation of opioid receptors inhibits adenylyl cyclase activity and thus reduces the formation of cAMP.[35,36] Opioid receptors also regulate Ca^{2+} and K^+ channel activity.[37] Recently, the three opioid receptor types have been cloned.[38] To date, the complementary DNA encoding one δ-,[39,40] one μ-,[41,42] and one κ-[43,44] receptor have been isolated and characterized, with no evidence for receptor subtypes. The three opioid receptors share a high degree of animo acid sequence similarity, with approximately 50% of the residues being identical.[38] Cloning of the opioid receptors has confirmed that these receptors are members of the family of G protein-linked receptors containing seven putative membrane-spanning regions. Their closest relatives in terms of amino acid sequence homology are the somatostatin receptors.

More than 20 peptides with opioid activity have been extracted and sequenced from brain, all containing the sequence Tyr-Gly-Gly-Phe as their N-terminus. Each of the opioid peptides arises from one of three distinct precursor peptides and each precursor is the unique product of one of three distinct genes: (1) proopiomelanocortin (POMC) which gives rise to β-endorphin, as well as to the nonopioid peptides adrenocorticotropin and melanocyte stimulating hormone,[45-47] (2) proenkephalin which gives rise to met- and leu-enkephalin, as well as to many longer enkephalin peptides,[48-51] and (3) prodynorphin

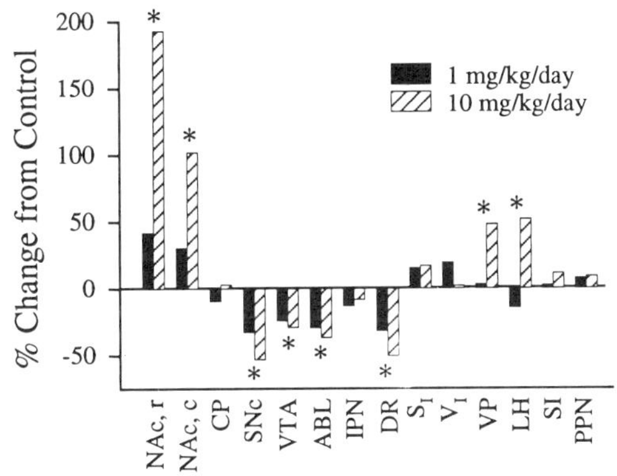

Figure 1. Effect of chronic cocaine administration on regional [^{3}H]naloxone binding in rat brain. NAc,r, nucleus accumbens, rostral; NAc,c, nucleus accumbens, caudal; CP, caudatoputamen; SNc, substantia nigra pars compacta; VTA, ventral tegmental area; ABL, basolateral amygdala nucleus; IPN, interpeduncular nucleus; DR, dorsal raphe; S_I, somatosensory cortex; V_I, visual cortex; VP, ventral pallidum; LH, lateral hypothalamus; SI, substantia innominata; PPN, pedunculopontine nucleus. * — $p \leq 0.05$. From Hammer et al., *NIDA Research Monograph,* 105, 184, 1991. With permission.

which gives rise to dynorphin A and B, each of which yield further active opioid peptides.[52-54] The endogenous opioid peptides are distributed widely throughout the brain and periphery. In the brain, they are thought to act as neurotransmitters or neuromodulators and have been implicated in mediating such functions as analgesia, reinforcement, temperature regulation, feeding and drinking behaviors, endocrine and immune regulation, and locomotion. In the periphery, opioid peptides modulate functions related to gastrointestinal, immune, cardiovascular, and endocrine systems.

II. COCAINE AND OPIOID RECEPTORS

Several investigators have observed that *in vivo* cocaine administration has profound effects on central opioid receptors. Hammer[55] studied the effects of chronic, continuous cocaine administration on regional opioid receptor density (Figure 1). In this study, cocaine (1 and 10 mg/kg/day) was administered to Sprague Dawley rats for 14 days by continuous subcutaneous infusion using osmotic minipumps, and the density of opioid receptors was measured by receptor autoradiography using [^{3}H]naloxone under μ-receptor-preferring conditions. Significant reductions of opioid receptor density were found in the ventral tegmental area, substantia nigra pars compacta, basolateral amygdala, and dorsal raphe in animals treated with 10 mg/kg/day of cocaine. In contrast, increases of opioid receptor density were seen in the nucleus accumbens, ventral pallidum, lateral hypothalamus, inferior colliculus, and medial geniculate. No other brain regions were affected. With the exception of the inferior colliculus and medial geniculate which are involved in auditory processes, the changes of opioid receptor expression were confined to brain regions associated either directly or indirectly with central reward mechanisms. μ-Receptor levels increased in mesolimbic terminal and reward output regions, namely the nucleus accumbens, ventral pallidum, and lateral hypothalamus, which are related directly to cocaine-induced reward,[6,9] without affecting the caudatoputamen. The ventral tegmental area, substantia nigra, and dorsal raphe, which contain the cell bodies of dopaminergic and serotonergic neurons that project to the forebrain, exhibited reductions of μ-receptor density. Cells of the basolateral amygdaloid nucleus project to the nucleus accumbens, and hence reduction of opioid activity in this area could alter limbic output to the nucleus accumbens. Therefore, chronic continuous cocaine exposure decreases μ-receptor density in regions which modulate neurochemical and functional projections to mesolimbic reward regions, while increasing μ-receptor density in the mesolimbic reward pathway.[55]

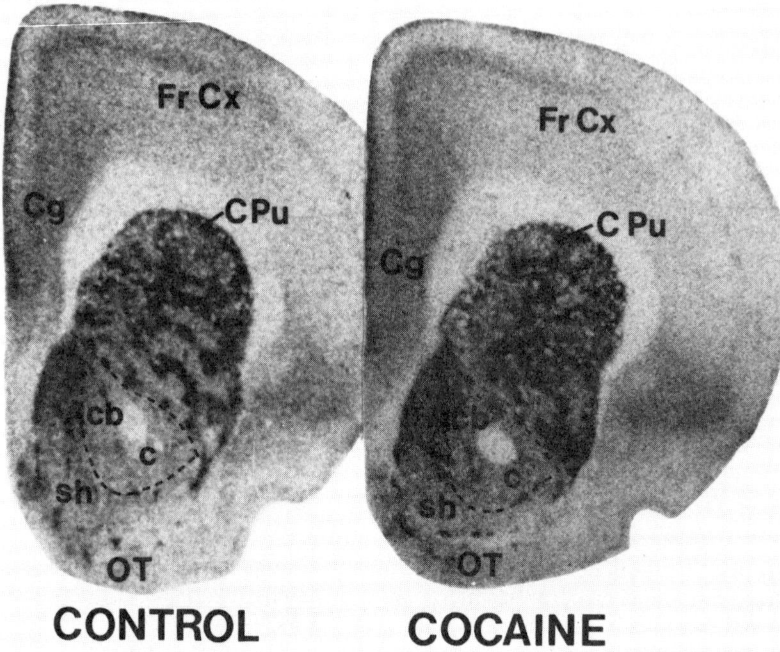

Figure 2. Autoradiographs of coronal sections of rat brain labeled with [^{3}H]DAMGO from a saline-injected control rat (left) and a cocaine-treated (30 mg/kg/day) rat (right). Cocaine treatment increased binding in the caudatoputamen, nucleus accumbens, and cingulate cortex. Acb, nucleus accumbens; c, nucleus accumbens core; sh, nucleus accumbens shell; CPu, caudatoputamen; Fr Cx, frontal cortex; Cg, cingulate cortex; OT, olfactory tubercle. From Unterwald et al., *Brain Res.,* 584, 316, 1992. With permission.

We have studied the effects of cocaine on opiate receptor density using an administration paradigm that mimics the pattern of human cocaine abuse.[56] Cocaine abusers often self-administer the drug repeatedly over a short time period (commonly referred to as a "binge"), most frequently in the early evening hours, followed by a period of abstinence. In our experiments, male Fischer rats were injected with cocaine (10 or 30 mg/kg/day) three times daily at 1-hr intervals at the beginning of the light (inactive) cycle for 14 days. Opioid receptors were measured 30 min after the last injection using quantitative receptor autoradiography under selective receptor labelling conditions. μ-Opioid receptors labeled with [^{3}H]D-Ala^2,N-Me-Phe^4,Gly-ol^5-enkephalin (DAMGO) were found to be upregulated in limbic and striatal regions following 14 days of binge cocaine administration. The density of μ-opioid receptors in the rostral aspect of the nucleus accumbens core and the caudatoputamen of animals receiving 30 mg/kg/day cocaine was higher than that of saline-injected control animals (Figure 2). The outer layer of the cingulate cortex also showed a striking upregulation of μ-receptors which was greatest in rostral sections of the cingulate cortex. The olfactory tubercle, septum, and hypothalamus exhibited smaller nonsignificant increases of μ-opioid receptors (Figure 3). At the level of the diencephalon, a significant upregulation of μ-opioid receptors was found in the basolateral amygdaloid nucleus. No significant change of μ-receptor labelling was found in any other brain region examined such as the hippocampus, thalamus, central gray matter, substantia nigra, ventral tegmental area, interpeduncular nucleus, or locus coeruleus. The time-course of these changes was examined by measuring μ-opioid receptors after 1, 2, 7, or

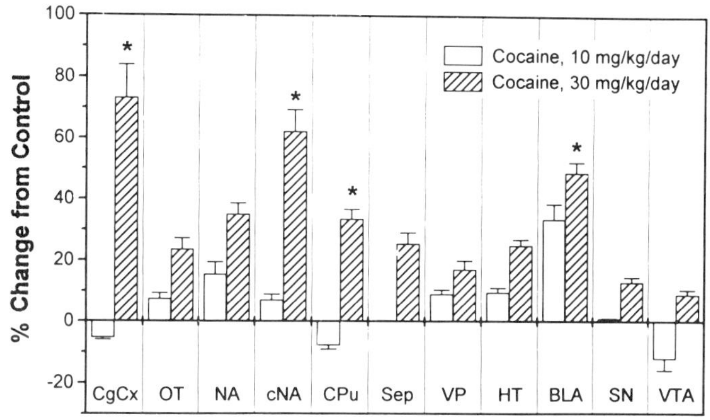

Figure 3. Effect of chronic binge cocaine administration on regional [^{3}H]DAMGO binding to rat brain. CgCx, cingulate cortex; OT, olfactory tubercle; NA, nucleus accumbens; cNA, nucleus accumbens core; CPu, caudatoputamen; Sep, septum; VP, ventral pallidum; HT, hypothalamus; BLA, basolateral amygdala; SN, substantia nigra; VTA, ventral tegmental area. $^*p \leq 0.05$. Data from Unterwald et al., *Brain Res.*, 584, 315, 1992.

14 days of binge cocaine administration. μ-Receptor density in the cingulate cortex and caudatoputamen increased gradually over time, becoming statistically significant by day 14 of cocaine administration. In contrast in the nucleus accumbens, significant μ-receptor upregulation was observed after 7 and 14 days of cocaine administration.[57]

We also have investigated the regulation of κ- and δ-opioid receptors during chronic binge cocaine administration.[58] The results demonstrate that the density of κ-opioid receptors in several brain regions was higher in animals injected with 45 mg/kg/day of cocaine for 14 days. Significant upregulation of κ-opioid receptors was observed in the rostral aspect of the cingulate cortex (124% of control), the rostral aspect of the caudatoputamen (122% of control), the caudal portion of the olfactory tubercle (126% of control), and the ventral tegmental area (131% of control). κ-Receptor density in the rostral nucleus accumbens was slightly higher in cocaine-treated animals (119% of control), although this increase was not statistically significant. No other brain region, including the frontal cortex, globus pallidus, hippocampus, thalamus, hypothalamus, amygdala, central gray matters, substantia nigra, interpeduncular nucleus, or locus coeruleus, showed increases of κ-opioid receptor density. In contrast to the upregulation of μ- and κ receptors, the number of δ-opioid receptors was not altered significantly following 14 days of binge cocaine administration in any brain region measured.[58]

Of interest is the finding that the brain regions containing the major dopaminergic pathways showed the greatest upregulation of μ- and κ-opioid receptors following cocaine treatment. Fibers of the nigrostriatal dopamine pathway terminate in the caudatoputamen; dopaminergic neurons of the mesocortical tract innervate the frontal cortex, cingulate cortex, and entorhinal cortex; and the mesolimbic dopamine pathway projects to the nucleus accumbens, olfactory tubercle, amygdaloid nuclei, and septum. Significant μ-receptor upregulation occurred in layer I of the rostral cingulate cortex, rostral caudatoputamen, rostral nucleus accumbens, and basolateral amygdala.[56] Significant κ-receptor upregulation occurred in the rostral cingulate cortex, rostral caudatoputamen, caudal olfactory tubercle, and ventral tegmental area.[58] These results suggest that dopaminergic activity, and its alteration by cocaine, can regulate the expression of μ- and κ-opioid receptors. The finding of greater μ-receptor upregulation in the rostral aspect of the nucleus accumbens and particularly in its core subdivision, could be related to the higher density of dopamine transporter sites in the rostral nucleus accumbens and in the core region as compared to the caudal accumbens and the shell subdivision, respectively.[59] Furthermore, a recent study has demonstrated that chronic administration of the dopamine D_2 receptor agonist quinpirole produces an upregulation of μ-, but not δ-receptors in the striatum.[60]

Results of a study by Ishizuka et al.[61] suggest that the effects of cocaine on opioid receptors are different in different strains of rats. In this study, 3 mg/kg of cocaine was administered once daily by bolus intravenous injection for 5 days to spontaneously hypertensive and Wistar-Kyoto rats. Opioid receptors were measured by [^{3}H]naloxone binding to membranes prepared from 6 different brain regions 24 hr after the last injection. In the Wistar-Kyoto rats, the B_{max} of [^{3}H]naloxone binding was significantly higher in the cortex (129% of control) and hippocampus (115% of control), lower in the striatum (86% of control for the low affinity site), and unchanged in the hypothalamus and medulla/pons following cocaine treatment. In contrast, cocaine decreased [^{3}H]naloxone binding in the hippocampus (84% of control), hypothalamus (76% of control), and striatum (58% and 69% of control for the high and low affinity sites respectively) in membranes from spontaneously hypertensive rats without affecting binding in the cortex or medulla/pons.

Gestational cocaine exposure also has been shown to increase opioid receptor binding in weanling offspring.[62] Cocaine was administered to dams once daily from E8 - E20 by subcutaneous injection at doses (10 to 40 mg/kg) that produce maternal plasma cocaine levels in the range of those reported in human cocaine users. Receptor density was measured in male pups on postnatal day 21 by *in vitro* receptor autoradiography using [^{3}H]naloxone under conditions of preferential μ-receptor labeling. Gestational cocaine exposure produced a dose-dependent increase of [^{3}H]naloxone binding in many brain regions, including dopaminergic terminal, limbic, and cortical areas. Specifically, [^{3}H]naloxone binding was significantly higher in weanlings exposed to the highest dose of cocaine in the medial prefrontal cortex, rostral olfactory tubercle, cingulate cortex, caudal nucleus accumbens, caudal caudatoputamen, basolateral amygdala, dorsomedial hypothalamus, mediodorsal thalamus, lateral habenula, CA1 and CA3 regions of the hippocampus, molecular layer of the dentate gyrus, agranular motor cortex, somatosensory cortex, and superficial entorhinal cortex. No alterations of neonatal or weanling body weights were observed, so it is unlikely that malnutrition was the cause of the observed effects.

The upregulation of μ-receptors produced by gestational cocaine exposure in dopaminergic terminal fields is consistent with a direct dopamine-opioid interaction at these sites. However, because many regions affected are not thought to be involved in cocaine reinforcement or to contain dopaminergic terminals, gestational cocaine exposure might produce more extensive, less specific effects on opioid systems than produced in adult animals. These data demonstrate that gestational cocaine exposure produces a long-term effect on opioid receptor binding, assessed 24 days after cessation of cocaine administration, and further support the hypothesis that dopaminergic activity can regulate opioid systems.

A recent study has investigated opioid receptor alteration in guinea pigs following cocaine administration.[63] Male Hartley guinea pigs were injected once daily with 40 mg/kg of cocaine i.p. for 7 days and opioid receptors were measured 24 hr after the last injection by homogenate binding assays. A significant reduction in the number of μ-opioid receptors as measured by [^{3}H]DAMGO binding was observed in the frontal cortex (68% of control), amygdala (71% of control), thalamus (67% of control), and hippocampus (69% of control) of cocaine-treated animals. No changes were detected in the hypothalamus, caudatoputamen, nucleus accumbens, cortex, or substantia nigra. The number of κ-opioid receptors as measured by [^{3}H]U69,593 binding was significantly higher in the cerebellum (136% of control) following cocaine administration. The cerebellum of the guinea pig contains high levels of κ-receptors, but few or no μ-, δ-, or dopamine receptors. No changes of κ-receptor number were found in any other brain area examined. Cocaine treatment had no effect on [^{3}H]D-Penicillamine^2, D-Penicillamine^5-enkephalin (DPDPE) binding to δ-receptors in any of the brain regions investigated. These findings indicate that repeated exposure of guinea pigs to cocaine induces distinct effects on μ-, δ-, and κ-opioid receptors.

III. COCAINE AND OPIOID RECEPTOR-MEDIATED SIGNAL TRANSDUCTION

As summarized above, chronic administration of cocaine using a variety of treatment paradigms alters opioid receptor number, although the location and direction of these changes varies with the particular procedure employed. Adenylyl cyclase activity has been used as a functional measure of opioid receptor-mediated signal transduction, as activation of opioid receptors inhibits adenylyl cyclase activity.[35,36] We have investigated the functional consequences of cocaine-induced opioid receptor upregulation by measuring adenylyl cyclase activity in the nucleus accumbens and rostral caudatoputamen of cocaine-treated rats.[64] Basal cyclase activity and the ability of selective μ- and δ-opioid receptor agonists to inhibit adenylyl cyclase activity was measured in rats after administration of cocaine or saline 3 times daily at 1-hr intervals for 14 days. This binge regimen of cocaine administration produced an upregulation of μ-, but not δ-, opioid receptors in the nucleus accumbens and rostral caudatoputamen,[56,58] as described in the previous section. Adenylyl cyclase activity was assessed by determining the amount of cAMP formed in tissue homogenates using a cAMP-binding protein.[65,66] The adenylyl cyclase activator, forskolin, was not used in this study because it is not known whether cocaine treatment changes the relationship between the maximum rate of adenylyl cyclase activity in the presence of forskolin and the actual rate of enzyme activity in the absence of this nonphysiological activator.

The results from this study demonstrate that cAMP production in the rostral caudatoputamen of cocaine-treated animals under basal conditions is significantly lower than that produced in the caudatoputamen of saline-injected animals. Adenylyl cyclase activity in the caudatoputamen of animals receiving 30 and 45 mg/kg/day was 52 ± 13% and 66 ± 3%, respectively, compared to saline-treated control animals. In contrast, basal adenylyl cyclase activity was unchanged in the nucleus accumbens following 14 days of binge cocaine administration. The selective μ-agonist DAMGO and the selective δ-agonist DPDPE each maximally inhibited approximately 25% and 30%, respectively, of basal adenylyl cyclase activity in the rostral caudatoputamen and nucleus accumbens of saline-injected control animals. Administration of cocaine attenuated the ability of DPDPE to inhibit adenylyl cyclase in both brain regions, but had no effect on the efficacy or potency of DAMGO for inhibiting adenylyl cyclase activity.[64] Figure 4 shows the inhibition of adenylyl cyclase activity by DPDPE (A) and by DAMGO (B) in the nucleus accumbens of saline- and cocaine-injected animals. The loss of δ-receptor-regulated cyclase activity was unexpected in light of the finding of μ-, but not δ-, receptor changes in animals treated with cocaine by the same regimen,[56,58] and suggests that chronic repeated cocaine administration results in a selective impairment of δ-opioid receptor-mediated effector function. Since the number of δ-binding sites does not appear to be altered following this regimen of cocaine administration, it is possible that exposure to cocaine results in an uncoupling of the δ-opioid receptor from its G protein.

The effect of chronic continuous cocaine administration on opioid-regulated adenylyl cyclase activity has been investigated by Izenwasser and Cox.[67] Basal cyclase activity and the ability of the selective μ- and δ-opioid receptor agonists, DAMGO and DPDPE, respectively, to inhibit adenylyl cyclase activity was measured in rats treated with 50 mg/kg/day cocaine or saline for 7 days by subcutaneously implanted osmotic minipumps. Adenylyl cyclase activity in the rostral caudatoputamen and nucleus accumbens was

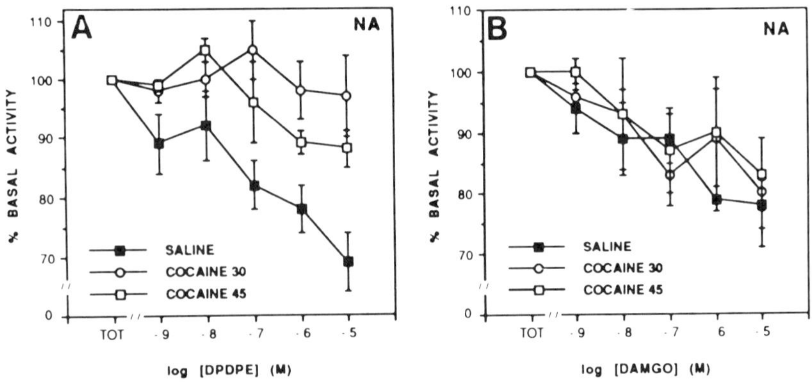

Figure 4. Inhibition of adenylyl cyclase activity by DPDPE (A) and DAMGO (B) in the nucleus accumbens of animals treated with saline, or 30 mg/kg/day or 45 mg/kg/day cocaine. Binge cocaine administration significantly attenuated the inhibition of adenylyl cyclase activity by DPDPE (A), but had no effect on inhibition by DAMGO (B). From Unterwald et al., *Synapse,* 15, 35-36, 1993. With permission.

assessed by determining the amount of cAMP formed in tissue homogenates using a cAMP-binding protein.[65,66] The results demonstrate that there is no change in basal adenylyl cyclase activity in the caudatoputamen or nucleus accumbens of animals treated with continuous cocaine as compared to those receiving saline. However, chronic continuous cocaine exposure increased the maximal inhibition of adenylyl cyclase activity by DAMGO, but not by DPDPE, in the nucleus accumbens. No change of opioid-regulated cyclase was observed in the caudatoputamen.[67] The results of these two studies[64,67] further demonstrate that the treatment paradigm influences the outcome of chronic cocaine exposure. Intermittent cocaine administration decreased basal cyclase activity in the caudatoputamen, which was not observed following continuous cocaine administration. Intermittent cocaine produced a loss of δ-receptor-mediated adenylyl cyclase inhibition in the caudatoputamen and nucleus accumbens, whereas continuous cocaine exposure selectively increased μ-receptor-mediated cyclase inhibition in the nucleus accumbens.

IV. COCAINE AND THE ENDOGENOUS OPIOID PEPTIDES

A. COCAINE AND ENKEPHALIN AND DYNORPHIN PEPTIDES

Several neuropeptide systems have been shown to be regulated by postsynaptic dopaminergic activity. Of the two major striatal efferent pathways, one projects to the globus pallidus and the other to the entopeduncular nucleus/substantia nigra.[68] Both of these pathways utilize GABA as a neurotransmitter,[69] but they differ in their neuropeptide and dopamine receptor content. Striatopallidal neurons contain enkephalin[70] and the mRNA encoding dopamine D_2 receptors,[71] whereas striatonigral neurons contain dynorphin,[72] substance P,[73] and D_1 receptor mRNA.[71] Dopamine differentially regulates the expression of these neuropeptides and the regulation appears to be dependent on the relative abundance of D_1 and D_2 dopamine receptors on striatonigral and striatopallidal neurons, respectively.[71] Dopamine has a tonic inhibitory influence on the synthesis of enkephalin in the striatum. Blockade of dopaminergic neurotransmission by chronic administration of dopamine receptor antagonists or by destruction of nigrostriatal neurons increases the levels of enkephalin and its mRNA.[71,74-80] Enkephalin biosynthesis appears

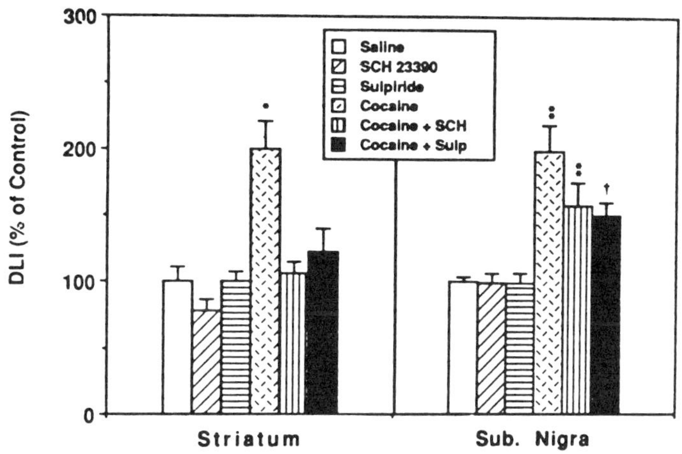

Figure 5. Effects of 30 mg/kg cocaine, 0.5 mg/kg SCH 23390 (D_1 antagonist), or 80 mg/kg sulpiride (D_2 antagonist) on dynorphin-like-immunoreactivity in the striatum and substantia nigra. Cocaine-induced increases of dynorphin were completely attenuated in the striatum and were partially attenuated in the substantia nigra by either SCH 23390 or sulpiride. $p \leq 0.5$, * cocaine vs. all other groups, ** cocaine vs. corresponding saline, + cocaine plus sulpiride vs. cocaine and saline. From Smiley et al., *J. Pharmacol. Exp. Ther.*, 253, 940, 1990. With permission.

to be maximally inhibited by dopamine under basal conditions because further stimulation of dopamine receptors has no effect on striatal enkephalin levels. D_2 receptors might mediate this inhibitory influence of dopamine on enkephalin biosynthesis,[60,71,81] but D_1 receptors also have been implicated.[82,83] Dynorphin biosynthesis is regulated by dopaminergic tone in a different manner. Enhancing dopaminergic neurotransmission by administration of dopamine receptor agonists increases tissue content of dynorphin and prodynorphin mRNA,[84-87] an effect that can be block by D_1 receptor antagonists.[83,86] D_2 receptors appear to play no role in the regulation of dynorphin.[83]

Several studies have investigated the effects of cocaine exposure on enkephalin and dynorphin peptides. Sivam[88] demonstrated that daily injections of 20 mg/kg/day cocaine for 4 days in rats significantly increased dynorphin A_{1-8}-immunoreactivity in the striatum and substantia nigra, which was maintained for at least 4 days after the last cocaine injection. No change of dynorphin was induced by acute cocaine administration, and the levels of met-enkephalin and substance P were not altered by acute or subchronic cocaine treatment. Co-administration of the D_1 receptor antagonist SCH 23390 or the D_2 receptor antagonist spiperone with cocaine attenuated the increase of striatal dynorphin concentration. These results demonstrate that repeated cocaine administration selectively increases striatonigral dynorphin A_{1-8} content.

Similarly, Smiley et al.[89] showed that 5 injections of 30 mg/kg cocaine at 6-hr intervals increased dynorphin A_{1-17}-immunoreactivity in the striatum, substantia nigra, and nucleus accumbens, but not in the hippocampus. These different brain regions exhibited unique time courses of induction of dynorphin expression. The increase of dynorphin occurred at 1 and 8 hr after the last cocaine injection in the striatum, at 1, 8, 24, and 48 hr after the last cocaine injection in the substantia nigra and only at 8 hr after the last cocaine injection in the nucleus accumbens. Co-administration of the D_1 antagonist SCH 23390 or the D_2 antagonist sulpiride with cocaine blocked the increase of dynorphin levels in the striatum, but only partially attenuated the increase in the substantia nigra (Figure 5). Administration of another dopamine uptake inhibitor GBR 12909, but not the serotonergic uptake inhibitor fluoxetine, similarly increased striatal and nigral dynorphin expression.

An increase in tissue content of neuropeptide could indicate either an increase or a decrease of peptide neurotransmission. An increase of tissue content could reflect an increase of peptide synthesis or, alternatively, an accumulation of the peptide in nerve terminals due to decreased release or turnover. Measurement of peptide gene expression can be more indicative of peptide synthesis and turnover. Spangler and co-workers[90] have observed that 45 mg/kg/day of cocaine administered using the binge paradigm (three injections daily at 1-hr intervals) increases prodynorphin mRNA content in the caudatoputamen. In this study, mRNA levels were quantified using a modified solution-hybridization assay.[91,92] Prodynorphin mRNA levels increased in the caudatoputamen (123% of control) following 14 days of cocaine administration, but were unchanged in the pituitary, hypothalamus, nucleus accumbens, or hippocampus, indicating a selective regulation of dynorphin gene expression in striatonigral neurons. Further time course analyses revealed that prodynorphin mRNA increased after 3 days of cocaine administration, which was the earliest time tested, and this increase persisted throughout the 14 days of cocaine treatment.[90]

In an earlier study, proenkephalin mRNA level was examined in rats treated with 30 mg/kg/day cocaine for 14 days by the same binge paradigm.[91] Using a solution hybridization assay, no significant change of enkephalin gene expression was observed in any brain region analyzed including the pituitary, frontal cortex, nucleus accumbens, caudatoputamen, hippocampus, hypothalamus, central grey, and cerebellum. Further, no change of enkephalin-immunoreactivity was found in any of these brain regions.[93] When shorter time points were examined, an increase (118% of control) of proenkephalin mRNA was observed in the caudatoputamen after 3 days (the earliest time measured) of binge cocaine treatment.[90] Earlier time points are currently being examined.

Striatal levels of prodynorphin mRNA have been measured in animals self-administering cocaine. Hurd and co-workers[94] used *in situ* hybridization to measure levels of mRNA encoding dynorphin, enkephalin, and substance P in the striatum of rats allowed to self-administer cocaine for 7 days. Significant increases of prodynorphin mRNA levels were observed in the dorsal caudate (119% of control) and nucleus accumbens (109% of control) following cocaine self-administration. The mRNA encoding substance P was also significantly higher in the dorsal caudate (115% of control) and nucleus accumbens (106% of control). Proenkephalin mRNA level showed a small but statistically significant increase in the nucleus accumbens (103% of control), but not in the dorsal caudate, after 7 days of cocaine self-administration. These findings demonstrate that cocaine self-administration produces an increase of dynorphin and substance P gene expression in striatonigral neurons.

Striatal prodynorphin gene expression in cocaine self-administering animals was also studied by Daunais and colleagues.[95] Rats self-administered cocaine for 2 to 6 weeks and were grouped according to their mean daily cocaine intake during the last week of self-administration. No significant change of prodynorphin mRNA was detected in the dorsal striatum (caudatoputamen) or nucleus accumbens in animals self-administering low or moderate doses of cocaine. In the high dose cocaine animals, however, a significant increase of prodynorphin mRNA level was found in the dorsal striatum, but not in the nucleus accumbens. The increase of prodynorphin mRNA was particularly apparent in the patch-like areas of the dorsal striatum, and appeared to be due to a large increase in the number of cells expressing the mRNA and a slight enhancement in the level of mRNA per cell. These investigators also observed an increase of dynorphin-immunoreactivity in striatal patches and in the shell of the nucleus accumbens[96] and an increase of striatal prodynorphin mRNA level[97] in rats injected with cocaine once daily for 10 days.

The effects of acute cocaine administration on the expression of dynorphin, enkephalin, and their mRNA are less consistent. Sivam[88] observed that 20 mg/kg acute cocaine had no effect on dynorphin- or enkephalin-immunoreactivity. Likewise, Daunais et al.[97]

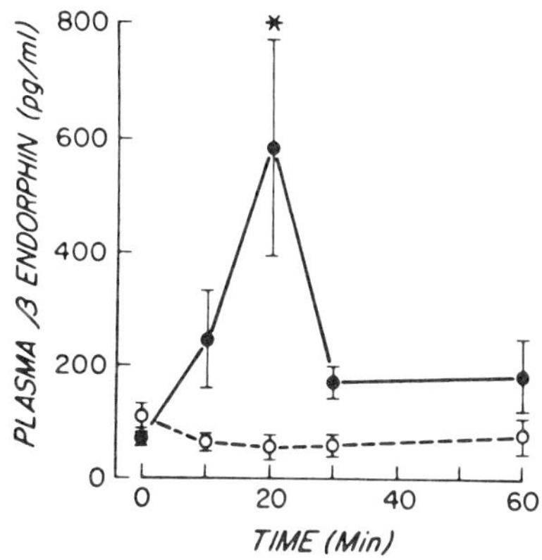

Figure 6. Time course of plasma β-endorphin concentrations after a 20 mg/kg cocaine (solid line) or saline (dashed line) injection in male rats. Cocaine administration increased plasma β-endorphin levels which reach a maximum 20 min after the injection. * $p \leq 0.01$. From Moldow and Fischman, *Peptides,* 8, 820, 1987. With permission.

reported no significant change of striatal prodynorphin mRNA content as measured by *in situ* hybridization 1 hr following a single i.p. injection of 10 to 30 mg/kg of cocaine. In contrast, Hurd and Herkenham[98] observed a significant increase (135% of control) of prodynorphin mRNA expression in the dorsolateral caudatoputamen 2 hr after a single i.p. injection of 30 mg/kg of cocaine as measured by *in situ* hybridization. In this study, acute cocaine administration also increased (135% of control) proenkephalin mRNA content in the dorsolateral caudatoputamen. No significant differences of either prodynorphin or proenkephalin gene expression were measured in the nucleus accumbens or ventromedial caudatoputamen.[98] In the amygdala, Cohen et al.[99] found a 3.5-fold increase in prodynorphin and a 2.5-fold increase in proenkephalin gene expression 75 min after a single i.v. injection of 10 mg/kg of cocaine using a Northern blot technique.

A possible mechanism of cocaine-induced dynorphin and enkephalin regulation involves nuclear transcription factors. The expression of immediate early genes is altered by cocaine administration. Two products of immediate early genes, c-*fos* and c-*jun,* interact with each other to form heterodimers that bind to Activator Protein 1 (AP-1) sites. Binding to AP-1 sites can activate or repress gene transcription.[100] Several neuropeptide genes including those of proenkephalin and prodynorphin contain AP-1 sites upstream of their promoter regions.[101-104] By this mechanism, induction of the expression of nuclear transcription factors can couple dopamine receptor activation to changes in postsynaptic gene expression.

Several studies have demonstrated that the levels of c-*fos* mRNA and its protein product, Fos, increased in the striatum, amygdala, and hippocampus following a single cocaine injection.[30,96,97,99,105,106] Activation was transient and readily returned to basal levels.[30,99,106] Pretreatment with a dopamine D_1 receptor antagonist completely suppressed cocaine-induced Fos activation in the striatum, whereas pretreatment with a D_2 receptor antagonist only partially diminished activation.[105,106] In the amygdala, hippocampus, and striatum, the levels of c-*fos* mRNA increased dramatically (6- to 11-fold) 45 min following treatment with 10 mg/kg cocaine i.v.[99] A concurrent increase of the mRNA encoding for c-*jun*, another nuclear transcription factor, was found in the amygdala, followed by increases of prodynorphin and proenkephalin mRNA.[99] The observed co-induction of c-*fos* and c-*jun* further supports the hypothesis that these two nuclear transcription factors might mediate the subsequent activation of proenkephalin and prodynorphin gene expression.

B. COCAINE AND β-ENDORPHIN

Both acute and chronic exposure to cocaine significantly alters plasma and tissue β-endorphin content. Following a single i.p. injection of 20 mg/kg cocaine in rats, plasma β-endorphin concentration rapidly and transiently increased (Figure 6).[107] Increased plasma β-endorphin content also was observed following chronic cocaine exposure.[108] Rats treated with 2.5 to 20 mg/kg cocaine once daily for 10 days had significantly higher β-endorphin plasma concentrations (approximately 3-fold) 24 hr after the last cocaine injection than those injected with saline. Significantly greater β-endorphin-immunoreactivity also was observed in the anterior and neurointermediate lobes of the pituitary in cocaine-treated rats, while no change of hypothalamic β-endorphin was found. The *in vitro* release of β-endorphin from the anterior pituitary and the neurointermediate pituitary was also assessed in this study. *In vitro* β-endorphin release from the anterior and neurointermediate lobes of the pituitary of rats chronically treated with cocaine was significantly greater than the release from the pituitary of control animals.[108] Administration of antiserum to corticotropin releasing factor (CRF) blocks the cocaine-induced stimulation of POMC release,[109] suggesting that CRF mediates the POMC response to cocaine.

Release of β-endorphin from the rat pituitary is directly inhibited by dopamine.[110-114] Although the acute effect of cocaine is a transient stimulation of dopaminergic neurotransmission, chronic cocaine exposure appears to produce an overall reduction of dopaminergic tone, possibly due to a relative dopamine depletion.[115-120] A reduction of dopaminergic activity then would contribute to increased β-endorphin release and plasma concentrations.

The effects of cocaine self-administration on β-endorphin serum and tissue concentration also have been investigated.[121,122] In these studies, rats were allowed to self-administer cocaine 6 hr daily for 5 days, and β-endorphin levels were measured immediately as well as 18 hr later. Plasma β-endorphin-immunoreactivity was higher immediately (123% of control) and 18 hr (129% of control) after the last cocaine self-administration session as compared to in control rats. A significant reduction of β-endorphin-immunoreactivity was found in the anterior pituitary (55% of control), nucleus accumbens (63% of control), rostral striatum (62% of control), and hippocampus (76% of control) 18 hr after the last cocaine injection, and in the septum (57% of control) immediately after the final cocaine self-administration. No significant changes in other brain regions including the neurointermediate lobe of the pituitary, amygdala, hypothalamus, thalamus, caudal striatum, mesencephalon, or medulla oblongata were observed at either time after cocaine self-administration. The authors hypothesize that the reduction of β-endorphin content selectively in the anterior parts of the limbic system might reflect the craving state of the animal.[121,122]

V. SUMMARY AND CONCLUSIONS

It is evident from the current literature that cocaine has profound effects on several components of the endogenous opioid system. Cocaine administered to a variety of species using many different treatment regimens alters the expression of opioid receptors. A consensus on the exact nature of these changes is still lacking, probably due to the different schedules of cocaine administration and the different strains and species of animals used in these studies. The mechanism underlying the regulation of opioid receptors by cocaine has not been elucidated, but activation of D_2 receptors might be involved. Opioid receptor-inhibition of adenylyl cyclase activity is altered following cocaine exposure and, once again, the schedule of cocaine administration appears to influence cocaine's effect on cyclase activity. Intermittent cocaine administration produces a loss of δ-receptor-mediated adenylyl cyclase inhibition in the caudatoputamen and nucleus accumbens, whereas continuous cocaine exposure selectively increases μ-receptor-mediated cyclase inhibition in the nucleus accumbens only. A variety of cocaine treatment regimens increase the expression of dynorphin and prodynorphin mRNA in striatonigral neurons. The activation of prodynorphin gene expression by cocaine might be mediated by the induction of nuclear transcription factors such as *c-fos* and c-*jun*. The effect of cocaine on enkephalin peptides is less consistent, but most evidence supports a lack of cocaine-induced enkephalin regulation. Both acute and chronic exposure to cocaine increases circulating levels of β-endorphin. Cocaine stimulation of β-endorphin release appears to be mediated by its activation of CRF.

The regulation of the endogenous opioid system by cocaine is clearly evident at cellular and molecular levels. The functional significance of this regulation is not presently understood, but behavioral studies indicate that opioid systems play a role in cocaine-induced behaviors. Further examination of the interaction of cocaine with the endogenous opioid system might provide insights into the cellular, molecular, and functional alterations that occur during cocaine exposure and that persist long after cocaine exposure has ceased. Changes in the endogenous opioid system might contribute to the perpetuation of cocaine abuse and/or to relapse.

ACKNOWLEDGMENTS

I wish to thank Dr. Mary Jeanne Kreek, head of the Laboratory of the Biology of Addictive Diseases at The Rockefeller University, for her active role in studies described in this chapter and for critical review of the manuscript. I am grateful to the other members of this Laboratory who have conducted or participated in these experiments including Dr. Rudolph Spangler, Dr. Andrea Branch, Dr. Yan Zhou, Dr. Ann Ho, Dr. Isabelle Maisonneuve, Mr. Joshua M. Rubenfeld, Ms. Jennifer Horne-King, and Mr. Christopher Maggos. This work was supported by USPHS Awards DA08267 (EMU) and P50 DA05130 (MJK), and by the Aaron Diamond Foundation (MJK).

REFERENCES

1. Ritz, M. C., Lamb, R. J., Goldberg, S. R., and Kuhar, M. J., Cocaine receptors on dopamine transporters are related to self-administration of cocaine, *Science,* 237, 1219, 1987.
2. Heikkila, R. E., Orlansky, H., and Cohen, G., Studies on the distinction between uptake inhibition and release of [^{3}H]dopamine in rat brain tissue slices, *Biochem. Pharmacol.,* 24, 847, 1975.
3. Kilty, J. E., Lorang, D., and Amara, S. G., Cloning and expression of a cocaine-sensitive rat dopamine transporter, *Science,* 254, 578, 1991.
4. Shimada, S., Kitayama, S., Lin, C. L., Patel, A., Nanthakumar, E., Gregor, P., Kuhar, K., and Uhl, G., Cloning and expression of a cocaine-sensitive dopamine transporter complementary DNA, *Science,* 254, 576, 1991.
5. Wise, R. A. and Bozarth, M. A., Brain reward circuitry: four circuit elements wired in apparent series, *Brain Res. Bull.,* 12, 203, 1984
6. Koob, G. F., Vaccarino, F. J., Amalric, M., and Swerdlow, N. R., Neural substrates for cocaine and opiate reinforcement, in *Cocaine: Clinical and Behavioral Aspects,* Fisher, S., Raskin, A., and Uhlenhuth, E. H., Eds., Oxford Press, New York, 1987, 80.
7. Roberts, D. C. S., Koob, G. F., Klonoff, P., and Fibiger, H. C., Extinction and recovery of cocaine self-administration following 6-hydroxydopamine lesions of the nucleus accumbens, *Pharmacol. Biochem. Behav.,* 12, 781, 1980.
8. Goeders, N. E. and Smith, J. E., Cortical dopaminergic involvement in cocaine reinforcement, *Science,* 221, 773, 1983.
9. Hubner, C. B. and Koob, G. F., The ventral pallidum plays a role in mediating cocaine and heroin self-administration in the rat, *Brain Res.,* 508, 20, 1990.
10. Kornetsky, C., Huston-Lyons, D., and Porrino, L. J., The role of the olfactory tubercle in the effects of cocaine, morphine and brain-stimulation reward, *Brain Res.,* 541, 75, 1991.
11. Khachaturian, H., Lewis, M. E., Hollt, V., and Watson, S. J., Telencephalic enkephalinergic systems in the rat brain., *J. Neurosci.,* 3, 844, 1983.
12. Fallon, J., H. and Leslie, F. M., Distribution of dynorphin and enkephalin peptides in the rat brain, *J. Comp. Neurol.,* 249, 293, 1986.

13. Mansour, A., Lewis, M. E., Khachaturian, H., Akil, H., and Watson, S. J., Pharmacological and anatomical evidence of selective μ, δ, and κ opioid receptor binding in rat brain, *Brain Res.*, 399, 69, 1986.
14. Tempel, A. and Zukin, R. S., Neuroanatomical patterns of the μ, δ, and κ opioid receptors of rat brain as determined by quantitative in vitro autoradiography, *Proc. Natl. Acad. Sci. USA,* 84, 4308, 1987.
15. Pollard, H., Llorens-Cortes, C., and Schwartz, J. C., Enkephalin receptors on dopaminergic neurons in rat striatum, *Nature,* 268, 745, 1977.
16. Reisine, T. D., Nagy, J. I., Beaumont, K., Fibiger, H. C., and Yamamura, H. I., The localization of receptor binding sites in the substantia nigra and striatum of the rat, *Brain Res.,* 177, 241, 1979.
17. Eghbali, M., Santoro, C., Paredes, W., Gardner, E. L., and Zukin, R. S., Visualization of multiple opioid-receptor types in rat striatum after specific mesencephalic lesions, *Proc. Natl. Acad. Sci. USA,* 84, 6582, 1987.
18. Unterwald, E. M., Tempel, A., Koob, G. F., and Zukin, R. S., Characterization of opioid receptor in rat nucleus accumbens following mesolimbic dopaminergic lesions, *Brain Res.,* 505, 111, 1989.
19. deWit, H. and Wise, R. A., Blockade of cocaine reinforcement in rats with the dopamine receptor blocker pimozide but not with the noradrenergic blockers phentolamine or phenoxybenzamine, *Can. J. Psychol.,* 31, 195, 1977.
20. Bain, G. T. and Kornetsky, C., Naloxone attenuation of the effect of cocaine on rewarding brain stimulation, *Life Sci.*, 40, 1119, 1987.
21. Houdi, A. A., Bardo, M. T., and Van Loon, G. R., Opioid mediation of cocaine-induced hyperactivity and reinforcement, *Brain Res.,* 497, 195, 1989.
22. Bilsky, E. J., Montegut, M. J., Delon, C. L., and Reid, L. D., Opioidergic modulation of cocaine conditioned place preferences, *Life Sci.,* 50, PL-85, 1992.
23. Corrigal, W. A. and Coen, K. M., Opiate antagonists reduce cocaine but not nicotine selfadministration, *Psychopharmacology,* 104, 167, 1991.
24. Mello, N. K., Kamien, J. B., Lukas, S. E., Drieze, J., and Mendelson, J. H., The effects of nalbuphine and butorphanol treatment on cocaine and food self-administration by rhesus monkeys, *Neuropsychopharmacology,* 8, 45, 1993.
25. Izenwasser, S. and Cox, B. M., Daily cocaine treatment produces a persistent reduction of [^{3}H]dopamine uptake in vitro in rat nucleus accumbens but not in striatum, *Brain Res.,* 531, 338, 1990.
26. Post, R. M., Intermittent versus continuous stimulation: effect of time interval on the development of sensitization or tolerance, *Life Sci.,* 26, 1275, 1980.
27. Izenwasser, S. and Cox, B. M., Inhibition of dopamine uptake by cocaine and nicotine: tolerance to chronic treatment, *Brain Res.,* 573, 119, 1992.
28. Reith, M. E. A., Benuck, M., and Lajtha, A., Cocaine dispostion in the brain after continuous or intermittent treatment and locomotor stimulation in mice, *J. Pharmacol. Exp. Ther.,* 243, 281, 1987.
29. Kreek, M. J., Factors modifying the pharmacological effectiveness of methadone, in *Research in the Treatment of Narcotic Addictions: State of the Art,* Cooper, J. R., Altman, F., Brown, B. S., and Czechowicz, D., Eds., NIDA Res. Monograph, Bethesda, MD, 1983, 444.
30. Torres, G. and Rivier, C., Differential effects of intermittent or continuous exposure to cocaine on the hypothalamic-pituitary-adrenal axis and c-fos expression, *Brain Res.,* 571, 204, 1992.
31. Pasternak, G. W. and Wood, P. J., Multiple mu opioid receptors, *Life Sci.,* 38, 1889, 1986.
32. Attali, B., Gouarderes, C., Mazarguil,, H., Audigier, Y., and Cros, J., Evidence for multiple kappa binding sites by use of opioid peptides in the guinea-pig lumbo-sacral spinal cord, *Neuropeptides,* 3, 53, 1982.
33. Rothman, R. B., Bowen, W. D., Bykov, V., Schumacher, U. K., Pert, C. B., Jacobson, A. E., Burke, T. R., Jr., and Rice, K. C., Preparation of rat brain membranes greatly enriched with either type-I-delta or type-II-delta opiate binding sites using site directed alkylating agents: evidence for a two-site allosteric model, *Neuropeptides,* 4, 201, 1984.
34. Zukin, R. S., Eghbali, M., Olive, D., Unterwald, E. M., and Temple, A., Characterization and visualization of rat and guinea pig brain kappa opioid receptors: evidence for κ_1 and κ_2 opioid receptors, *Proc. Natl. Acad. Sci. USA,* 85, 4061, 1988.
35. Klee, W. A., Sharma, S. K., and Nirenberg, M., Opiate receptors as regulators of adenylate cyclase, *Life Sci.,* 16, 1869, 1975.
36. Childers, S. R., Opiate-inhibited adenylate cyclase in rat brain membranes depleted of G_s-stimulated adenylate cyclase, *J. Neurochem.,* 50, 543, 1988.

37. DiChiara, G. and North, R. A., Neurobiology of opiate abuse, *Trends Pharmacol. Sci.*, 13, 185, 1992.
38. Reisine, T. and Bell, G. I., Molecular biology of opioid receptors, *Trends Neurosci.*, 16, 506, 1993.
39. Evans, C. J., Keith, D. S., Jr., Morrison, H., Magendzo, K., and Edwards, R. H., Cloning of a delta opioid receptor by functional expression, *Science*, 258, 1952, 1992.
40. Kieffer, B. L., Befort, K., Gaveriaux-Ruff, C., and Hirth, C. G., The delta-opioid receptor: isolation of a cDNA by expression cloning and pharmacological characterization, *Proc. Natl. Acad. Sci. USA*, 89, 12048, 1992.
41. Chen, Y., Mestek, A., Liu, J., Hurley, J. A., and Yu, L., Molecular cloning and fuctional expression of a mu opioid receptor from rat brain, *Mol. Pharmacol.*, 44, 8, 1993.
42. Wang, J. B., Imai, Y., Eppler, C. M., Gregor, P., Spivak, C., and Uhl, G. R., Mu-opiate receptor: cDNA cloning and expression, *Proc. Natl. Acad. Sci. USA*, 90, 10230, 1993.
43. Yasuda, K., Raynor, K., Kong, H., Breder, C. D., Takeda, J., Reisine, T., and Bell, G. I., Cloning and functional comparison of kappa and delta opioid receptors from mouse brain, *Proc. Natl. Acad. Sci. USA*, 90, 6736, 1993.
44. Minami, M., Toya, T., Katao, Y., Maekawa, K., Nakamura, S., Onogi, T., Kaneko, S., and Satoh, M., Cloning and expression of a cDNA for the rat kappa-opioid receptors, *FEBS*, 329, 291, 1993.
45. Loh, H. H., Tseng, L. F., Wei, E., and Li, C. H., Beta-endorphin is a potent analgesic agent, *Proc. Natl. Acad. Sci. USA*, 73, 2895, 1976.
46. Roberts, J. L. and Herbert, E., Characterization of a common precursor to corticotropin and beta-lipotropin: indentification of beta-lipotropin peptides and their arrangement relative to corticotropin in the precursor synthesis in a cell-free system, *Proc. Natl. Acad. Sci. USA*, 74, 5300, 1977.
47. Kita, T., Inoue, A., Nakanashi, S., and Numa, S., Purification and characterization of the messenger RNA coding for bovine corticotropin/beta-lipotropin precursor, *Eur. J. Biochem.*, 93, 213, 1979.
48. Watson, S. J., Akil, H., Richard, C. W., and Barchas, J. D., Evidence for two separate opiate peptide neuronal systems, *Nature*, 275, 226, 1978.
49. Kimura, S., Lewis, R. V., Stern, A. S., Rossier, J., Stein, S., and Udenfriend, S., Probable precursors of [Leu]enkephalin and [Met]enkephalin in adrenal medulla: peptides of 3-5 kilodaltons, *Proc. Natl. Acad. Sci. USA*, 77, 1681, 1980.
50. Gubler, U., Kilpartick, D. L., Seeburg, P. H., Gage, L. P., and Udenfriend, S., Detection and partial characterization of proenkephalin mRNA, *Proc. Natl. Acad. Sci. USA*, 78, 5484, 1981.
51. Howells, R. D., Kilpatrick, D. L., Bhatt, R. S., Monahan, J. J., Poonian, H., and Udenfriend, S., Molecular cloning and sequence determination of rat preproenkephalin in cDNA: sensitive probe for studying transcriptional changes in rat tissues, *Proc. Natl. Acad. Sci. USA*, 81, 7651, 1984.
52. Cox, B. M., Opheim, K. E., Teschmacher, H., and Goldstein, A., A peptide-like substance from pituitary that acts like morphine. 2. Purification and properties, *Life Sci.*, 16, 1777, 1975.
53. Goldstein, A., Fischli, W., Lowney, L., Hunkapiller, M., and Hood, L., Porcine pituitary dynorphin: complete amino acid sequence of the biologically active heptadecapeptide, *Proc. Natl. Acad. Sci. USA*, 78, 7219, 1981.
54. Kakidani, H., Furutani, Y., Takahashi, H., Noda, M., Morimoto, Y., Hirose, T., Asai, M., Inayama, S., Nakanishi, S., and Numa, S., Cloning and sequence analysis of cDNA for porcine beta-neo-endorphine/dynorphin precursor, *Nature*, 298, 245, 1982.
55. Hammer, R. P., Jr., Cocaine alters opiate receptor binding in critical brain reward regions, *Synapse*, 3, 55, 1989.
56. Unterwald, E. M., Horne-King, J., and Kreek, M. J., Chronic cocaine alters brain mu opioid receptors, *Brain Res.*, 584, 314, 1992.
57. Unterwald, E. M., Rubenfeld, J. M., and Kreek, M. J., Time course of cocaine-induced alterations in opioid and dopamine receptors and transporter sites, *NIDA Research Monographs*, in press.
58. Unterwald, E. M., Rubenfeld, J. M., and Kreek, M. J., Repeated cocaine administration upregulates κ and μ, but not δ, opioid receptors, *NeuroReport*, 5, 1613, 1994.
59. Graybiel, A. M. and Moratalla, R., Dopamine uptake sites in the striatum are distributed differentially in striosome and matrix compartments, *Proc. Natl. Acad. Sci. USA*, 86, 9020, 1989.
60. Chen, J. F., Aloyo, V. J., and Weiss, B., Continuous treatment with the D2 dopamine receptor agonist quinpirole decreases D2 dopamine receptors, D2 dopamine receptor messenger RNA and proenkephlain messenger RNA, and increases mu opioid receptors in mouse striatum, *Neuroscience*, 54, 669, 1993.

61. Ishizuka, Y., Rockhold, R. W., Hoskins, B., and Ho, I. K., Cocaine-induced changes in ^{3}H-naloxone binding in brain membranes isolated from spontaneously hypertensive and Wistar-Kyoto rats, *Life Sci.,* 43, 2275, 1988.
62. Clow, D. W., Hammer, R. P., Jr., Kirstein, C. L., and Spear, L. P., Gestational cocaine exposure increases opiate receptor binding in weanling offspring, *Devel. Brain Res.,* 59, 179, 1991.
63. Itzhak, Y., Differential regulation of brain opioid receptors following repeated cocaine administration to guinea pigs, *Drug Alc. Dep.,* 33, 53, 1993.
64. Unterwald, E. M., Cox, B. M., Kreek, M. J., Cote, T. E., and Izenwasser, S., Chronic repeated cocaine administration alters basal and opioid-regulated adenylyl cyclase activity, *Synapse,* 15, 33, 1993.
65. Frey, E. A. and Kebabian, J. W., A mu-opiate receptor in 7315c tumor tissue mediates inhibition of immunoreactive prolactin release and adenylate cyclase activity, *Endocrinology,* 115, 1797, 1984.
66. Puttfarcken, P. S., Werling, L. L., and Cox, B. M., Effects of chronic morphine exposure on opioid inhibition of adenylyl cyclase in 7315c cell membranes: a useful model for the study of tolerance at mu opioid receptors, *Mol. Pharmacol.,* 33, 520, 1988.
67. Izenwasser, S. and Cox, B., M., Effects of continuous cocaine administration on opioid-regulated adenylyl cyclase activity in rat nucleus accumbens, *Soc. Neurosci. Abst.,* 18, 546, 1992.
68. Kawaguchi, Y., Wilson, C. J., and Emson, P., Projection subtypes of rat neostriatal matrix cells revealed by intracellular injection of biocytin, *J. Neurosci.,* 10, 3421, 1990.
69. Chevalier, G., Vacher, S., Deniau, J. M., and Desban, M., Dis-inhibition as a basic process in the expression of striatal function. I. The striato-nigral influence on tecto-spinal/tecto-diencephalin neurons, *Brain Res.,* 334, 215, 1985.
70. Haber, S. N. and Nauta, W. J. H., Ramifications of the globus pallidus in the rat as demonstrated by patterns of immunohistochemistry, *Neuroscience,* 9, 245, 1983.
71. Gerfen, C. R., Engber, T. M., Mahan, L. C., Susel, Z., Chase, T. N., Monsma, F. J., Jr., and Sibley, D. R., D_1 and D_2 dopamine receptor-regulated gene expression of striatonigral and striatopallidal neurons, *Science,* 250, 1429, 1990.
72. Vincent, S. R., Hokfelt, T., Christensson, I., and Terenius, L., Immunohistochemical evidence for a dynorphin immunoreactive striatonigral pathway, *Eur. J. Pharmacol.,* 85, 251, 1982.
73. Brownstein, M. J., Mroz, E. A., Tappaz, M. L., and Leeman, S. E., On the origin of substance P and glutamic acid decarboxylase (GAD) in the substantia nigra, *Brain Res.,* 135, 2315, 1977.
74. Hong, J. S., Yang, H.-Y. T., Gillin, J. C., Diguillio, A. M., Fratta, W., and Costa, E., Chronic treatment with haloperidol accelerates the biosynthesis of enkephalins in rat brain, *Brain Res.,* 160, 192, 1979.
75. Thal, L. J., Sharpless, N. S., Hirshhorn, I. D., Horowitz, S. G., and Makman, M. H., Striatal met-enkephalin concentration increases following nigrostriatal denervation, *Biochem. Pharmacol.,* 44, 3297, 1983.
76. Tang, F., Costa, E., and Schwartz, J. P., Increase of proenkephalin mRNA and enkephalin content of rat striatum after daily injection of haloperidol for 2 to 3 weeks, *Proc. Natl. Acad. Sci. USA,* 80, 3841, 1983.
77. Mocchetti, I., Schwartz, J. P., and Costa, E., Use of mRNA hybridization and radioimmunoassay to study mechanisms of drug-induced accumulation of enkephalins in rat brain structures, *Mol. Pharmacol.,* 28, 86, 1985.
78. Young, S. W., Bonner, T. I., and Brann, M. R., Mesencephalic dopamine neurons regulate the expression of neuropeptide mRNAs in the rat forebrain, *Proc. Natl. Acad. Sci. USA,* 83, 9827, 1986.
79. Morris, B. J., Herz, A., and Hollt, V., Localization of striatal opioid gene expression, and its modulation by the mesostriatal dopamine pathway: an in situ hybridization study, *J. Mol. Neurosci.,* 1, 9, 1989.
80. Abood, M. E., Eberwine, J. H., Erdelyi, E., and Evans, C. J., Regulation of both preproenkephalin mRNA and its derived opioids by haloperidol—a method for measurement of peptides and mRNA in the same tissue extract, *Mol. Brain Res.,* 8, 243, 1990.
81. Morris, B. J., Hollt, V., and Herz, A., Dopaminergic regulation of striatal proenkephalin mRNA and prodynorphin mRNA: contrasting effects of D1 and D2 antagonists, *Neuroscience,* 25, 525, 1988.
82. Mocchetti, I., Naranjo, J. R., and Costa, E., Regulation of enkephalin turnover in rats receiving antagonists of specific dopamine receptor subtypes, *J. Pharmacol. Exp. Ther.,* 241, 1120, 1987.

83. Jiang, H.-K., McGinty, J. F., and Hong, J. S., Differential modulation of striatonigral dynorphin and enkephalin by dopamine receptor subtypes, *Brain Res.*, 507, 57, 1990.
84. Peterson, M. R. and Robertson, H. A., Effect of dopaminergic agents on levels of dynorphin 1-8 in rat striatum, *Prog. Neuro-Psychopharmacol. Biol. Psychiat.*, 8, 725, 1984.
85. Li, S. J., Sivam, S. P., and Hong, J. S., Regulation of the concentration of dynorphin A(1-8) in the striatonigral pathway by the dopaminergic system, *Brain Res.*, 398, 390, 1986.
86. Hanson, G. R., Merchant, K. M., Letter, A. A., Bush, L., and Gibb, J. W., Methamphetamine-induced changes in the striatonigral dynorphin system: role of D-1 and D-2 receptors, *Eur. J. Pharmacol.*, 144, 245, 1987.
87. Nylander, I. and Terenius, L. H., Dopamine receptors mediate alterations in striato-nigral dynorphin and substance P pathways, *Neuropharmacology*, 26, 1295, 1987.
88. Sivam, S. P., Cocaine selectively increases striatonigral dynorphin levels by a dopaminergic mechanism, *J. Pharmacol. Exp. Ther.*, 250, 818, 1989.
89. Smiley, P. L., Johnson, M., Bush, L., Gibb, J. W., and Hanson, G. R., Effects of cocaine on extrapyramidal and limbic dynorphin systems, *J. Pharmacol. Exp. Ther.*, 253, 938, 1990.
90. Spangler, R., Unterwald, E. M., and Kreek, M. J., 'Binge' cocaine administration induces a sustained increase of prodynorphin mRNA in rat caudate-putamen, *Mol. Brain Res.*, 19, 323, 1993.
91. Branch, A. D., Unterwald, E. M., Lee, S. E., and Kreek, M. J., Quantitation of preproenkephalin mRNA levels in brain regions from male Fischer rats following chronic cocaine treatment using a recently developed solution hybridization assay, *Mol. Brain Res.*, 14, 231, 1992.
92. Hellman, W., Suzuki, F., Ohkubo, H., Nakanishi, S., Ludwig, G., and Ganten, D., Angiotensinogen gene expression in extra-hepatic rat tissues: application of a solution hybridization assay, *N.-S. Arch. Pharmacol.*, 338, 327, 1988.
93. Unterwald, E. M., Horne, J., Nyberg, F., Terenius, L., and Kreek, M. J., Regional levels of met-enkephalin-arg^6-phe^7 in rat brain following chronic cocaine administration, *NIDA Res. Monogr.*, 119, 372, 1992.
94. Hurd, Y. L., Brown, E. E., Finlay, J. M., Fibiger, H. C., and Gerfen, C. R., Cocaine self-administration differentially alters mRNA expression of striatal peptides, *Mol. Brain Res.*, 13, 165, 1992.
95. Daunais, J. B., Roberts, D. C. S., and McGinty, J. F., Cocaine self-administration increases preprodynorphin, but not c-fos, mRNA in rat striatum, *Neuroreport*, 4, 543, 1993.
96. Daunais, J. B., Helton, T. E., Bohler, W. T., and McGinty, J. F., Cocaine-induced elevation of dynorphin, fos, and PKC immunoreactivity in rat striatum and hippocampus, *Soc. Neurosci. Abstr.*, 17, 357, 1991.
97. Daunais, J. B., Bohler, W. T., and McGinty, J. F., Acute and chronic cocaine administration: dose response effects of ZIF268, c-fos, preproenkephalin, and preprodynorphin mRNAs in rat brain, *Soc. Neurosci. Abstr.*, 18, 1432, 1992.
98. Hurd, Y. L. and Herkenham, M., Influence of a single injection of cocaine, amphetamine or GBR 12909 on mRNA expression of striatal neuropeptides, *Mol. Brain Res.*, 16, 97, 1992.
99. Cohen, B. M., Van Nguyen, T., and Hyman, S. E., Cocaine-induced changes in gene expression in rat brain, *NIDA Res. Monogr.*, 105, 175, 1991.
100. Chiu, R., Boyle, W. J., Meek, J., Smeal, T., Hunter, T., and Karin, M., The c-fos protein interacts with c-Jun/AP-1 to stimulate the transcription of AP-1 responsive genes, *Cell*, 54, 541, 1988.
101. Hyman, S. E., Comb, M., Lin, Y. S., Pearlberg, J., Green, M. R., and Goodman, H. M., A common Trans-acting factor is involved in transcriptional regulation of neurotransmitter ques by cAMP, *Mol. Cell. Biol.*, 8, 4225, 1988.
102. Comb, M., Mermod, N., Hyman, S. E., Pearlberg, J., Ross, M. E., and Goodman, H. M., Proteins bound at adjacent DNA elements act synergistically to regulate human proenkephalin cAMP inducible transcription, *EMBO J.*, 7, 3793, 1988.
103. Sonnenberg, J. L., Rauscher, F. J., III, Morgan, J. I., and Curran, T., Regulation of proenkephalin by fos and jun, *Science*, 246, 1622, 1989.
104. Naranjo, J. R., Mellstrom, B., Achaval, M., and Sassone-Corsi, P., Molecular pathways of pain: Fos/jun-mediated activation of a noncanonical AP-1 site in the prodynorphin gene, *Neuron*, 6, 607, 1991.
105. Graybiel, A. M., Moratalla, R., and Robertson, H. A., Amphetamine and cocaine induced drug-specific activation of the c-fos gene in striosome-matrix compartments and limbic subdivisions of the striatum, *Proc. Natl. Acad. Sci. USA*, 87, 6912, 1990.

106. Young, S. T., Porrino, L. J., and Iadarola, M. J., Cocaine induces striatal c-fos-immunoreactive proteins via dopaminergic D1 receptors, *Proc. Natl. Acad. Sci. USA,* 88, 1291, 1991.
107. Moldow, R. L. and Fischman, A. J., Cocaine induced section of ACTH, beta-endorphin, and corticosterone, *Peptides,* 8, 819, 1987.
108. Forman, L. J. and Estilow, S., Cocaine influences beta-endorphin levels and release, *Life Sci.,* 43, 309, 1988.
109. Rivier, C. and Vale, W., Cocaine stimulates adrenocorticotropin (ACTH) secretion through a corticotropin-releasing factor (CRF)-mediated mechanism, *Brain Res.,* 422, 403, 1987.
110. Bower, A., Hadley, M. E., and Hruby, V. J., Biogenic amines and control of melanophore stimulating hormone release, *Science,* 184, 70, 1974.
111. Munemura, M., Eskay, R. L., and Kebabian, J. W., Release of alpha-melanocyte-stimulating hormone from dispersed cells of the intermediate lobe of the rat pituitary gland: involvement of catecholamines and adenosine 3',5'-monophosphate, *Endocrinology,* 106, 1795, 1980.
112. Vermes, I., Mulder, G. H., Smelik, P. G., and Tilders, F. J. H., Differential control of beta-endorphin/beta-lipotrophin secretion from anterior and intermediate lobes of the rat pituitary gland in vitro, *Life Sci.,* 27, 1761, 1980.
113. Farah, J. M., Sapun-Malcolm, D., and Mueller, G. P., Dopaminergic inhibition of pituitary beta-endorphin-like immunoreactivity secretion in the rat, *Endocrinology,* 110, 657, 1982.
114. Jackson, S. and Lowrey, P. J., Secretion of pro-opiocortin peptides from isolated perfused rat pars intermedia cells, *Neuroendocrinology,* 37, 248, 1983.
115. Parsons, L. H., Smith, A. A., and Justice, J. B., Jr., Basal extracellular dopamine is decreased in the rat nucleus accumbens during abstinence from chronic cocaine, *Synapse,* 9, 60, 1991.
116. Robertson, M. W., Leslie, C. A., and Bennett, J. P., Jr., Apparent synaptic dopamine deficiency induced by withdrawal from chronic cocaine treatment, *Brain Res.,* 538, 337, 1991.
117. Rossetti, Z. L., Hmaidan, Y., and Gessa, G. L., Marked inhibition of mesolimbic dopamine release: a common feature of ethanol, morphine, cocaine and amphetamine abstinence in rats, *Eur. J. Pharmacol.,* 221, 227, 1992.
118. Maisonneuve, I. M., Ho, A., and Kreek, M. J., Dopaminergic recovery and subsequent response to "relapse" following chronic cocaine "binge", *Soc. Neurosci. Abstr.,* 19, 1849, 1993.
119. Maisonneuve, I. M. and Kreek, M. J., Chronic administration of a cocaine "binge" alters basal extracellular levels in male rats: an in vivo microdialysis study, *J. Pharmacol. Exp. Ther.,* 272, 652, 1995.
120. Imperato, A., Mele, A., Scrocco, M. G., and Puglisi-Allerga, S., Chronic cocaine alters limbic extracelluar dopamine. Neurochemical basis for addiction, *Eur. J. Pharmacol.,* 212, 299, 1992.
121. Sweep, C. G. J., Van Ree, J. M., and Wiegant, V. M., Characterization of beta-endorphin-immunoreactivity in limbic brain structures of rats self-administering heroin or cocaine, *Neuropeptides,* 12, 229, 1988.
122. Sweep, C. G. J., Wiegant, V. M., deVry, J., and Van Ree, J. M., Beta-endorphin in brain limbic structures as neurochemical correlate of psychic dependence on drugs, *Life Sci.,* 44, 1133, 1989.

Chapter 10

Neurochemistry of Cocaine Withdrawal

Friedbert Weiss, Loren H. Parsons, and Athina Markou

CONTENTS

I. INTRODUCTION

Abstinence from chronic cocaine abuse in humans is associated with a variety of psychiatric symptoms such as depression, insomnia, anergia, anxiety, and cocaine craving.[1-5] Although clinical descriptions of the cocaine abstinence syndrome differ as to the time course and persistence of these symptoms,[1-3,6] withdrawal distress is severe at the outset and intense feelings of craving, in particular, are thought to trigger relapse and sustain continued abuse and dependence.

Animal research directed at elucidating the changes in brain function that underlie the cocaine withdrawal syndrome has taken several directions. While the common objective of these efforts has been to examine the neurochemical correlates of cocaine withdrawal, different methodologies and durations of exposure to cocaine prior to withdrawal have been employed. The most widely used approach has been to assess brain neurotransmitter function during a period of abstinence following daily, intermittent cocaine treatments. In an attempt to mimic more closely human cocaine abuse behavior that proceeds in extended "binges", a second approach has utilized continuous administration regimens. This approach has been extended more recently to the self-administration model using animals given unlimited access to intravenous cocaine.

The great majority of investigations of synaptic function following termination of repeated or chronic cocaine treatment have focused on changes in central dopamine and serotonin systems. Cocaine has a similar potency for inhibiting the uptake of dopamine and serotonin[7] and increases the extracellular concentrations of both dopamine and serotonin in the nucleus accumbens, prefrontal cortex, ventral tegmental area, and the dorsal raphe nucleus.[8-10] It has become clear over the past decade that chronic cocaine administration can modify the function and activity of central dopamine and serotonin systems at the synaptic, cellular, and molecular levels. It is the objective of this review to discuss, in particular, the changes in extracellular neurotransmitter concentrations associated with cocaine abstinence. These neurochemical alterations will be examined

1-8493-8311-0/95/$0.00+$.50

further in relation to behavioral signs of withdrawal and their possible role in the cocaine withdrawal syndrome.

II. MONOAMINERGIC FUNCTION DURING WITHDRAWAL FROM CHRONIC COCAINE

A. EFFECTS OF ABSTINENCE AFTER INTERMITTENT EXPOSURE

1. Dopamine

A substantial body of evidence implicates the mesocorticolimbic dopamine system in the mediation of the acute reinforcing effects of cocaine.[11-13] For example, destruction of mesolimbic dopamine neurons by 6-hydroxydopamine abolishes responding for cocaine,[12,13] locally administered selective dopamine D_1 and D_2 antagonists attenuate the reinforcing efficacy of cocaine in self-administering rats,[14-17] and rats given access to intravenous cocaine regulate their drug intake in a manner that elevates and maintains stable extracellular levels of dopamine in the nucleus accumbens.[10,18-21] The critical involvement of dopaminergic neurotransmission in the nucleus accumbens in the acute reinforcing effects of cocaine suggests that this circuitry also may play a role in dependence and withdrawal. A conceptual framework for this putative mechanism was formulated by Dackis and Gold[22] in the form of the "dopamine depletion hypothesis", which stipulates that the dysphoric aspects of cocaine withdrawal are the result of a deficiency in dopamine availability due to excessive synaptic metabolism associated with sustained overstimulation of mesolimbic dopamine neurons during chronic cocaine use.

Most investigations of postcocaine consequences on brain dopamine (and serotonin) systems have employed treatment regimens involving daily bolus injections over periods typically ranging from 1 to 3 weeks. Consistent with an impairment of dopamine function, termination of cocaine administration after such treatments is often associated with decreased levels and activity of the rate-limiting enzyme tyrosine hydroxylase in the nucleus accumbens[23,24] and reduced dopamine synthesis as measured by accumulation of the dopamine precursor 3,4-dihydroxyphenylalanine (DOPA) following inhibition of DOPA decarboxylase.[25-27] In one instance the reduction of dopamine biosynthesis was reported to persist for up to 60 days.[26] However, reduced synthesis is not consistently observed.[28,29] Moreover, the effects of alterations of dopamine synthesis on neurotransmitter stores remain unclear because significant and long lasting postcocaine changes of striatal or accumbal dopamine content are not observed,[29-33] even after aggressive, high-dose or long-term[31] treatment regimens. The discrepancies among these reports of cocaine-induced alterations of dopamine synthesis could be related to differences in experimental procedures such as cocaine dose, treatment duration, and the time elapsed since termination of drug treatments, a possibility that seems to warrant further systematic study in light of evidence that extracellular dopamine levels often vary considerably as a function of such factors. Although the failure to observe any changes in tissue dopamine content in animals that had been exposed to repeated cocaine treatment[29-31,33] is in apparent disagreement with several reports of reduced transmitter synthesis,[25-27] it is possible that small but functionally relevant changes in dopamine content or turnover affect limited or specific pools that are not detected easily by traditional post mortem tissue assays.

Intracranial microdialysis results suggest that extracellular dopamine levels are altered profoundly by chronic cocaine in contrast to the lack of effect on whole tissue dopamine content. While earlier investigations, which were concerned primarily with the effects of repeated cocaine pretreatment on the dopaminergic response to a subsequent cocaine challenge, did not observe significant postcocaine changes in the efflux of dopamine or its metabolites in the nucleus accumbens,[34,35] more recent systematic examinations of basal dopamine function reveal an *elevation* of dopamine efflux or extracellular turnover

Table 1. Apparent Dependence of Changes in Extracellular Dopamine Levels in the Nucleus Accumbens During Cocaine Abstinence on Treatment Duration, Cocaine Dose, and Time of Postcocaine Testing

Cocaine Treatment		Abstinence		
Duration (days)	**Daily Dose**	**Time Postcocaine**	**Effect**	**Reference**
3	20 mg/kg; IP	2 d	increase	Heidbreder &
		10 d	no change	Shippenberg[43]
4	10 mg/kg; IP	1 d	no change	Segal & Kuczenski[35]
5	30 mg/kg; IP	7 d	increase[a]	Kalivas et al.[27]
		20 d	no change	
1-5	2 × 10 mg/kg	12 hr	increase	Imperato et al.[39]
6-9		12 hr	decrease	
9		7 d	decrease	
10	10 or 30 mg/kg; IP	1 d	increase	Weiss et al.[38]
		3 d	increase	
		7 d	no change	
10	20 mg/kg; IP	1 d	increase[a]	Parsons et al.[36,40]
		10 d	decrease	
10-14	30 mg/kg (self-	1 d	increase[b]	Weiss et al.[19]
	administration; IV)	1 d	increase[b]	Parsons et al.[10]
16	2 × 15 mg/kg; IP	1 d	no change	Rosetti et al.[42]
		3 d	decrease	
		5 d	decrease	
		10 d	decrease	
		15 d	no change	
18	30 mg/kg; IP	7 d	decrease	Robertson et al.[41]

[a] Data reflect changes in the "*in vivo* recovery" of dopamine.

[b] The 30 mg/kg dose represents an average dose self-administered in daily sessions with access limited to 3 hr/day.

during the early postcocaine stage,[36,37] followed by a deficiency of extracellular dopamine at a later stage.[19,36,38-43] As summarized in Table 1, the direction of change of extracellular dopamine availability during the postcocaine period appears to depend on the dose, duration, and frequency of cocaine pretreatment as well as on the interval between termination of intermittent drug exposure and testing. Pretreatment with a single daily injection of cocaine or daily limited-access self-administration over 3 to 10 consecutive days significantly increases dopamine level in microdialysates when measured 1 to 3 days after the last cocaine injection.[19,36,38,39,43] Under these conditions dopamine efflux returned to control levels 7 days postcocaine.[38] In contrast, basal extracellular dopamine level in the nucleus accumbens were *decreased* 10 days postcocaine[36] with the same or similar pretreatments that increased dialyate dopamine levels[38,43] or enhanced extracellular turnover[36,37] during the first 3 postcocaine days. Similarly, dialysate dopamine concentration in the ventral striatum measured daily 12 hr after the last of 2 cocaine injections was elevated during the first 4 days of a 9-day treatment period, while deficiencies of basal dopamine efflux emerged after 5 days of exposure to cocaine and persisted when the drug was discontinued.[39] Suppressed basal dopamine efflux has been reported to occur as early as 3[42] and 7[41] days following cessation of longer (16- to 18-day) intermittent treatment periods involving high doses, or two daily injections of cocaine. Together, these findings suggest that daily, intermittent cocaine administration produces biphasic alterations of basal postcocaine dopamine release: an initial adaptive stage of dopaminergic "hyperactivity" characterized by elevated extracellular dopamine concentrations or "turnover"[10,19,27,36,38,40,43] (which is augmented further by a supersensitivity of postsynaptic dopamine D_1 receptors[44]), and a subsequent dopaminergic "deficiency" stage whose onset is accelerated by higher cocaine doses, increased frequency of administration, or longer intermittent treatment durations (Table 1).

Several lines of evidence implicate autoregulatory changes of mesolimbic dopamine neurons as a possible mechanism underlying the biphasic postcocaine profile of basal dopamine efflux in the nucleus accumbens. The increase of basal dopaminergic tone during the early postcocaine stage is associated with an increase of dopamine recovery by *in vivo* microdialysis[9,36,37] (Table 1) suggesting an upregulation of both release and uptake processes period.[9,36,45] The *in vivo* recovery of dopamine can be manipulated by administration of dopamine D_2 receptor agonists and antagonists implicating an interaction between D_2 receptors and transporter activity.[40,46,47] Thus, enhanced *in vivo* recovery during the early stages of cocaine withdrawal is likely to be the result of an alteration of dopamine transporter regulation by the D_2 receptor, resulting in enhanced dopamine uptake. This view is supported by the finding that pretreatment with the D_2 antagonist pimozide prior to daily cocaine injections significantly attenuates the cocaine-induced enhancement of *in vivo* dopamine recovery by microdialysis.[40] The increase of postcocaine dopamine uptake is paralleled by presynaptic alterations indicative of enhanced release processes. Repeated cocaine administration induces a transient subsensitivity of ventral tegmental (A10) somatodendritic, impulse-regulating autoreceptors as evidenced by increases of both the firing rate and the number of spontaneously active ventral tegmental dopamine neurons,[48,49] increases of extracellular dopamine level,[50] and reduced attenuation of *d*-amphetamine-stimulated dopamine release[27] in the ventral tegmental area. The desensitization of somatodendritic autoreceptors appears to involve an N-methyl-D-aspartate (NMDA) receptor-mediated cascade of second-messenger events that results in reduced G protein coupling of the D_2 autoreceptor to the potassium channel.[51-53] This autoreceptor subsensitivity and associated molecular changes such as altered G protein ADP-ribosylation in the ventral tegmental area occur during the early postcocaine stage but are no longer apparent 3 to 4 days after termination of intermittent cocaine treatment.[48,52]

The transient derangement of autoinhibitory control, thus, parallels the time course of postcocaine elevation of basal dopamine efflux from the limbic terminal areas of these midbrain dopamine neurons. Similarly, the number of spontaneously active A10 dopamine neurons is reduced significantly when measured 10 to 14 days after discontinuation of daily cocaine administration[54] corresponding to the time of emergence of basal accumbal dopamine efflux deficiencies.[36,39,41] The decrease of A10 neuronal activity is accompanied by a significant attenuation of potassium-stimulated dopamine efflux from slice preparations of the ventromedial mesencephalon.[55] In contrast to the hyperactivity during the early postcocaine stage, however, the mechanisms underlying the dopaminergic hypoactivity are less well understood and might involve alterations of afferent input[54] or increased somatodendritic dopamine receptor sensitivity although the latter mechanism has been confirmed only in nigral dopamine neurons after a continuous cocaine treatment procedure to date.[56]

2. Serotonin

A growing body of literature indicates a modulatory role for serotonin in the reinforcing effects of psychostimulant drugs[57-63] suggesting that serotonin could be an important substrate for certain aspects of the reinforcing and/or dependence-inducing properties of cocaine. The neuroanatomical connectivity between dopaminergic and serotonergic systems[64,65] as well as electrophysiological[66,67] and neurochemical[68-72] evidence suggest that serotonin might interact with mesocorticolimbic dopamine transmission at both the cell body and the terminal level. Changes of serotonin function produced by repeated cocaine exposure might, therefore, modify dopaminergic neurotransmission and indirectly participate in the cocaine withdrawal syndrome. Moreover, the role of serotonin in psychiatric symptoms of depression,[73] anxiety,[74] and insomnia,[75] suggests that a dysfunction of serotonin neurotransmission during cocaine abstinence could play a *direct* role in these cocaine withdrawal-associated symptoms.

Examinations of the effects of repeated cocaine administration on serotonin function to date have been largely limited to the early (first 48 hr) postcocaine period after intermittent administration. Under these conditions serotonin synthesis in the nucleus accumbens, caudate nucleus, and medial prefrontal cortex appears to remain unaffected as measured by the accumulation of the serotonin precursor 5-hydroxytryptophan following synthesis inhibition with NSD-1015.[28] As with dopamine, repeated cocaine exposure has been reported to inhibit the activity of the rate-limiting enzyme tryptophan hydroxylase, and to decrease serotonin tissue levels and metabolism in the septum and caudate nucleus,[76,77] but reduced tissue levels of serotonin or its metabolite 5-hydroxyindoleacetic acid are not observed consistently in basal forebrain regions.[31,33,78]

In contrast to A10 somatodendritic dopamine autoreceptors that exhibit subsensitivity early after discontinuation of repeated cocaine treatment, 5-HT_{1A} autoreceptors in the dorsal raphe nucleus develop supersensitivity[79-81] although the basal firing rate of dorsal raphe neurons is not affected.[82] The relative density of 5-HT_{1A} receptors is unaltered in several brain regions after repeated cocaine exposure with the exception of the central medial amygdala where the density of 5-HT_{1A} receptors is decreased.[79] In addition, intermittent cocaine treatment results in enhanced density of serotonin uptake sites in several brain areas including the prefrontal cortex and dorsal raphe nucleus.[79] The findings of increased density of serotonin uptake sites and supersensitivity of 5-HT_1 receptors are of interest because they suggest that postcocaine *extracellular* levels of serotonin may be decreased. Recent microdialysis studies have provided tentative support for this hypothesis. Decreased baseline dialysate levels of serotonin were observed in the nucleus accumbens, ventral tegmental area, and dorsal raphe nucleus 24 hr after cessation of 10 days of noncontingent cocaine treatment.[9] Although these reductions only revealed a trend and were not statistically reliable, serotonin levels decreased significantly below baseline levels in rats withdrawn from extended access to intravenous cocaine.[10]

3. Summary and Implications

Repeated exposure to cocaine with intermittent administration regimens produces alteration of both dopamine and serotonin function that persists into or becomes manifest during the postcocaine period. These modifications of serotonergic neurotransmission appear to be confined largely to increases in the density of uptake sites and enhanced sensitivity of the 5-HT_{1A} receptors with little change in the biosynthesis rate and whole-tissue neurotransmitter level. These changes in 5-HT_{1A} receptor sensitivity and transporter density have been implicated in the development of behavioral sensitization to cocaine,[79] but their significance with regard to symptoms of cocaine withdrawal or their relevance for cocaine-seeking behavior still is poorly understood. As pointed out above, postcocaine alterations of serotonergic neurotransmission have been examined only during the early abstinence period. Whether more significant (or biphasic) alterations occur at later periods after termination of intermittent cocaine treatments remains to be determined.

By comparison with serotonin, profound postcocaine changes of mesocorticolimbic dopamine neurotransmission that proceed in a biphasic manner are evident at the extracellular and autoregulatory level. Based on early clinical observations describing the cocaine withdrawal syndrome in humans as a three-stage process proceeding from "crash" to "withdrawal" to "extinction",[2] it has been proposed that the dopaminergic hypoactivity in animal studies corresponds to the late middle phase of human cocaine withdrawal that is characterized by anhedonia, anergia, anxiety, and craving. While this interpretation would be in agreement with the dopamine depletion hypothesis[22] several considerations complicate this issue. Recent descriptions of cocaine withdrawal characteristics in humans[1,3,6] and other clinical findings[83] differ from the three-stage process identified by Gawin and Kleber[2] and place relevant symptoms that may result from dopamine depletion or hypoactivity at a much earlier time than previously thought, particularly in inpatient populations.[1,6] Thus, intense depression and craving were greatest on the first day of withdrawal and these symptoms declined linearly over 4 weeks.[1,6] The sequence of withdrawal symptoms in these clinical studies[1,3,83] is difficult to reconcile with the animal data which have revealed elevated extracellular dopamine levels and enhanced postsynaptic dopamine receptor sensitivity[44] in the mesolimbic circuitry during the early postcocaine stage, and dopaminergic deficiencies at a time when clinical symptoms begin to dissipate. Moreover, short-term, intermittent cocaine administration regimens as employed in most of the animal work do not produce reliable overt symptoms of dependence and withdrawal in pigeons,[84] rats,[37,85-87] and monkeys[88] as measured by disruptions in operant behavior, altered brain stimulation reward thresholds, or depression of spontaneous motor activity. Similarly, there is little evidence that limited, recreational abuse of cocaine ("chipping") produces symptoms of dependence or withdrawal in humans.[2,89-92] It is well known that intermittent cocaine treatments produce strong behavioral sensitization to subsequent psychostimulant challenges (see Parsons et al.[93] for review). However, in view of the above observations, the behavioral significance of both the early hyperactivity and subsequent deficiency in extracellular dopamine per se is unclear and remains to be established.

The lack of reliable behavioral correlates of cocaine withdrawal as well as the discrepancy between the time course of clinical symptoms and the biphasic changes in dopamine function in animal studies make it difficult to evaluate the significance of the alterations of dopamine function associated with intermittent treatment regimens for the cocaine withdrawal syndrome. However, it seems likely that these changes play an important role in enhanced susceptibility to more severe abuse, and ultimately, in the development of dependence. Specifically, the postcocaine dopaminergic hyperactivity has been implicated in the initiation of behavioral sensitization,[48,49,53] cocaine psychosis,[38,49] and cocaine craving.[94] The apparent dopamine "hyperfunction" during the early postcocaine stage seems inconsistent with a dopaminergic involvement in clinical early withdrawal symptoms. Recent findings in cocaine self-administering rats suggest that a relationship between changes in extracellular dopamine levels and behavioral signs of withdrawal can, nonetheless, be demonstrated if dopamine function during withdrawal is examined with respect to adaptations that occur during chronic exposure.[10,19,85]

B. EFFECTS OF WITHDRAWAL FROM SUSTAINED AND CONTINUOUS EXPOSURE TO COCAINE

A further consideration for the evaluation of neurochemical correlates of cocaine withdrawal is that the typical single daily injection procedure may not adequately model the characteristics of high-dose, chronic human cocaine abuse. Cocaine abuse in humans frequently proceeds in repetitive episodes or "binges" lasting from 12 hr to several days during which the drug is readministered continuously.[89] It is believed that this pattern of cocaine abuse, in particular, is associated with severe withdrawal symptoms and high rates of relapse.[2,89] Although plasma cocaine levels fluctuate during the course of "cocaine binges," the pattern of continuous readministration of the drug will maintain rather sustained elevations of plasma cocaine concentrations over extended periods of time, a pharmacological condition that is not accomplished by intermittent treatment designs. Indeed, the literature suggests that when administration regimens are employed that approximate this condition by increasing the frequency of daily injections[95,96] or by continuous subcutaneous or intravenous infusion,[85,97-99] behavioral signs of withdrawal are more readily observed than with single-dose, daily treatments.[37,84,86-88]

Studies on the neurochemical consequences of cocaine withdrawal that incorporate features of this type of human cocaine abuse have employed continuous infusion procedures via osmotic minipumps or intravenous self-administration with unlimited access to the drug. In spite of their relevance as alternative models of cocaine dependence and withdrawal, continuous infusion and unlimited-access self-administration have not been employed as extensively as intermittent procedures and there is, as yet, a dearth of neurochemical data generated by these models.

1. Continuous Administration Models

The most prominent difference between the outcomes of continuous and intermittent administration procedures occurs at the behavioral level in that the former induces tolerance while the latter results in sensitization of motor behaviors in response to cocaine challenge after cessation of chronic treatments.[81,100,101] The behavioral tolerance is apparent both during early (1 day) and intermediate (7 days) stages of cocaine abstinence[81,100] suggesting that in contrast to the biphasic alterations seen with intermittent injections, suppressive effects on mesolimbic dopamine neurotransmission prevail at all stages of cocaine withdrawal after continuous administration.

Some support for this position comes from the finding that continuous administration of cocaine via an osmotic minipump produces tolerance to the inhibition of [^{3}H]dopamine uptake by cocaine in the striatum and nucleus accumbens[102] while daily administration can enhance the effects of cocaine on dopamine uptake.[103,104] Similarly, cocaine-stimulated [^{3}H]dopamine efflux from striatal slices assessed after 7 days of abstinence was reduced significantly in rats pretreated with continuous infusion but increased in animals that received intermittent injections,[105] while no postcocaine changes in amphetamine-stimulated [^{3}H]dopamine release were observed with intermittent cocaine administration.[104] Nevertheless, as with intermittent injections,[29-31,33,106] continuous intravenous infusion of cocaine over 21 days failed to alter dopamine tissue levels in the terminal regions of the mesolimbic and mesostriatal dopamine projections.[31]

Unfortunately, sufficient data mapping the profile of postcocaine changes of dopamine function over time are not available to substantiate the hypothesis that impairments in dopaminergic neurotransmission prevail at all stages of withdrawal after continuous treatment. In fact, while electrophysiological, behavioral, and pharmacological data generated with this model indicate that nigral D_2 dopamine autoreceptors are supersensitive after 7 days of abstinence,[56,107] they are apparently subsensitive to the inhibitory effects of apomorphine on the first day of withdrawal.[56] Although the sensitivity of ventral tegmental dopamine neurons has not been examined in these studies, it appears likely that biphasic autoregulatory alterations occur in this neuronal population during withdrawal from continuous cocaine as with intermittent cocaine treatments. Thus, in spite of the observations of behavioral tolerance to the locomotor-stimulatory effects of cocaine at both early and later withdrawal stages (see Chapter 6),[81,100,101] the available data do not rule out impaired autoinhibitory control of mesolimbic dopamine neurons and elevated extracellular neurotransmitter levels during the early postcocaine stage after continuous administration.

2. Withdrawal Effects of Unlimited-Access Self-Administration

Like many human abusers, laboratory animals including monkeys[108-110] and rats[10,19,85,99] will exhibit "binge" patterns of cocaine self-administration when given unlimited access to cocaine. In humans, such sustained cocaine-taking episodes are followed by a severe withdrawal syndrome with depression, anergia, and craving as prominent symptoms.[1-3,111]

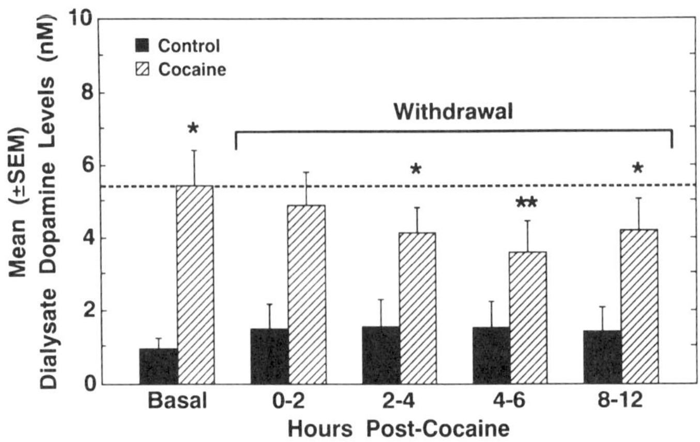

Figure 1. Dopamine concentrations in microdialysates collected from the nucleus accumbens before unlimited access to cocaine (Basal) and during cocaine withdrawal. The duration of self-administration episodes varied across animals (see Figure 2B). Dopamine release was significantly suppressed between 2 and 12 hr postcocaine (* $p \leq 0.05$; ** $p \leq 0.01$). Data are compared to drug-naive rats habituated to the self-administration chamber (Control). Note that the cocaine group had undergone 10 to 15 days of daily 3-hr limited-access self-administration training. The presession basal dopamine efflux in these animals was significantly higher than in drug-naive control rats ($p < 0.02$). Modified from Weiss et al., *Brain Res.*, 593, 314, 1992. With permission.

In animals, termination of access to cocaine after unrestricted intravenous self-administration produces behavioral disruptions believed to be indicative of dependence and withdrawal[85,97,98] In fact, a syndrome of postcocaine "anhedonia" that may model the dysphoric aspects of human cocaine withdrawal has been described in rats.[85,99] When tested for their brain stimulation reward thresholds before and after access to cocaine, such animals showed considerably elevated brain stimulation reward thresholds after 12 to 48 hr of continuous intravenous self-administration. The reward deficits increased in parallel with increases in the duration of the self-administration episode. The reduced reinforcing efficacy of electrical brain stimulation during cocaine withdrawal is thought to reflect a diminished ability to experience rewarding stimulation and seems to involve impairment of dopamine neurotransmission as suggested by the dose-dependent reversibility of this phenomenon by pretreatment with the dopamine D_2 agonist, bromocriptine.[99]

a. Dopamine

Two studies have examined dopamine release in the nucleus accumbens of behaving rats after cessation of self-administration under conditions of unlimited access.[10,19] Withdrawal from cocaine was associated with rapid, marked reductions of basal dopamine release below levels recorded prior to the start of the self-administration episode. Maximal inhibition of dopamine efflux was observed within 2 to 4 hr postcocaine and extracellular dopamine levels remained depressed throughout a 12-hr monitoring period (Figure 1). As with brain stimulation reward,[85] this effect was "dose-dependent" such that the degree of dopamine deficiency increased as a function of the duration of the preceding self-administration episode (Figure 2).

The time course of this dopamine "deficiency" resembles the rapid onset of withdrawal in humans, and in particular, the symptoms of depression and craving that have been reported to be most intense during the initial 24 hr of abstinence.[1,3] Similarly, the onset and duration of the neurochemical changes correspond closely to the behavioral manifestations of withdrawal in animal studies. Maximal attenuation of dopamine release

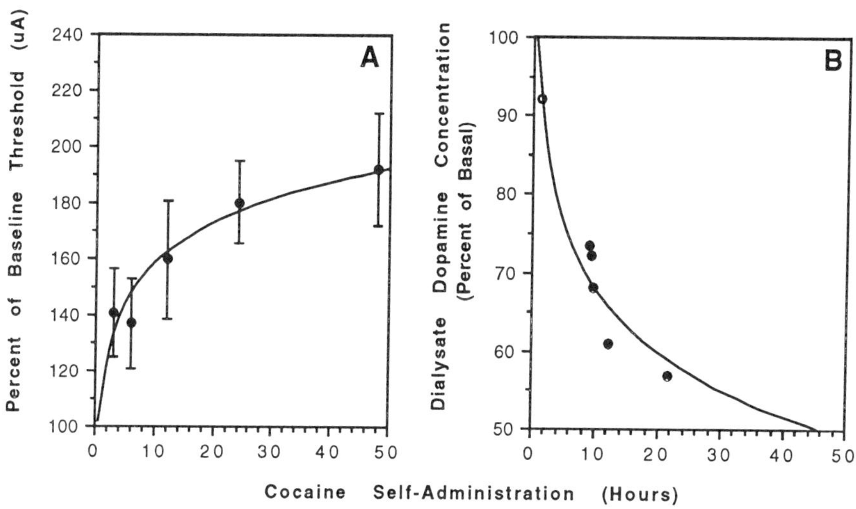

Figure 2. A: Average reward thresholds for lateral hypothalamic brain stimulation obtained between 1 and 6 hr after termination of 3, 6, 12, 24, or 48 hours of intravenous cocaine self-administration. Data are expressed as percent increases over baseline thresholds measured before access to cocaine. **B:** Dopamine concentration in dialysate collected from the nucleus accumbens of individual rats between 4 and 6 hr after cessation of unlimited-access cocaine self-administration. Dopamine concentrations are represented as percent of basal values determined before the start of self-administration sessions. (Open circle: Cocaine exposure terminated after 1 hr. Data not included in Figure 1.) Elevation of intracranial self-stimulation reward threshold and inhibition of basal dopamine release were both significantly correlated with the duration of cocaine self-administration prior to withdrawal. Modified from Markou and Koob, *Neuropsychopharmacology,* 4, 17, 1991 (A) and Weiss et al., *Brain Res.,* 593, 314, 1992 (B). With permission.

and brain stimulation reward[85] both occurred between 1 and 6 hr postcocaine. Also, like the inhibition of basal dopamine release, the elevation of brain stimulation reward thresholds was correlated significantly with the duration of cocaine self-administration prior to withdrawal (Figure 2). Taken together, these data support the hypothesis that some impairment of mesolimbic dopamine neurotransmission underlies the cocaine withdrawal syndrome as measured by intracranial self-stimulation thresholds, and suggest that unlimited-access self-administration provides a promising animal model of cocaine dependence and withdrawal.

Interestingly, both unlimited-access studies revealed a difference in basal dopamine levels between rats previously trained to self-administer cocaine and drug-naive control animals (Figures 1 and 3). Basal dopamine efflux in cocaine-experienced (i.e., self-administering) animals was *elevated* substantially relative to drug-naive control animals before unlimited access to cocaine, and remained at higher levels even during the withdrawal period in spite of the significant reductions with respect to their own presession basal values (Figure 1). This suggests an "adaptation" of mesolimbic dopamine function induced by repeated cocaine exposure during the limited-access self-administration training period. As reviewed above, daily injections of cocaine induce a subsensitivity of A10 somatodendritic dopamine autoreceptors, and an elevation of extracellular dopamine concentration during the early abstinence phase.[10,19,39,49] Thus, the adaptive changes leading to the relative deficiency of extracellular dopamine during cocaine withdrawal in the unlimited-access model seem to be superimposed on a state of functional hyperactivity of mesolimbic dopamine neurons. Whether repetitive cocaine binges would lead to an

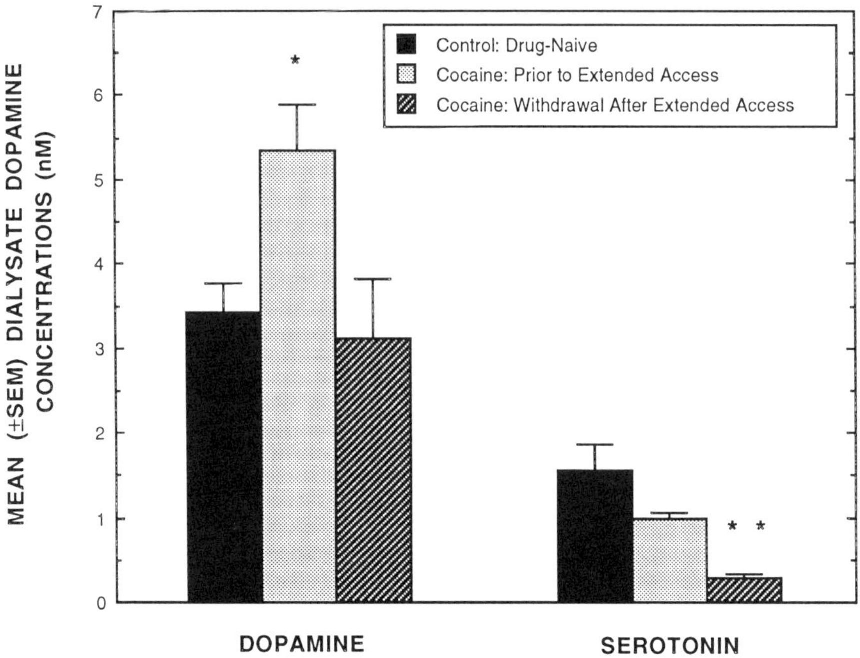

Figure 3. Basal concentration of dopamine and serotonin in dialysates from the nucleus accumbens of rats measured 24 hr after completion of a 10-day cocaine self-administration training period with access limited to 3 hr/day, and 6 hr after termination of an extended (12 hr) self-administration session. As in the experiment shown in Figure 1, daily limited-access to cocaine produced a marked enhancement in basal dopamine efflux relative to drug-naive controls. Cocaine withdrawal after extended access was associated with a significant decrease of dopamine release compared to the level recorded before session onset but not with respect to basal release in drug-naive control animals. In contrast, a decrease of serotonin efflux with respect to drug-naive controls was evident already at the end of the limited-access self-administration training period. Basal serotonin output was further and significantly suppressed during cocaine withdrawal after unlimited access. (* $p \leq 0.05$, significantly different from drug-naive and cocaine withdrawal groups; ** $p \leq 0.05$, significantly different from levels prior to extended access and drug-naive controls.)

eventual reversal of this "hyperactivity" and an absolute, rather than relative decrease in synaptic dopamine levels during the early withdrawal stage is unclear at present. However, this appears unlikely based on the finding that nigral dopamine autoreceptors are subsensitive during the early withdrawal phase, even after 2 weeks of continuous cocaine infusion.[56]

The mechanisms underlying the withdrawal-associated, relative suppression of dopamine release remain largely unclear. Intermittent cocaine treatment does not reduce dopamine tissue content in the nucleus accumbens.[29-31,33] Depletion of intraneuronal storage pools would also seem counterintuitive since basal dopamine efflux during withdrawal is *increased* relative to the level in cocaine-naive animals. Nonetheless, a partial depletion of intraneuronal dopamine after sustained self-administration cannot be ruled out. With a partial depletion, the amount of releasable transmitter might be insufficient to maintain the high basal dopamine output induced by daily cocaine exposure, a condition that would contribute to the relative deficiency in extracellular dopamine during withdrawal. Tentative support for this possibility comes from the finding that intravenous cocaine, administered in a manner mimicking the pattern of self-administration by rats in an extended-access situation enhances tyrosine hydroxylase activity and tyrosine hydroxylase mRNA content in the A9 and A10 cell body regions, but not in their striatal and limbic

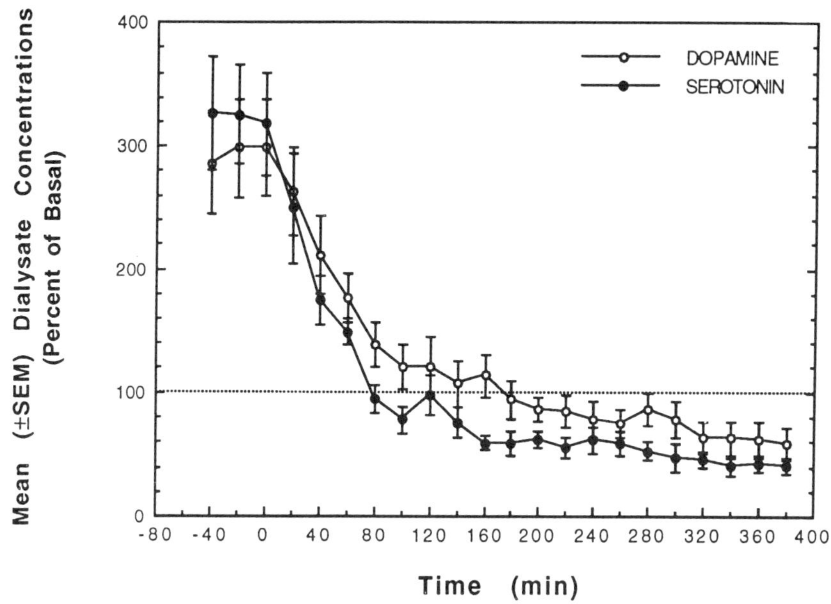

Figure 4. Illustration of the decrease of dopamine and serotonin efflux after termination of self-administration sessions (Time "0") in rats given 12 hr of access to intravenous cocaine. Note the greater and more rapid decrease of serotonin relative to dopamine levels. The concentration of both neurotransmitters reached minimum at approximately 6 hr postcocaine (380 min) and beyond (data not shown). The suppression of dopamine efflux at that time was 40% below baseline levels and is similar to that shown in Figure 2B after a 12-hr self-administration episode. Serotonin level decreased 60% below pre-session basal efflux, and 74% below basal concentration of drug-naive rats.

terminal fields.[112] Based on the finding that neurofilament gene expression is reduced by chronic cocaine[113] one might hypothesize that while the expression of the biosynthetic enzyme is increased in the cell body regions, impairments of axonal transport could prevent increases in dopamine terminal areas. The resulting interference with increased demand for biosynthesis would seem in agreement with the possibility of a partial depletion of releasable transmitter pools in dopamine terminal areas.

An additional or alternative possible mechanism for the suppression of basal dopamine release might be the development of supersensitivity of release-regulating, terminal dopamine D_2 autoreceptors. Increased sensitivity of these receptors could attenuate or retard accumulation of additional extracellular dopamine by cocaine in spite of the concurrent subsensitivity of somatodendritic dopamine autoreceptors. Terminal autoreceptor supersensitivity has been demonstrated in nigral neurons after continuous cocaine infusion, albeit at 1 week postcocaine.[107] Supersensitivity of release regulating autoreceptors also has been implicated as a possible mechanism mediating the reduction of basal dopamine release during the late cocaine abstinence stage after intermittent cocaine[41] although direct evidence for this hypothesis is scarce.[114,115]

b. Serotonin

Withdrawal from cocaine after unlimited-access self-administration also had profound effects on extracellular serotonin concentrations in the nucleus accumbens. While only trends toward a suppression of basal serotonin efflux were apparent in rats after 24 hr of abstinence from intermittent cocaine treatment[9] or daily limited-access self-administration, basal dialysate serotonin concentrations were decreased substantially in animals given 12-hr access to intravenous cocaine (Figure 4).[10] Under these conditions basal serotonin efflux measured by quantitative microdialysis methods was decreased by more than 50% as early as 6 to 10 hr postcocaine.[10] This finding is consistent with the supersensitivity of 5-HT_{1A} receptors and increased density of serotonin uptake sites seen after intermittent cocaine administration,[79-81] but suggests that marked extracellular consequences of these presynaptic changes become evident only after extended periods of exposure to cocaine. Whether the emergence of the extracellular serotonin deficiency also involves some depletion of intraneuronal transmitter stores remains to be determined in appropriate animal models.

The suppression of basal serotonin release greatly exceeded that of dopamine (Figures 3 and 4) and thus implicates a possible prominent role for this neurotransmitter in the cocaine withdrawal symptomatology. Decreased serotonergic transmission is thought to be associated with depression,[73] panic disorder,[116] insomnia,[75] impulsiveness, and aggression.[117,118] Thus, a decrease of extracellular serotonin might contribute directly to many aspects of the cocaine withdrawal syndrome. Indeed, clinical studies suggest that nearly all of these symptoms including depression,[1,3,5] anger and hostility,[1] and insomnia[1,3-5] are present in patients on the first day of withdrawal.

While the deficiency of extracellular serotonin might play a direct role in behavioral manifestations of cocaine withdrawal, it also might exert indirect effects by modifying mesolimbic dopamine neurotransmission. Of interest for this hypothesis is the observation that the decline of extracellular dopamine concentration during cocaine withdrawal was preceded by a drop in serotonin level, which also progressed more rapidly and greatly exceeded that of the catecholamine (Figure 3).[119] Several laboratories have demonstrated that serotonin has "facilitatory" effects on the release of dopamine in the nucleus accumbens after local application[68,71] and can enhance accumbal dopamine release via activation of ventral tegmental[120] or dorsal raphe neurons.[69] These effects are reversible by serotonergic antagonists[68-71] suggesting a receptor-mediated mechanism rather than a transmitter exchange via the dopamine uptake carrier. On the basis of these findings and the temporal profile of decline and degree of postcocaine inhibition of serotonin release, one can speculate that the withdrawal-associated reduction of extracellular dopamine is a consequence of diminished facilitatory serotonergic activity and, consequently, is secondary to an impairment of serotonin release. In support of this possibility is the observation that a substantial sensitization develops during the withdrawal period to the stimulatory effect of serotonin on dopamine release in the nucleus accumbens.[10] The dopaminergic response to locally applied serotonin showed a fivefold increase and appears to represent a compensatory adaptation without which dopamine efflux might show an even greater deficiency. This effect seems specific to cocaine withdrawal after prolonged access, since it did not occur when daily access to cocaine was limited to 3 hr.[10]

In spite of the general support by these data for a role of serotonin in the decrease of basal dopamine efflux during cocaine withdrawal, the interactions among the 2 neurotransmitters are likely to be considerably more complex. In contrast to the excitatory influence on dopamine activity discussed above, inhibitory actions have also been widely reported.[121,122] More importantly, a recent report described 2 populations of dopaminergic neurons in the nucleus accumbens that respond differently to iontophoretically applied serotonin.[119] While a large population of neurons are inhibited by serotonin, a smaller population are stimulated leading to the hypothesis that more than one serotonin receptor-mediated response exists in nucleus accumbens neurons. Alterations in the magnitude of these responses or the proportion of neurons responding with excitation or inhibition following chronic or extended access to cocaine could alter the overall effect of serotonin on dopaminergic output. Thus it is possible, albeit difficult to assess, that serotonin contributes in some fashion both to the overall elevation of dopaminergic tone during the early abstinence phase as well as to the relative suppression of basal dopamine efflux during withdrawal in the unlimited-access model.

3. Summary and Implications

The changes of extracellular levels of dopamine and serotonin after unlimited or prolonged access to intravenous cocaine suggest a role for both transmitters in the cocaine withdrawal syndrome. A marked decrease in extracellular serotonin concentration that was specific to the early withdrawal stage points toward a more important involvement of this neurotransmitter in symptoms of cocaine withdrawal than suggested by data generated with intermittent administration.

With regard to the role of dopamine, the behavioral and neurochemical data obtained with the unlimited-access self-administration model seem to reconcile to a certain extent the apparent discrepancy between the behavioral deficits in clinical and animal studies and the dopaminergic "hyperactivity" during early stages of cocaine abstinence after intermittent administration regimens. The finding that the deficit of brain stimulation reward after extended access to intravenous cocaine[85,99] appears to be associated not with an absolute, but a relative suppression in dopamine efflux that is superimposed upon an overall elevation of basal dopamine output suggests that dopamine-dependent reward systems might adapt in response to sustained enhancement of dopamine activity after prolonged exposure to cocaine, and eventually require increased synaptic dopamine levels for the maintenance of the "normal" hedonic state. In this sense—namely a condition of "relative" suppression of synaptic dopamine availability in an otherwise hyperactive system underlying the withdrawal symptoms of anergia, craving, and depression—the stipulations of the dopamine depletion hypothesis[22] would still seem to hold. Finally, these results illustrate the need to consider adaptations of the "baseline" state of neurotransmitter and biochemical systems that might develop during chronic cocaine exposure for the analysis and evaluation of the behavioral significance of withdrawal-associated alterations in these systems.

REFERENCES

1. Weddington, W. W., Brown, B. S., Haertzen, C. A., Cone, E. J., Dax, E. M., Herning, R. I., and Michaelson, B. S., Changes in mood, craving, and sleep during short-term abstinence reported by male cocaine addicts, *Arch. Gen. Psychiatry,* 47, 861, 1990.
2. Gawin, F. H. and Kleber, H. D., Abstinence symptomatology and psychiatric diagnosis in cocaine abusers, *Arch. Gen. Psychiatry,* 43, 107, 1986.
3. Miller, N. S., Summers, G. L., and Gold, M. S., Cocaine dependence: alcohol and other drug dependence and withdrawal characteristics, *J. Addict. Dis.,* 12, 25, 1993.
4. Kowatch, R. A., Schnoll, S. S., Knisely, J. S., Green, D., and Elswick, R. K., Electroencephalographic sleep and mood during cocaine withdrawal, *J. Addict. Dis.,* 11, 21, 1992.
5. Watson, R., Bakos, L., Compton, P., and Gawin, F., Cocaine use and withdrawal: the effect on sleep and mood, *Am. J. Drug Alcohol Abuse,* 18, 21, 1992.
6. Satel, S. L., Price, L. H., Palumbo, J. M., McDougle, C. J., Krystal, J. H., Gawin, F., Charney, D. S., Heninger, G. R., and Kleber, H. D., Clinical phenomenology and neurobiology of cocaine abstinence: a prospective inpatient study, *Am. J. Psychiatry,* 148, 1712, 1991.
7. Koe, B. K., Molecular geometry of inhibitors of the uptake of catecholamines and serotonin in synaptosomal preparations of the rat brain, *J. Pharmacol. Exp. Ther.,* 199, 649, 1976.
8. Moghaddam, B. and Bunney, B. S., Differential effects of cocaine on extracellular dopamine levels in rat medial prefrontal cortex and nucleus accumbens: comparison to amphetamine, *Synapse,* 4, 156, 1989.
9. Parsons, L. H. and Justice, J. B., Jr., Serotonin and dopamine sensitization in the nucleus accumbens, ventral tegmental area, and dorsal raphe nucleus following repeated cocaine administration, *J. Neurochem.,* 61, 1611, 1993.
10. Parsons, L. H., Koob, G. F., and Weiss, F., Extracellular serotonin and dopamine in the nucleus accumbens following unlimited-access to cocaine self-administration, *Soc. Neurosci. Abstr.,* 19, 1856, 1993.

11. Roberts, D., Corcoran, M. E., and Fibiger, H. C., On the role of ascending catecholaminergic systems in intravenous self-administration of cocaine, *Pharmacol. Biochem. Behav.*, 6, 615, 1977.
12. Roberts, D., Koob, G. F., Klonoff, P., and Fibiger, H. C., Extinction and recovery of cocaine self-administration following 6-hydroxydopamine lesions of the nucleus accumbens, *Pharmacol. Biochem. Behav.*, 12, 781, 1980.
13. Roberts, D. and Koob, G. F., Disruption of cocaine self-administration following 6-hydroxydopamine lesions of the ventral tegmental area in rats, *Pharmacol. Biochem. Behav.*, 17, 901, 1982.
14. Phillips, A. G., Broekkamp, C. L., and Fibiger, H. C., Strategies for studying the neurochemical substrates of drug reinforcement in rodents, *Prog. Neuropsychopharmacol. Biol. Psychiatry*, 7, 585, 1983.
15. Maldonado, R., Robledo, P., Chover, A. J., Caine, B., and Koob, G. F., D_1 Dopamine receptors in the nucleus accumbens modulate cocaine self-administration in the rat, *Pharmacol. Biochem. Behav.*, 45, 239, 1993.
16. MacGregor, A. and Roberts, D. C. S., Dopaminergic antagonism within the nucleus accumbens or the amygdala produces differential effects on intravenous cocaine self-administration under fixed and progressive ratio schedules of reinforcement, *Brain Res.*, 624, 245, 1993.
17. Caine, S. B., Heinrichs, S. C., Williams, C. E., Coffin, V. L., and Koob, G. F., Effects of the dopamine D-1 antagonist SCH23390 microinjected into the accumbens, amygdala, or striatum on cocaine self-administration in the rat, *Brain Res.*, in press, 1995.
18. Weiss, F., Hurd, Y. L., Ungerstedt, U., Markou, A., Plotsky, P. M., and Koob, G. F., Neurochemical correlates of cocaine and ethanol self-administration, in *The Neurobiology of Drug and Alcohol Addiction*, Kalivas, P. W. and Samson, H. H., Eds., *Ann. NY. Acad. Sci.*, Vol. 654, 1992, 220.
19. Weiss, F., Markou, A., Lorang, M. T., and Koob, G. F., Basal extracellular dopamine levels in the nucleus acumbens are decreased during cocaine withdrawal after unlimited-access self-administration, *Brain Res.*, 593, 314, 1992.
20. Pettit, H. O. and Justice, J. B., Jr., Dopamine in the nucleus accumbens during cocaine self-administration as studied by in vivo microdialysis, *Pharmacol. Biochem. Behav.*, 34, 899, 1989.
21. Pettit, H. O. and Justice, J. B., Jr., Effect of dose on cocaine self-administration behavior and dopamine levels in the nucleus accumbens, *Brain Res.*, 539, 94, 1991.
22. Dackis, C. A. and Gold, M. S., New concepts in cocaine addiction: the dopamine depletion hypothesis, *Neurosci. Biobehav. Rev.*, 9, 469, 1985.
23. Trulson, M. E., Joe, J. C., Babb, S., and Raese, J. D., Chronic cocaine administration depletes tyrosin hydroxylase immunoreactivity in the meso-limbic dopamine system in the rat, *Brain Res. Bull.*, 19, 39, 1987.
24. Beitner-Johnson, D. and Nestler, E. J., Morphine and cocaine exert common chronic actions on tyrosine hydroxylase in dopaminergic brain reward regions, *J. Neurochem.*, 57, 344, 1991.
25. Brock, J. W., Ng, J. P., and Justice, J. B., Jr., Effect of chronic cocaine on dopamine synthesis in the nucleus accumbens as determined by microdialysis perfusion with NSD-1015, *Neurosci. Lett.*, 117, 234, 1990.
26. Trulson, M. E. and Ulissey, M. J., Chronic cocaine administration decreases dopamine synthesis rate and increases [^{3}H]spiroperidol binding in rat brain, *Brain Res. Bull.*, 19, 35, 1987.
27. Kalivas, P. W., Duffy, P., DuMars, L. A., and Skinner, C., Behavioral and neurochemical effects of acute and daily cocaine administration in rats, *J. Pharmacol. Exp. Ther.*, 245, 485, 1988.
28. Bauman, M. H., Raley, T. J., Partilla, J. S., and Rothman, R. B., Biosynthesis of dopamine and serotonin in the rat brain after repeated cocaine injections: a microdissection mapping study, *Synapse*, 14, 40, 1993.
29. Hurd, Y. L., Weiss, F., Koob, G. F., Andén, N.-E., and Ungerstedt, U., Reinforcing effects of cocaine are dissociated from dopamine release in rat nucleus accumbens: an in vivo microdialysis study, *Brain Res.*, 489, 199, 1989.
30. Karoum, F., Suddath, R. L., and Wyatt, R. J., Chronic cocaine and rat brain catecholamines: long-term reduction in hypothalamic and frontal cortex metabolism, *Eur. J. Pharmacol.*, 186, 1, 1990.
31. Kleven, M. S., Woolverton, W. L., and Seiden, L. S., Lack of long-term monoamine depletions following repeated or continuous exposure to cocaine, *Brain. Res. Bull.*, 21, 233, 1988.
32. Taylor, D. and Ho, B. T., Effects of short- and long-term treatment with cocaine on rat brain typtophane hydroxylase, *Res. Commun. Chem. Pathol. Pharmacol.*, 15, 805, 1976.
33. Yeh, S. Y. and De Souza, E. B., Lack of neurochemical evidence for neurotoxic effects of repeated cocaine administration in rats on brain monoamine neurons, *Drug Alcohol Depend.*, 27, 51, 1991.

34. Pettit, H. O., Pan, H. T., Parsons, L. H., and Justice, J. B., Jr., Extracellular concentrations of cocaine and dopamine are enhanced during chronic cocaine administration, *J. Neurochem.,* 55, 798, 1990.
35. Segal, D. S. and Kuczenski, R., Repeated cocaine administration induces behavioral sensitization and corresponding decreased extracellular dopamine responses in caudate and accumbens, *Brain Res.,* 577, 351, 1992.
36. Parsons, L. H., Smith, A. D., and Justice, J. B., Jr., Basal extracellular dopamine is decreased in the rat nucleus accumbens during abstinence from chronic cocaine, *Synapse,* 9, 60, 1991.
37. Kalivas, P. W. and Duffy, P., Time course of extracellular dopamine and behavioral sensitization to cocaine. I. Dopamine axon terminals, *J. Neurosci.,* 13, 266, 1993.
38. Weiss, F., Paulus, M. P., Lorang, M. T., and Koob, G. F., Increases in extracellular dopamine in the nucleus accumbens by cocaine are inversely related to basal levels: effects of acute and chronic administration, *J. Neurosci.,* 12, 4372, 1992.
39. Imperato, A., Mele, A., Scrocco, M. G., and Puglisi-Allegra, S., Chronic cocaine alters extracellular dopamine, neurochemical basis for addiction, *Eur. J. Pharmacol.,* 212, 299, 1992. 299-300
40. Parsons, L. H., Schad, C. A., and Justice, J. B., Jr., Co-administration of the D_2 antagonist pimozide inhibits up-regulation of dopamine release and uptake induced by repeated cocaine, *J. Neurochem.,* 60, 376, 1993.
41. Robertson, M. W., Leslie, C. A., and Bennet, J. P., Jr., Apparent synaptic deficiency induced by withdrawal from chronic cocaine treatment, *Brain Res.,* 538, 337, 1991.
42. Rosetti, Z. L., Hmaidan, Y., and Gessa, G. L., Marked inhibition of mesolimbic dopamine release: a common feature of ethanol, morphine, cocaine and amphetamine abstinence in rats, *Eur. J. Pharmacol.,* 221, 227, 1992.
43. Heidbreder, C. and Shippenberg, T. S., High basal dopamine tone in the nucleus accumbens blunts the response profile to repeated cocaine exposure. In: Monitoring Molecules in Neuroscience: Proceedings of the 6th International Conference on In Vivo Methods, Louilot, A., Durkin, T., Spampinato, U., and Cadar, M., Eds., pp. 153-154.
44. Henry, D. J. and White, F. J., Repeated cocaine administration causes persistent enhancement of D1 dopamine receptor sensitivity within the rat nucleus accumbens, *J. Pharmacol. Exp. Ther.,* 258, 882, 1991.
45. Ng, J. P., Hubert, G. W., and Justice, J. B., Jr., Increased stimulated release and uptake of dopamine in the nucleus accumbens after repeated cocaine administration as measured by in vivo voltammetry, *J. Neurochem.,* 56, 1485, 1991.
46. Smith, A. D., Characterization of quantitative microdialysis and its application in studying receptor regulation of the dopamine transporter, Doctoral Dissertation, Emory University, Department of Chemistry, Atlanta, GA, 1994.
47. Meiergerd, S. M., Terrell, A. P., and Schenk, J. O., D_2 receptors may modulate the function of the striatal transporter for dopamine: kinetic evidence from studies in vitro and in vivo, *J. Neurochem.,* 61, 764, 1993.
48. Ackerman, J. M. and White, F. J., A10 somatodendritic dopamine autoreceptor sensitivity following withdrawal from repeated cocaine treatment, *Neurosci. Lett.,* 117, 181, 1990.
49. Henry, D. J., Greene, M. A., and White, F. J., Electrophysiological effects of cocaine in the mesoaccumbens dopamine system: repeated administration, *J. Pharmacol. Exp. Ther.,* 251, 833, 1989.
50. Kalivas, P. W. and Duffy, P., Time course of extracellular dopamine and behavioral sensitization to cocaine. II. Dopamine perikarya, *J. Neurosci.,* 13, 276, 1993.
51. Nestler, E. J., Terwilliger, R. Z., Walker, J. R., Sevarino, K. A., and Duman, R. S., Chronic cocaine treatment decreases levels of the G-protein subunits Gi_a and Go_a in discrete regions of the rats brain, *J. Neurochem.,* 55, 1079, 1990.
52. Striplin, C. D. and Kalivas, P. W., Correlation between behavioral sensitization to cocaine and G-protein ADP-ribosylation in the ventral tegmental area, *Brain Res.,* 579, 181, 1992.
53. Kalivas, P. W., Interactions between dopamine and excitatory amino acids in behavioral sensitization to psychostimulants, *Drug Alcohol Depend.,* in press, 1994.
54. Ackerman, J. M. and White, F., Decreased activity of rat A10 dopamine neurons following withdrawal from repeated cocaine, *Eur. J. Pharmacol.,* 218, 171, 1992.
55. Kalivas, P. W. and Duffy, P., Effects of daily cocaine and morphine treatment on somatodendritic and terminal field dopamine release, *J. Neurochem.,* 50, 1498, 1988.
56. Zhang, H., Lee, T. H., and Ellinwood, E. H., Jr., The progressive changes of neuronal activities of the nigral dopaminergic neurons upon withdrawal from continuous infusion of cocaine, *Brain Res.,* 594, 315, 1992. 315-318

57. Smith, F. L., Yu, D. S., Smith, D. G., Leceese, A. P., and Lyness, W. H., Dietary tryptophan supplements attentuate amphetamine self-administration in the rat, *Pharmacol. Biochem. Behav.*, 25, 849, 1986.
58. Leccese, A. P. and Lyness, W. H., The effects of putative-5-hydroxytryptamine receptor active agents on d-amphetamine self-administration in controls and rats, *Brain Res.*, 303, 153, 1984.
59. Carroll, M. E., Lac, S. T., Asencio, M., and Kragh, R., Fluoxetine reduces intravenous cocaine self-administration in rats, *Pharmacol. Biochem. Behav.*, 35, 237, 1990.
60. Loh, E. A. and Roberts, D., Break-points on a progressive ratio schedule reinforced by intravenous cocaine increase following depletion of forebrain serotonin, *Psychopharmacology*, 101, 262, 1990.
61. Richardson, N. R. and Roberts, D. C. S., Fluoxetine pretreatment reduces breaking points on a progressive ratio schedule reinforced by intravenous cocaine self-administration, *Life Sci.*, 49, 833, 1991.
62. Yu, S. L., Smith, F. L., Smith, D. G., and Lyness, W. H., Fluoxetine-induced attenuation of amphetamine self-administration in rats, *Life Sci.*, 39, 1383, 1986.
63. Carroll, M. E., Lac, S. T., Asencio, M., and Kragh, R., Intravenous cocaine self-administration in rats is reduced by dietary L-tryptophan, *Psychopharmacology*, 100, 293, 1990.
64. Hervé, D., Pickel, V. M., Joh, T. H., and Beaudet, A., Serotonin axon terminals in the ventral tegmental area of the rat: fine structure and synaptic input to dopaminergic neurons, *Brain Res.*, 435, 71, 1987.
65. Vertes, R. P., A PHA-L analysis of ascending projections of the dorsal raphé nucleus in the rat, *J. Comp. Neurol.*, 313, 643, 1991.
66. Sinton, C. M. and Fallon, S. L., Electrophysiological evidence for a functional differentiation between subtypes of of the 5-HT_1 receptor, *Eur. J. Pharmacol.*, 157, 173, 1988.
67. Sorenson, S. M., Humphreys, T. M., and Palfreyman, M. G., Effect of acute and chronic MDL 73,174EF, a 5-HT_3 receptor antagonist, on A9 and A10 dopamine neurons, *Eur. J. Pharmacol.*, 115, 1989.
68. Chen, J., van Praag, H. M., and Gardner, E. L., Activation of 5-HT_3 receptors by 1-phenylbiguanide increases dopamine release in the nucleus accumbens, *Brain. Res.*, 543, 354, 1991.
69. Yoshimoto, K. and McBride, W. J., Regulation of nucleus accumbens dopamine release by the dorsal raphe nucleus in the rat, *Neurochem. Res.*, 17, 401, 1992.
70. Jiang, L. H., Ashby, C. R., Jr., Kasser, R. J., and Wang, R. Y., The effects of intraventricular administration of the 5-HT_3 receptor agonist 2-methylserotonin on the release of dopamine in the nucleus accumbens: an in vivo chronocoulometric study, *Brain Res.*, 513, 156, 1990.
71. Parsons, L. H. and Justice, J. R., Perfusate serotonin increases extracellular dopamine in the nucleus accumbens as measured by in vivo microdialysis, *Brain Res.*, 606, 195, 1993.
72. Sorenson, S. M., Humphreys, T. M., Taylor, V. L., and Schmidt, C. J., 5-HT_2 receptor antagonists reverse amphetamine-induced slowing of dopaminergic neurons by interfering with stimulated dopamine synthesis, *J. Pharmacol. Exp. Ther.*, 260, 872, 1992.
73. Meltzer, H. Y. and Lowy, M. T., The serotonin hypothesis of depression, in *Psychopharmacology: The Third Generation of Progress*, Raven Press, New York, 1987, 513.
74. Charney, D. S., Woods, S. W., Krystal, J. H., and Heninger, G. R., Serotonin function and human anxiety disorders, *Ann. N.Y. Acad. Sci.*, 600, 558, 1990.
75. Nicholson, A. N. and Pascoe, P. A., Studies on the modulation of the sleep-wakefulness continuum in man by fluoxetine, a 5-HT uptake inhibitor, *Neuropharmacology*, 27, 597, 1988.
76. Taylor, D. and Ho, B. T., Neurochemical effects of cocaine following acute and repeated injection, *J. Neurosci. Res.*, 3, 95, 1977.
77. Roy, S. N., Bhattacharyya, A. K., Pradhan, S., and Pradhan, S. N., Behavioral and neurochemical effects of repeated administration of cocaine in rats, *Neuropharmacology*, 17, 559, 1978.
78. Pradhan, S., Roy, S. N., and Pradhan, S. N., Correlation of behavioral and neurochemical effects of acute administration of cocaine in rat, *Life Sci.*, 22, 1737, 1987.
79. Cunningham, K. A. and Paris, J. M., Serotonin neurotransmission in cocaine sensitization, in *The Neurobiology of Drug and Alcohol Addiction*, Kalivas, P. W. and Samson, H. H., Eds., *Ann. NY Acad. Sci.*, 1992, 117.
80. Darmani, N. A., Martin, B. R., and Glennon, R. A., Repeated administration of low doses of cocaine enhances the sensitivity of 5-HT_2 receptor function, *Pharmacol. Biochem. Behav.*, 41, 519, 1992.

81. King, G. R., Joyner, C., Lee, T., and Ellinwood, E. H., Jr., Withdrawal from continuous of intermittent cocaine: effects of NAN-190 on cocaine-induced locomotion, *Pharmacol. Biochem. Behav.*, 44, 253, 1993.
82. Cunningham, K. A., Paris, J. M., and Goeders, N. E., Chronic cocaine enhances serotonin autoregulation and serotonin uptake binding, *Synapse*, 11, 112, 1992.
83. Volkow, N. D., Fowler, J. S., Wolf, A. P., Hitzemann, R., Dewey, S., Bendriem, B., Alpert, R., and Hoff, A., Changes in glucose metabolism in cocaine dependence and withdrawal, *Am. J. Psychiatry*, 148, 621, 1991.
84. Clark, R. D. and Poling, A., Effects of cocaine administration and withdrawal on the performance of pigeons under a fixed-consecutive-number schedule with and without an external discriminative stimulus, *Behav. Pharmacol.*, 491, 1990.
85. Markou, A. and Koob, G. F., Post cocaine anhedonia: an animal model of cocaine withdrawal, *Neuropsychopharmacology*, 4, 17, 1991.
86. Frank, R. A., Martz, S., and Pommering, T., The efect of chronic cocaine on self-stimulation train duration thresholds, *Pharmacol. Biochem. Behav.*, 29, 755, 1988.
87. Frank, R. A., Manderscheid, P. Z., Panicker, S., Williams, H. P., and Kokoris, D., Cocaine euphoria, dysphoria, and tolerance assessed using drug-induced changes in brain-stimulation reward, *Pharmacol. Biochem. Behav.*, 771, 1992.
88. Branch, M. N. and Sizemore, G. M., Behavioral tolerance to cocaine in squirrel monkeys: acute and chronic effects on complex operant behavior, *Pharmacol. Biochem. Behav.*, 30, 737, 1988.
89. Gawin, F. H. and Ellinwood, E. H., Jr., Cocaine and other stimulants. Actions, abuse, and treatment, *N. Engl. J. Med.*, 318, 1173, 1988.
90. Kleber, H. D. and Gawin, F. H., The spectrum of cocaine abuse and its treatment, *J. Clin. Psychiatry*, 45, 18, 1984.
91. Resnick, R. and Schuyten-Resnick, E., Clinical aspects of cocaine: assessment of cocaine abuse behavior in man, in *Cocaine*, CRC Press Inc., Boca Raton, 1977,
92. Van Dyke, C. and Byck, R., Cocaine, *Sci. Amer.*, 28, 108, 1982.
93. Kalivas, P. W. and Stewart, J., Dopamine tranmission in drug- and stress-induced behavioral sensitization, *Brain Res. Rev.*, 16, 223, 1991.
94. Robinson, T. E. and Berridge, K. C., The neural basis of drug craving: an incentive-sensitization theory of addiction, *Brain Res. Rev.*, 18, 247, 1993.
95. Wood, D. M. and Lal, H., Anxiogenic properties of cocaine withdrawal, *Life Sci.*, 41, 1431, 1987.
96. Kokkinidis, L. and McCarter, B. D., Postcocaine depression and sensitization of brain stimulation reward: analysis of reinforcement and performance effects, *Pharmacol. Biochem. Behav.*, 36, 463, 1990.
97. Kleven, M. S. and Woolverton, W. L., Effects of continuous cocaine administration on schedule-controlled behavior in rhesus monkeys, *Behav. Pharmacol.*, 2, 471, 1991.
98. Woolverton, W. L. and Kleven, M. S., Evidence for cocaine dependence in monkeys following a prolonged period of exposure, *Psychopharmacology*, 94, 288, 1988.
99. Markou, A. and Koob, G. F., Bromocriptine reverses the elevation in intracranial self-stimulation reward thresholds observed in a rat model of cocaine withdrawal, *Neuropsychopharmacology*, 7, 213, 1992.
100. King, G. R., Joyner, C., Lee, T., Kuhn, C., and Ellinwood, E. H., Jr., Intermittent and continuous cocaine administration: residual behavioral states during withdrawal, *Pharmacol. Biochem. Behav.*, 43, 243, 1992.
101. Reith, M. E. A., Benuck, M., and Lajtha, A., Cocaine disposition in the brain after continuous or intermittent treatment and locomotor stimulation in mice, *J. Pharmacol. Exp. Ther.*, 243, 281, 1987.
102. Izenwasser, S. and Cox, B. M., Inhibition of dopamine uptake by cocaine and nicotine: tolerance to chronic treatments, *Brain Res.*, 573, 119, 1992.
103. Izenwasser, S. and Cox, B. M., Daily cocaine treatment produces a persistent reduction of [^{3}H]dopamine uptake in rat nucleus accumbens but not in striatum, *Brain Res.*, 531, 303, 1990.
104. Yi, S.-Y. and Johnson, K. J., Effects of acute and chronic administration of cocaine on striatal uptake, compartmentalization and release of [^{3}H] dopamine, *Neuropharmacology*, 29, 475, 1990.
105. King, G. R., Kuhn, C., and Ellinwood, E. H. J., Dopamine efflux during withdrawal from continuous or intermittent cocaine, *Psychopharmacology*, 111, 179, 1993.
106. Kalivas, P. W. and Duffy, P., The effect of acute and daily cocaine treatment on extracellular dopamine in the nucleus accumbens, *Synapse*, 5, 48, 1990.

107. King, G. R., Ellinwood, E. H., Jr., Silvia, C., Joyner, C. M., Xue, Z., Caron, M. G., and Lee, T. H., Withdrawal from continuous or intermittent cocaine; I. Changes in D_2 receptor function, *J. Pharmacol. Exp. Ther.*, in press, 1994.
108. Bozarth, M. A. and Wise, R. A., Toxicity associated with long-term intravenous heroin and cocaine self-administration in the rat, *J. Am. Med. Assoc.*, 254, 81, 1985.
109. Deneau, G. A., Yanagita, T., and Seevers, M. H., Self-administration of psychoactive substances by the monkeys, *Psychopharmacology,* 16, 30, 1969.
110. Johanson, C. E., Balster, R. L., and Bonese, K., Self-administration of psychomotor stimulant drugs: the effects of unlimited access, *Pharmacol. Biochem. Behav.*, 4, 45, 1976.
111. Ellinwood, E. H. and Petrie, W. M., Dependence on amphetamine, cocaine and other stimulants, in *Drug Abuse: Clinical and Basic Aspects,* C.V. Mosby, St. Louis, 1977, 248.
112. Vrana, S. L., Vrana, K. E., Koves, T. R., Smith, J. E., and Dworkin, S. I., Chronic cocaine administration increases CNS tyrosine hydroxylase enzyme activityand mRNA levels and tryptophan hydroxylase enzyme activity levels, *J. Neurochem.*, 61, 2262, 1993.
113. Beitner-Johnson, D., Guitart, X., and Nestler, E., Neurofilament proteins and the mesolimbic dopamine system: common regulation by chronic morphine and chronic cocaine in the rat ventral tegmental area, *J. Neurosci.,* 12, 2165, 1992.
114. Dwoskin, L. P., Peris, J., Yasuda, R. P., Philpott, K., and Zahniser, N. R., Repeated cocaine administration results in supersensitivity of striatal D-2 dopamine autoreceptors to pergolide, *Life Sci.,* 42, 255, 1988.
115. Weiss, F., Hurd, Y. L., Ungerstedt, U., and Koob, G. F., Repeated cocaine administration potentiates the hypomotility effects of apomorphine, *Soc. Neurosci. Abstr.,* 15, 250, 1989.
116. Charney, D. S. and Heninger, G. R., Serotonin function in panic disorders: the effect of intravenous tryptophan in healthy subjects and patients with panic disorder before and during alprazolam treatment, *Arch. Gen. Psychiatry,* 43, 1059, 1986.
117. Soubrie, P., Reconciling the role of central serotonin neurons in human and animal behavior, *Behav. Brain Sci.,* 9, 319, 1986.
118. Coccaro, E. F., Srever, L. J., Klar, H., Maurer, K., Rubenstein, K., Cooper, T. B., Mohs, R. C., and Davis, K. L., Serotonergic studies in affective and personality disorder patients: correlates with behavioral aggression and impulsivity, *Arch. Gen. Psychiatry,* 46, 587, 1989.
119. White, F. J., Hu, X. T., and Henry, D. J., Electrophysiological effects of cocaine in the rat nucleus accumbens: microiontophoretic studies, *J. Pharmacol. Exp. Ther.*, 266, 1075, 1993.
120. Guan, X. M. and McBride, W. J., Serotonin microinfusion into the ventral tegmental area increases accumbens dopamine metabolism, *Brain Res. Bull.*, 23, 541, 1989.
121. Ugedo, L., Grenhoff, J., and Svensson, T. H., Ritanserin, a 5-HT_2 receptor antagonist, activates midbrain dopamine neurons by blocking serotonergic inhibition, *Psychopharmacology*, 98, 45, 1989.
122. Hervé, D., Simon, H., Blanc, G., Kisoprawski, A., LeMoal, M., Glowinski, J., and Tassin, J. P., Increased utilization of dopamine in the nucleus accumbens but not in the cerebral cortex after dorsal raphe lesions in the rat, *Neurosci. Lett.*, 15, 127, 1979.

Chapter 11

Chronic Cocaine Action on the Dopamine Transporter

Nancy R. Zahniser, Greg A. Gerhardt, and Wayne A. Cass

CONTENTS

I. INTRODUCTION

Cocaine exerts its effects in the nervous system by two distinct mechanisms. The more potent effects of cocaine are due to its ability to inhibit neuronal monoamine transporters, whereas the less potent local anesthetic effects are due to the ability of cocaine to block voltage-gated Na^+ channels.[1-6] The monoamine transporters are critically important because they remove dopamine (DA), serotonin (5-hydroxytryptamine; 5-HT) and norepinephrine (NE) from the synaptic cleft and thereby represent the primary mechanisms for terminating the actions of these three neurotransmitters. Na^+ and Cl^- are co-transported back into neurons along with DA, 5-HT, or NE, the electrochemical gradients for these ions providing the energy for monoamine transport. It is now established that the monoamine transporters belong to a large family of Na^+/Cl^--dependent transporters.[7-9] However, molecular cloning also has demonstrated that the DA, 5-HT, and NE transporters are distinct molecules.[10-18] Cocaine inhibits the uptake of [^{3}H]DA, [^{3}H]5-HT and [^{3}H]NE by these transporters expressed in cell lines with affinities in the high nanomolar to low micromolar range. Cocaine inhibits the uptake activity of all three transporters in synaptosomes prepared from appropriate rat brain regions with similar high nanomolar affinities.[19]

Therefore, cocaine potentially can cause effects due to increased synaptic concentrations of DA, 5-HT, and NE. However, there are a number of reasons that most studies to date have focused on the interaction of cocaine with brain DA systems. The psychomotor stimulant effects produced by drugs such as amphetamine and cocaine are attributed largely to their ability to increase synaptic DA concentrations in the limbic forebrain and

1-8493-8311-0/95/$0.00+$.50

basal ganglia.[20-22] It also has been established that the mesocorticolimbic DA system plays a critical role in reward and drug reinforcement.[23,24] Furthermore, the powerful reinforcing properties of cocaine have been shown to be related to its ability to inhibit the DA transporter.[25]

When cocaine is administered chronically, it can produce behavioral sensitization (see Chapter 6) and/or tolerance (see Chapter 3).[19-21] Locomotor activity and stereotyped behaviors, such as head bobbing and rearing, are typical rodent responses to psychomotor stimulants such as cocaine. With repeated, intermittent administration of a constant dose of cocaine, the magnitude of these elicited behaviors often becomes augmented. This phenomenon is known as "behavioral sensitization". One hallmark of behavioral sensitization is that it is long lasting. Augmented locomotor and stereotyped behaviors can be elicited by a challenge dose of cocaine weeks to months after the repeated drug administration has been discontinued. These characteristics are in contrast to those associated with the diminished locomotor and stereotyped behaviors typical of behavioral tolerance. Tolerance is observed during continuous administration of cocaine, such as with chronic infusions or pellet implantations. It is also apparent in response to a cocaine challenge for a few days to a week after continuous administration has been discontinued; however, tolerance usually dissipates relatively rapidly thereafter.

A number of possible mechanisms involving changes in DA systems have been suggested as the basis for stimulant-induced sensitization and tolerance.[31-34] In this regard, studies of chronic amphetamine administration have provided much insight into likely mechanisms. Although some regulation of DA receptors occurs with chronic psychomotor stimulant administration, few of these changes are persistent enough to explain sensitization. The most consistent observation with chronic amphetamine administration has been an alteration of extracellular brain DA concentration.[35-42] The results of studies of chronic cocaine administration (discussed below) are in agreement with this observation. With most psychomotor stimulants, there is no evidence that pharmacokinetic changes account for behavioral sensitization. However, unique to cocaine, the question of changes in drug disposition has arisen. Many studies have administered cocaine repeatedly by the intraperitoneal (i.p.) route. Increased brain levels of cocaine occur in response to this treatment paradigm.[43-47] However, we recently have found that repeated intravenous (i.v.) cocaine administration produces behavioral sensitization in the absence of any changes of brain cocaine levels.[48] This observation, along with those of Justice and colleagues,[45,46] suggests that the increased brain cocaine levels observed following repeated i.p. cocaine administration might contribute to, but certainly cannot completely account for, behavioral sensitization. Thus, the increased extracellular DA concentrations likely reflect a pharmacodynamic change. The fact that cocaine binds to the DA transporter and transiently inhibits its activity raises the possibilities that (1) the DA transporter is regulated in response to chronic cocaine administration, and (2) this contributes to the different behavioral changes observed in response to intermittent or continuous chronic cocaine exposure. This chapter will review the current evidence for and against these possibilities.

II. INTERMITTENT COCAINE ADMINISTRATION AND REGULATION OF THE DOPAMINE TRANSPORTER

A. RELEVANCE OF SENSITIZATION TO MENTAL HEALTH AND DRUG ABUSE

Chronic use of psychomotor stimulants such as amphetamine and cocaine by humans can lead to progressive increases in paranoia and anxiety culminating in paranoid psychosis or panic disorder.[49] These effects can be severe and long lasting. Persons who have abused psychomotor stimulants remain vulnerable to reactivation of stimulant-induced paranoia and panic attacks even after years of abstinence. The augmentation of behavioral responsiveness and the ability to manifest these behavioral changes with stimulant drug administration long after discontinuation of the repeated drug use are both reminiscent of behavioral sensitization. Thus, there is considerable interest in elucidating the mechanism(s) underlying the persistent nature of behavioral sensitization in animals because this may help in understanding the bases for both drug-induced and idiopathic psychoses in humans.[33] Sensitization also may play a role in drug addiction by increasing the strength of the drug-related associations and the risk of subsequent addiction.[50] Preexposure of rats to cocaine sensitizes them to the rewarding effects of the drug.[51] Furthermore, it has been suggested that sensitization to the reinforcing effects of psychomotor stimulants may account for the relative ease with which addicts relapse into drug abuse after a period of abstinence.[52]

B. ANIMAL STUDIES

1. *In Vivo* Extracellular Dopamine Concentrations

Several studies using *in vivo* microdialysis have shown that extracellular DA concentrations are enhanced to a greater extent in response to a cocaine challenge in the striatum and nucleus accumbens (NAc) of rats that have been treated repeatedly with cocaine than with saline.[53-58] Importantly, this relative increase of extracellular DA concentration occurs concurrently with an enhanced behavioral responsiveness characteristic of sensitization. However, some studies have observed decreased, rather than increased, levels of extracellular DA in rat brain in response to a cocaine challenge.[58-61] In most of these latter studies, rats were sensitized behaviorally but were withdrawn from repeated cocaine administration for a relatively short period (1 to 2 days) before the challenge dose was administered. Kalivas and Duffy have suggested that greater augmentation of extracellular DA is consistently observed provided that the challenge dose is administered after a relatively long withdrawal period (>4 days).[58] Furthermore, the longer the drug treatment period and the higher the dose administered, the longer the withdrawal period required. Nonetheless, the fact that behavioral sensitization is observed following a short withdrawal period dissociates the changes in extracellular DA concentration and behavior during the period immediately following discontinuation of repeated cocaine administration. During withdrawal from repeated cocaine administration, basal extracellular DA concentration measured in the absence of a challenge is depressed.[57,62-64] Similarly, neuronal activity and responsivity of DA systems appear to be blunted.[65,66] Thus, the augmentation of extracellular DA in response to a cocaine challenge represents a long-lasting change which is not observed at short withdrawal times. Increased extracellular DA concentration could be due to augmented release and/or diminished DA uptake. However, the fact that cocaine challenge following repeated administration can produce augmented dopamine response suggests that the ability of cocaine to antagonize the DA transporter has been altered. This is consistent with the idea that DA transporter function is regulated in response to repeated cocaine administration, which contributes to the persistence of the behavioral changes.

2. *In Vitro* Dopamine Release

Local changes in the DA terminal region likely contribute to the increased extracellular DA concentration observed *in vivo* following withdrawal from repeated cocaine administration because DA release also is increased in *in vitro* slice preparations from the striatum and NAc. When cocaine is included in the superfusion buffer to inhibit the DA transporter, increased K^+-stimulated release of endogenous DA is observed in slices taken from DA terminal regions (e.g., the striatum and NAc), but not from cell body regions.[67] Amphetamine-stimulated [^{3}H]DA release from striatal slices also is increased following repeated cocaine administration; this increase is not apparent until days after repeated drug administration has terminated.[68,69] Amphetamine induces DA release via reversal of the DA transporter.[70,71] Thus, these findings further support the suggestion that repeated cocaine administration regulates the activity of the DA transporter.

3. Dopamine Transporter

Cocaine increases extracellular DA by blocking the neuronal DA transporter. Therefore, repeated cocaine administration could produce an augmented extracellular DA response because (1) there are fewer DA transporters for cocaine to inhibit and/or (2) the affinity of the DA transporter for cocaine is increased. The first suggestion comes from the observation that cocaine produces more potent *in vivo* effects in brain regions in which there are fewer DA transporters to inhibit (i.e., NAc vs. dorsal striatum).[72,73] The limits of this dose-response relationship should be noted, however. For example, cocaine is less effective in frontal cortex than in NAc, suggesting that when the number of DA transporters is extremely low, such as in frontal cortex, the effectiveness of cocaine may be diminished.[74] Most studies have examined either *in vitro* [^{3}H]DA uptake or radioligand binding assays to assess the state of the DA transporter. In uptake assays, the maximal velocity (V_{max}) and the apparent substrate affinity (K_m), as well as the affinity for cocaine inhibition (K_i), can be determined. In binding assays, the number of transporters (B_{max}), the binding affinity of the radioligand (K_d), and the binding affinity of cocaine (K_i) can be determined. The duration of cocaine treatment and withdrawal, the dose of cocaine administered, the administration of a challenge dose following withdrawal, and the brain region(s) investigated are all potentially important variables to take into account when comparing results from different studies.

Despite varied treatment protocols, the results of experiments which examined tissues from striatum and cell body regions have been generally in agreement. Most often, no changes in striatal [^{3}H]DA uptake or DA transporter binding have been observed following repeated cocaine administration.[69,75-86] These studies were conducted using several species including rat, mouse, and rabbit. In the cell body regions (i.e., the substantia nigra (SN) and ventral tegmental area (VTA)), [^{3}H]DA uptake and DA transporter binding were also unchanged following repeated cocaine administration in rats.[78,79,84] However, an increased V_{max} for [^{3}H]DA uptake was observed in rat striatum when a challenge dose of cocaine was administered after a long withdrawal period.[87] Similarly, following a long treatment period (1 to 3 weeks) and a short withdrawal period (12 hr), the densities of [^{3}H]BTCP and [^{3}H]cocaine binding sites were increased in rat striatum.[88,89] In rhesus monkey striatum, but not in NAc or prefrontal cortex, a selective, persistent down-regulation of DA transporter binding sites ([^{3}H]GBR 12935) has been observed.[90] In this latter study, a relatively high dose of cocaine (3 to 4 mg/kg, i.m., 4 times daily) was administered for 2 weeks and tissues were examined 2 weeks later. Decreased *in vivo* [^{3}H]WIN 35,428 binding in rat striatum also has been observed 1 and 10 days after repeated i.v., but not i.p., cocaine administration.[91] Thus, some limited evidence supports the presence of fewer DA transporters in the striatum and in cell body regions following repeated cocaine administration.

In limbic regions of rat brain, changes induced by repeated cocaine administration have been more evident. Following a short withdrawal period, [^{3}H]DA uptake V_{max} and DA transporter B_{max} are increased in rat NAc.[75,85] Similar results have been observed in NAc using electrical stimulation and *in vivo* voltammetry to measure the clearance rate of endogenous DA.[92] These effects did not persist upon longer withdrawal. However, other studies have observed no alteration of binding to the DA transporter in either NAc or frontal cortex shortly after cessation of repeated cocaine administration.[78,82] We also observed no change of [^{3}H]mazindol binding in the NAc after a long withdrawal period;

however, a cocaine challenge was administered shortly before the *in vitro* assay was conducted, perhaps making it more similar to a brief withdrawal study.[81] In contrast, following relatively long periods of withdrawal (10 days to 12 weeks), selective downregulation of DA transporter binding sites has been observed in both rat NAc and frontal cortex,[78,82,86,91] and the dependence of these results on the length of the withdrawal period has been clearly documented in these studies. Reductions in [^{3}H]DA uptake, due to decreased transporter affinity, also have been observed in the NAc 1 day after 3 days of cocaine administration.[77] However, one study reported no change of DA transporter binding sites in the NAc following a long withdrawal period.[79] In this study a higher dose of cocaine (35 mg/kg, i.p.) was administered than in any of the other studies. The presence of an inverted U-shaped dose-response curve for the ability of repeated cocaine administration to induce changes of the striatal DA transporter[88] could underlie these results.

Interestingly, DA transporter gene expression in rat SN, measured with a nuclease protection assay, is downregulated only after short withdrawal periods following repeated cocaine administration.[93] However, a long-lasting decrease of DA transporter mRNA has been observed using *in situ* hybridization histochemistry in the VTA, but not in the SN, 10 days after cessation of repeated cocaine administration,[94] which is consistent with earlier reports of a long-lasting downregulation of DA transporter binding in the NAc.[78,86,91] In summary, repeated intermittent cocaine administration causes a transient increase of DA uptake and DA transporter binding shortly after cessation of treatment. However, following relatively long withdrawal periods, DA uptake and DA transporter binding are reduced. Regulation of the DA transporter appears more likely to occur in limbic regions of rat brain. These findings generally are consistent with those from *in vivo* microdialysis studies. Furthermore, they suggest that fewer DA transporters may be available in the NAc and frontal cortex for cocaine to inhibit, thus contributing to increased synaptic concentrations of DA and behavioral sensitization.

The second possible explanation for augmented extracellular DA is that the affinity of the DA transporter for cocaine is increased following repeated treatment. Izenwasser and Cox reported that cocaine was a more potent inhibitor of [^{3}H]DA uptake in NAc following withdrawal from repeated cocaine administration.[49] However, we observed that the affinity of cocaine was unchanged when measured against [^{3}H]mazindol binding in either the striatum or NAc following withdrawal from repeated cocaine administration and subsequent cocaine challenge.[81]

In order to examine further the possible functional changes in the DA transporter in the intact animal, we have used a novel *in vivo* voltammetric approach. A micropipette was glued in place approximately 300 μm distant from a calibrated electrochemical electrode, and the electrode-pipette assembly was lowered stereotaxically into the dorsal striatum and NAc of an anesthetized rat. Calibrated amounts of DA were pressure-ejected from these assemblies once every 5 min; the amount of DA ejected was adjusted so that the signal amplitudes in the two brain regions would be approximately equal (Figure 1). This amount was always higher (+33 to 100%) in the dorsal striatum than in the NAc. DA clearance was examined by analyzing the disappearance of this locally applied, exogenous DA. This approach eliminates any contribution from the release of endogenous DA and confines the analysis to those processes responsible for clearing DA from the extracellular milieu. In regions containing DA neurons, such as the striatum and NAc, neither metabolism nor diffusion appeared to contribute significantly to the disappearance of the exogenous DA.[95] Therefore, DA clearance must reflect the functional activity of the DA transporter. Furthermore, by monitoring and controlling the amount of DA locally applied in the brain, this technique allows for direct comparison between DA clearance in different brain regions. Using this approach, we observed that both local administration and acute systemic administration of cocaine affected *in vivo* DA clearance in the NAc

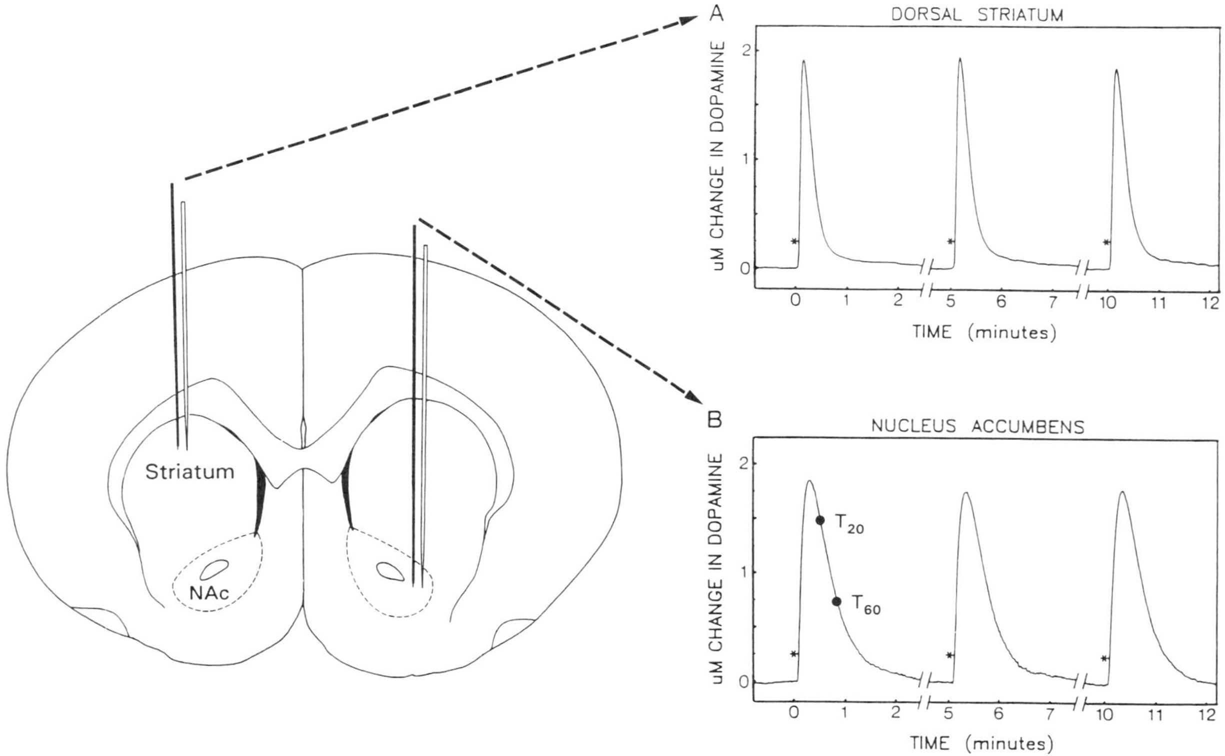

Figure 1. Representative placements of the *in vivo* electrochemical electrode/micropipette assemblies in the rat dorsal striatum and NAc and representative signals recorded in response to local application of exogenous DA in these two brain regions. The coronal brain section illustrated is located at a level approximately 1.5 mm anterior to bregma. The graphs show simultaneous recordings made in the dorsal striatum (A; 7.4 pmoles of DA ejected/application) and NAc (B; 5.0 pmoles of DA ejected/application) of a urethane-anesthetized rat to three consecutive applications of DA (*) pressure-ejected at 5-min intervals. Only the oxidation signals, which were converted directly to micromolar concentration using calibration factors determined in *vitro,* are shown. Also shown are the T_{20} and T_{60} time points used to calculate clearance rate. Adapted from Cass, W. A., Gerhardt, G. A., Mayfield, R. D., Curella, P. and Zahniser, N. R., *J. Neurochem.,* 59, 259, 1992; Cass, W. A., Gerhardt, G. A., Gillespie, K., Curella, P., Mayfield, R. D., and Zahniser, N. R., *J. Neurochem.,* 61, 273, 1993; Paxinos, G. and Watson, C., The Rat Brain in

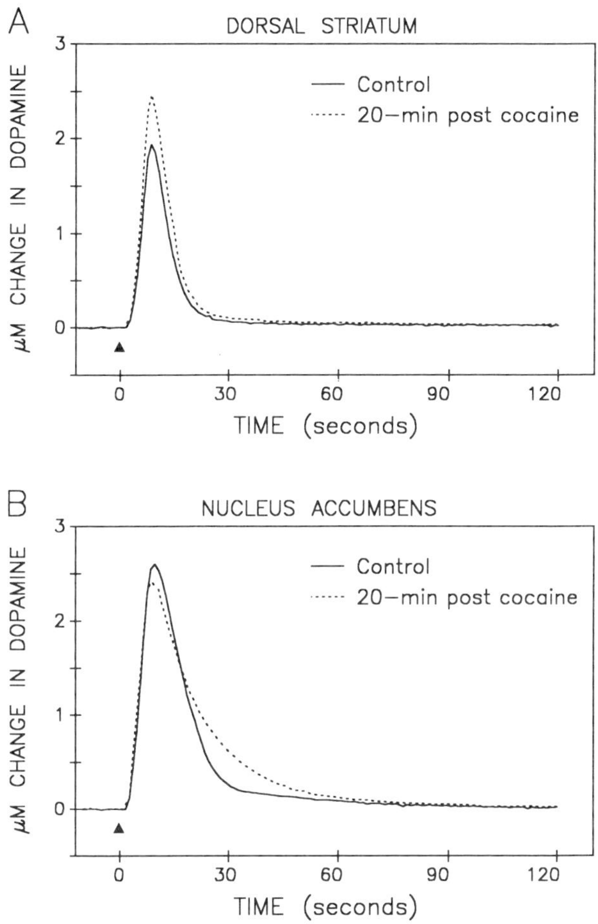

Figure 2. Representative data summarizing the changes observed in *in vivo* electrochemical signals induced by a cocaine challenge following withdrawal from repeated cocaine administration. The signals were recorded simultaneously in the rat dorsal striatum (A) and NAc (B) in response to local application of exogenous DA before (control) and 20 min after (20-min postcocaine) a cocaine challenge (10 mg/kg, i.p.). Arrowheads indicate the point at which DA was applied. The rat had been treated with cocaine (10 mg/kg, i.p.) once daily for 1 week and then was drug free for another week. In the dorsal striatum, cocaine challenge increased the amplitude of the signal, had no significant effect on the time course of the signal and increased the clearance rate of DA (slope of the line between the T_{20} and T_{60} time points indicated in Figure 1). In contrast in the NAc, cocaine challenge produced no significant effect on the amplitude of the signal, increased the time course of the signal and decreased the clearance rate of DA. (Adapted from Cass, W. A., Gerhardt, G. A., Gillespie, K., Curella, P., Mayfield, R. D., and Zahniser, N. R., *J. Neurochem.*, 61, 273, 1993. With permission.)

at a lower cocaine concentration/dose than in the dorsal striatum.[73,95] These results suggest that the NAc is more sensitive to the effects of DA uptake blockers such as cocaine. This observation may be explained by the fact that the density of DA transporter binding sites in the NAc is 60% lower than in the dorsal striatum.[73,96,97]

Following 1 week of withdrawal from repeated cocaine administration (10 mg/kg i.p. once daily for 7 days), a 10 mg/kg cocaine challenge increased the amplitude, but not the time course, of the DA signal in the dorsal striatum (Figure 2). In contrast, the signal

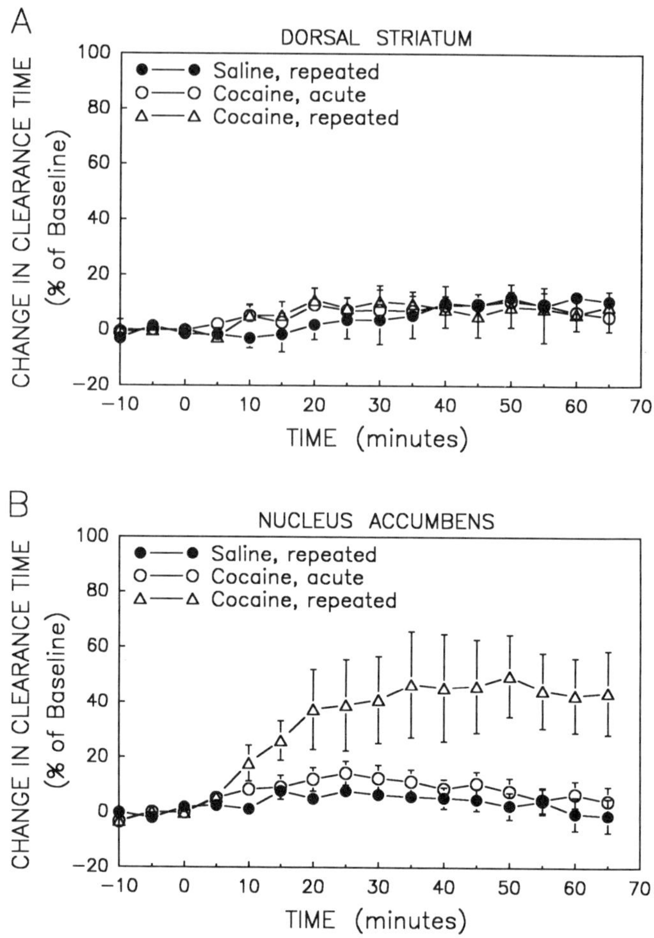

Figure 3. Increased DA clearance time in the rat NAc (B), but not dorsal striatum (A), in response to a cocaine challenge following withdrawal from repeated cocaine administration. The repeated saline group received once-daily saline injections (i.p.) for one week and then was untreated for another week prior to receiving a cocaine injection at time zero. The acute cocaine group was untreated prior to receiving a cocaine injection at time zero. The repeated cocaine group received once-daily cocaine injections for 1 week and then was drug free for another week before receiving a cocaine injection at time zero. All of the cocaine injections were administered at a dose of 10 mg/kg (i.p.). Clearance times were calculated as the time it took for each electrochemical oxidation signal to decay by 80% from its maximal value and are expressed as the percentage of change from baseline, where baseline was defined as the average clearance time at the –5 and 0 time points. Mean values ± SEM are shown for N = 4-6 rats. Two-factor ANOVA with repeated measures for the time factor was used to analyze the data. Dorsal striatum, all three groups ($F_{2,12} = 0.02$, $p \geq 0.05$). NAc, all three groups ($F_{2,12} = 4.16$, $p \leq 0.05$). Further analyses indicated that the repeated cocaine group was significantly different from both of the other groups ($p \leq 0.05$).

amplitude was unchanged in the NAc, whereas the time course increased (Figure 2). Therefore, more time was required to clear DA from the extracellular space in response to a cocaine challenge selectively in the NAc, suggesting that cocaine was either more efficacious or more potent in this brain region (Figure 3). The diminished DA clearance in the NAc in response to a 10 mg/kg cocaine challenge was similar to that observed in response to an acute injection of 20 mg/kg cocaine.[81] No change was observed in the NAc of rats withdrawn from repeated saline treatment or rats treated acutely with the same dose of cocaine (Figure 3). This change in DA clearance may not be sufficient to produce behavioral sensitization because it was observed in the NAc of all rats treated repeatedly with cocaine in response to a cocaine challenge, whereas not all of these rats exhibited behavioral sensitization.[81] In a preliminary report using different mouse lines, Miller and colleagues also suggest that changes in the DA transporter and behavioral sensitization may not be directly related.[98] Nonetheless, the preferential change in the ability of cocaine to inhibit the DA transporter in the NAc likely contributes to the increased behavioral responsiveness observed in response to a cocaine challenge. These results also support the idea that repeated cocaine administration induces long-lasting changes in the interactions of cocaine with the DA transporter and its subsequent activity.

III. CONTINUOUS COCAINE ADMINISTRATION AND REGULATION OF THE DOPAMINE TRANSPORTER

Unlike repeated intermittent administration, chronic infusion of cocaine leads to tolerance of behaviors such as cocaine-induced locomotor activity and stereotypy.[44,99,100] The results of two studies suggest that, consistent with this behavioral tolerance, the ability of cocaine to inhibit DA uptake *in vitro* is blunted following chronic infusion of cocaine. The first study showed that [^{3}H] DA uptake is unaltered, whereas the ability of cocaine to inhibit [^{3}H]DA uptake is diminished in both the rat NAc and striatum following chronic cocaine infusion.[101] These results suggest that the cocaine binding site, but not the functional activity of the DA transporter, is altered by chronic cocaine infusion. It is interesting to note that, as previously discussed, this group observed an augmented cocaine response that was restricted to the NAc following repeated intermittent cocaine administration.[77] The second report showed that chronic cocaine infusion diminished the ability of cocaine to increase the spontaneous efflux of endogenous DA from rat striatal slices.[102] In this case, the effect was shown to be long-lasting; rats were drug-free for 1 week prior *in vitro* assay. Additionally, K^+-evoked [^{3}H]DA release is decreased selectively in NAc slices taken from rats that had been infused with cocaine for 1 to 2 weeks.[103] These results suggest that behavioral tolerance may be associated with the diminished ability of cocaine to inhibit the DA transporter, as well as changes in stimulation-evoked DA release.

IV. COCAINE REGULATION OF THE HUMAN DOPAMINE TRANSPORTER

Two recent studies have reported increased numbers of DA transporter binding sites in postmortem brain samples from subjects who had histories of chronic cocaine abuse. In the first study, an increased number of [^{3}H]WIN 35,428 binding sites was observed in the caudate, putamen, and NAc of chronic cocaine abusers compared to control subjects.[104] All cocaine abusers had detectable levels of cocaine present in their brains at the time of death. In the second study, the number of [^{3}H]WIN 35,428 binding sites increased in the putamen of subjects who had chronically used cocaine and died from a cocaine overdose.[105] These results are consistent with those from animal studies showing increased DA transporter binding sites when there is little or no withdrawal period from repeated intermittent cocaine exposure. However, another study reported a maximal decrease of DA transporter binding sites, in the caudate/putamen of persons with a history of cocaine abuse and measurable cocaine levels at the time of death.[106] Again, these observations support the idea that chronic cocaine exposure can regulate the DA transporter. The circumstances under which up- or downregulation occurs still remain to be defined. However, all of these human studies concur that the DA transporter is regulated in basal ganglia, as well as in limbic regions.

V. COCAINE REGULATION OF THE SEROTONIN AND NOREPINEPHRINE TRANSPORTERS

Several recent studies also suggest that the 5-HT transporter may be regulated by repeated, intermittent cocaine administration. Parsons and Justice observed a greater augmentation of the extracellular concentration of 5-HT in the NAc, VTA, and dorsal raphe nucleus (DRN) in response to a cocaine challenge 1 day after 10 days of repeated cocaine treatment.[107] Interestingly, the 5-HT response was more sensitized in the cell body region (DRN) than in the terminal field (NAc). In the same animals, extracellular DA increased to a greater extent in response to cocaine challenge in the NAc and DRN than in the VTA. Thus, the opposite was true for the mesolimbic DA system (i.e., the terminal field was relatively more sensitized than the cell body region). The greater increase of 5-HT in the VTA and DRN induced by repeated cocaine administration might contribute to the augmented release of DA in the NAc.[107] Cunningham reported that cocaine sensitization is associated with increased [^{3}H]imipramine binding to the 5-HT transporter in the DRN and medial prefrontal cortex, but not in the NAc or striatum (see Chapter 8).[108] The changes in the DRN, but not in the NAc, might reflect the relatively greater increase of extracellular 5-HT in the cell body region as reported by Parsons and Justice. These changes were measured within one day after termination of repeated cocaine administration, so it is not known whether or not they are persistent. However, Benmansour and colleagues observed no evidence for persistent regulation of either the 5-HT transporter (as measured by [^{3}H]cyanoimipramine binding) or the NE transporter [as measured by [^{3}H]nisoxetine binding 1 week after 10 days of high dose cocaine treatment (35 mg/kg, i.p.)].[79] Also, chronic cocaine infusion (without a withdrawal period) does not alter [^{3}H]5-HT uptake or cocaine inhibition of uptake in the rat striatum or NAc.[101]

VI. FUTURE DIRECTIONS

A. RELATIONSHIP BETWEEN TRANSPORTER REGULATION AND BEHAVIORAL CHANGES

Evaluation of the functional significance of cocaine-induced transporter regulation and identification of the molecular mechanisms involved in this regulation represent two important avenues for future research. It is now feasible to carry out *in vivo* electrochemical recording in unanesthetized, behaving rats. This approach, when used to measure DA clearance, will permit concurrent measurements of behavior and DA transporter function on a second-to-second time scale during cocaine administration and withdrawal. Measurement of simultaneous changes in behavior and DA clearance will greatly increase the power to assess the relationship between these measures. Additional studies are also required to investigate cocaine-induced regulation of the DA transporter in human brain. *In vivo* imaging techniques should permit the measurement of DA transporter binding in the living human brain and would confirm the relevance of animal studies in understanding the basis for changes in brain function induced by cocaine. It is understandable that the primary focus of most studies to date has been on understanding regulation of the DA transporter. However, studies of 5-HT and NE transporter binding sites and function following repeated cocaine administration are also necessary to elucidate the potential functional contribution of regulating these transporters.

B. MOLECULAR MECHANISMS FOR TRANSPORTER REGULATION

There are numerous molecular mechanisms by which chronic cocaine administration could potentially regulate the DA transporter.[9,109] A preliminary report showed that cocaine administration for 3 days produced a long-lasting change in the Na^+-dependency of *in vitro* DA uptake in the rat striatum, as measured with an *in vitro* rotating disk electroanalytical technique.[110] DA transport was inhibited to a greater degree when the Na^+ concentration was reduced. Such a change could be due to a persistent alteration in the Na^+ binding site(s) on the DA transporter. Multi-affinity binding states for cocaine and its analogs (WIN 35,428 and RTI-55) have been shown to exist;[111] however, no chronic cocaine studies to date have addressed the possibility that DA transporter regulation involves a shift in the proportion of high and low affinity states for cocaine binding.

The increased level of synaptic DA produced by cocaine treatment can activate both the D_1 and D_2 DA receptor families. Recently, D_2 receptors have been shown to regulate the activity of the DA transporter; acute exposure to D_2 receptor agonists increases DA transport whereas D_2 receptor antagonists diminish it.[112–114] Furthermore, repeated co-administration of the D_2 receptor antagonist pimozide along with cocaine blocks the cocaine-induced increase of DA transporter activity.[112] Thus, coincident activation of D_2 receptors may facilitate regulation of the DA transporter.

D_1-like receptors increase cAMP production and also might increase phosphoinositide hydrolysis whereas D_2-like receptors inhibit cAMP production, activate K^+ channels, and inhibit Ca^{2+} channels.[115,116] Increasing intracellular cAMP has been shown to stimulate [^{3}H]DA uptake in rat hypothalamic cells.[117] The DA transporter contains several consensus sequences for phosphorylation by protein kinases, including cAMP-dependent protein kinase (PKA).[8] In addition, chronic cocaine administration regulates the cAMP pathway/and phosphorylation of cAMP response element binding (CREB) protein, as well as immunoreactivity of CREB protein and binding activity of the cAMP response element (CRE) in the NAc.[109] Phosphorylated CREB protein binds to the CRE in promoter regions of several genes to induce their transcription.

Cocaine-induced changes in other neurotransmitter systems also could regulate the DA transporter. For example, increased glutamate activation of NMDA receptors can induce short-term regulation. Such activation can alter the Na^+ electrochemical gradient in DA neurons and thereby induce reverse DA transport and increased extracellular concentrations of DA.[118] Activation of NMDA receptors also leads to the production of the diffusible messenger nitric oxide (NO). Compounds that generate NO have been reported to decrease the V_{max} and increase K_m for [^{3}H]DA uptake in rat striatal synaptosomes.[119,120] However, exposure to NO does not directly affect binding to the DA transporter, suggesting that their site of action is distinct from the DA recognition site. Such a change could involve a specific antagonist recognition site(s), the Na^+ binding–site(s) and/or the Cl^- binding site(s).

In addition to direct phosphorylation of the transporter molecule, phosphorylation could also trigger more long lasting effects by altering transporter gene expression, mRNA stability, protein turnover, protein conformation, accessory proteins, etc. There is recent evidence that chronic cocaine can alter several families of transcription factors including immediate early genes.[109] These transcription factors are thought to provide a liaison between short-term changes in neuronal activity and long-term changes in gene expression.[121] Chronic cocaine desensitizes the responsiveness of several immediate early genes such as c-*fos*, c-*jun*, and *zif*268.[122–124] However, it leads to a persistent upregulation of AP-1 binding,[123] which is associated with the induction of apparently novel *fos*-like proteins (see Chapter 14).[109] Whether the expression of the DA transporter could be regulated in this manner remains to be demonstrated.

VII. SUMMARY

There is increasing evidence that the DA transporter can be regulated in response to chronic cocaine. During chronic cocaine administration and at short times after cessation of drug treatment, a transient increase in DA transporter binding and DA uptake occur.

These changes suggest that DA would be cleared more quickly from the synaptic cleft and are consistent with the observation that behavioral tolerance is induced by continuous cocaine administration. However, days to weeks after cessation of repeated intermittent cocaine administration, DA transporter binding and uptake are reduced. Regulation of the DA transporter has most often been observed in limbic regions of the rat brain. This long-lasting regulation is consistent with the observed increase of extracellular DA concentration induced in the rat brain by cocaine challenge following withdrawal from repeated cocaine administration. These changes in DA have been linked to behavioral sensitization and, thus, suggest that altered DA transporter activity contributes to the behavioral changes induced by intermittent cocaine exposure.

ACKNOWLEDGMENTS

We thank Drs. S. G. Amara, K. J. Buck, D. L. Cameron, T. Nguyen, and M. S. Sonders, for their helpful discussions and comments. N.R.Z. is the recipient of USPHS Research Scientist Development Award DA 00174.

REFERENCES

1. Ross, S. B. and Renyi, A. L., Inhibition of the uptake of tritiated catecholamines by antidepressant and related agents, *Eur. J. Pharmacol.*, 2, 181, 1967.
2. Iversen, L. L., Uptake mechanisms for neurotransmitter amines, *Biochem. Pharmacol.*, 23, 1927, 1974.
3. Koe, B. K., Molecular geometry of inhibitors of the uptake of catecholamines and serotonin in synaptosomal preparations of rat brain, *J. Pharmacol. Exp. Ther.*, 199, 649, 1976.
4. Reith, M. E. A., Meisler, B. E., Sershen, H., and Lajtha, A., Structural requirements of cocaine congeners to interact with DA and serotonin uptake sites in mouse brain and to induce stereotyped behavior, *Biochem. Pharmacol.*, 35, 1123, 1976.
5. Yasuda, R. P., Zahniser, N. R., and Dunwiddie, T. V., Electrophysiological effects of cocaine in the rat hippocampus in vitro, *Neurosci. Lett.*, 45, 199, 1984.
6. Post, R. M., Weiss, S. B., and Pert, A., Sensitization and kindling effects of chronic cocaine administration, in *Cocaine: Pharmacology, Physiology, and Clinical Strategies,* Lakoski, J. M., Galloway, M. P., and White, F. J., Eds., CRC Press, Boca Raton, 1991, 115.
7. Uhl, G. R. and Hartig, P. R., Transporter explosion: update on uptake, *Trends Pharmacol. Sci.*, 13, 421, 1992.
8. Giros, B. and Caron, M. G., Molecular characterization of the dopamine transporter, *Trends Pharmacol. Sci.*, 14, 43, 1993.
9. Amara, S. G. and Kuhar, M. J., Neurotransmitter transporters: recent progress, *Annu. Rev. Neurosci.*, 16, 73, 1993.
10. Pacholczyk, T., Blakely, R. D., and Amara, S. G., Expression cloning of a cocaine- and antidepressant-sensitive human noradrenaline transporter, *Nature,* 350, 350, 1991.
11. Shimada, S., Kitayama, S., Lin, C.-L., Patel, A., Nanthakumar, E., Gregor, P., Kuhar, M., and Uhl, G., Cloning and expression of a cocaine-sensitive dopamine transporter complementary DNA, *Science,* 254, 576, 1991.
12. Kilty, J. E., Lorang, D., and Amara, S. G., Cloning and expression of a cocaine-sensitve rat dopamine transporter, *Science,* 254, 578, 1991.
13. Hoffman, B. J., Mezey, E., and Brownstein, M. J., Cloning of a serotonin transporter affected by antidepressants, *Science,* 254, 579, 1991.
14. Blakely, R. D., Berson, H. E., Fremeau, R. T., Jr., Caron, M. G., Peek, M. M., Prince, H. K., and Bradley, C. C., Cloning and expression of a functional serotonin transporter from rat brain, *Nature,* 354, 66, 1991.
15. Giros, B., Mestikawy, S. E., Bertrand, L., and Caron, M. G., Cloning and functional characterization of a cocaine-sensitive dopamine transporter, *FEBS Lett.*, 295,149, 1991.
16. Usdin, T. B., Mezey, E., Chen, C., Brownstein, M. J., and Hoffman, B. J., Cloning of the cocaine-sensitive bovine dopamine transporter, *Proc. Natl. Acad. Sci. USA,* 88, 11168, 1991.

17. Giros, B., El Mestikawy, S., Godinot, N., Zheng, K., Han, H., Yang-Feng, T., and Caron, M. G., Cloning, pharmacological characterization, and chromosome assignment of the human dopamine transporter, *Mol. Pharmacol.*, 42, 383, 1992.
18. Vandenbergh, D. J., Persico, A. M., and Uhl, G. R., A human dopamine transporter cDNA predicts reduced glycosylation, displays a novel repetitive element and provides racially-dimorphic *Taq* I RFLPs, *Mol. Brain Res.*, 15, 161, 1992.
19. Hyttel, J., Citalopram—pharmacological profile of a specific serotonin uptake inhibitor with antidepressant activity, *Prog. Neuro-Psychopharmacol. Biol. Psychiatry*, 6, 277, 1982.
20. Kelley, P. H. and Iversen, S. D., Selective 6-OHDA-induced destruction of mesolimbic dopamine neurons: abolition of psychostimulant-induced locomotor activity in rats, *Eur. J. Pharmacol.*, 40, 45, 1975.
21. Costall, B. and Naylor, R. J., Mesolimbic and extrapyramidal sites for the mediation of sterotyped behavior patterns and hyperactivity by amphetamine and apomorphine in the rat, in *Advances in Behavioral Biology, Vol. 21: Cocaine and Other Stimulants*, Ellinwood, E. H. and Kilbey, M. M., Eds., Plenum Press, New York, 1977, 47.
22. Delfs, J. M., Schreiber, L., and Kelley, A. E., Microinjection of cocaine into the nucleus accumbens elicits locomotor activation in the rat, *J. Neurosci.*, 10, 303, 1990.
23. Wise, R. A. and Bozarth, M. A., A psychomotor stimulant theory of addiction, *Psychol. Rev.*, 94, 469, 1987.
24. Koob, G., Drugs of abuse: anatomy, pharmacology, and function of reward pathways, *Trends Pharmacol. Sci.*, 13, 177, 1992.
25. Ritz, M. C., Lamb, R. J., Goldberg, S. R., and Kuhar, M. J., Cocaine receptors on dopamine transporters are related to self-administration of cocaine, *Science*, 237, 1219, 1987.
26. Post, R. M. and Contel, N. R., Human and animal studies of cocaine: implications for development of behavioral pathology, in *Stimulants: Neurochemical, Behavioral, and Clinical Perspectives*, Creese, I., Ed., Raven Press, New York, 1983, 169.
27. Post, R. M. and Weiss, S. B., Sensitization and kindling: implications for the evolution of psychiatric symptomatology, in *Sensitization in the Nervous System*, Kalivas, P. W. and Barnes, C. D., Eds., Telford Press, Caldwell, 1988, 257.
28. Goudie, A. J. and Emmett-Oglesby, M. W., *Psychoactive Drugs: Tolerance and Sensitization*, Humana Press, NJ, 1989.
29. Johanson, C.-E. and Fischman, M. W., The pharmacology of cocaine related to its abuse, *Pharmacol. Rev.*, 41, 3, 1989.
30. Stewart, J. and Badiani, A., Tolerance and sensitization to the behavioral effects of drugs, *Behav. Pharmacol.*, 4, 289, 1993.
31. Robinson, T. E. and Becker, J. B., Enduring changes in brain and behavior produced by chronic amphetamine administration: a review and evaluation of animal models of amphetamine psychosis, *Brain Res. Rev.*, 11, 157, 1986.
32. Kuczenski, R. and Segal, D. S., Psychomotor stimulant-induced sensitization: behavioral and neurochemical correlates, in *Sensitization in the Nervous System*, Kalivas, P. W. and Barnes, C. D., Eds., Telford Press, Caldwell, 1988, 175.
33. Kalivas, P. W. and Stewart, J., Dopamine transmission in the initiation and expression of drug- and stress-induced sensitization of motor activity, *Brain Res. Rev.*, 16, 223, 1991.
34. Zahniser, N. R. and Peris, J., Neurochemical mechanisms of cocaine-induced sensitization, in *Cocaine: Pharmacology, Physiology, and Clinical Strategies*, Lakoski, J. M., Galloway, M. P., and White, F. J., Eds., CRC Press, Boca Raton, 1991, 229.
35. Castaneda, E., Becker, J. B., and Robinson, T. E., The long-term effects of repeated amphetamine treatment *in vivo* on amphetamine, KCl and electrical stimulation evoked striatal dopamine release *in vitro*, *Life Sci.*, 42, 2447, 1988.
36. Robinson, T. E., Jurson, P. A., Bennett, J. A., and Bentgen, K. M., Persistent sensitization of dopamine neurotransmission in ventral striatum (nucleus accumbens) produced by prior experience with (+)amphetamine: a microdialysis study in freely moving rats, *Brain Res.*, 462, 211, 1988.
37. Kuczenski, R. and Segal, D., Concomitant characterization of behavioral and striatal neurotransmitter response to amphetamine using *in vivo* microdialysis, *J. Neurosci.*, 9, 2051, 1989.
38. Kuczenski, R., Segal, D. S., and Aizenstein, M. L., Amphetamine, cocaine, and fencamfamine: relationship between locomotor and stereotypy response profiles and caudate and accumbens dopamine dynamics, *J. Neurosci.*, 11, 2703, 1991.

39. Patrick, S. L., Thompson, T. L., Walker, J. M., and Patrick, R. L., Concomitant sensitization of amphetamine-induced behavioral stimulation and *in vivo* dopamine release from rat caudate nucleus, *Brain Res.*, 538, 343, 1991.
40. Paulson, P. E., Camp, D. M., and Robinson, T. E., Time course of transient behavioral depression and persistent behavioral sensitization in relation to regional brain monamine concentrations during amphetamine withdrawal in rats, *Psychopharmacology,* 103, 480, 1991.
41. Segal, D. S. and Kuczenski, R., *In vivo* microdialysis reveals a diminished amphetamine-induced dopamine response corresponding to behavioral sensitzation produced by repeated amphetamine pretreatment, *Brain Res.*, 521, 330, 1992.
42. Vezina, P., Amphetamine injected into the ventral tegmental area sensitizes the nucleus accumbens dopaminergic response to systemic amphetamine: an *in vivo* microdialysis study in the rat, *Brain Res.*, 605, 332, 1993.
43. Estevez, V. S., Ho, B. T., and Englert, L. F., Metabolism correlates of cocaine-induced stereotypy in rats, *Pharmacol. Biochem. Behav.*, 10, 267, 1979.
44. Reith, M. E. A., Benuck, M., and Lajtha, A., Cocaine disposition in the brain after continuous or intermittent treatment and locomotor stimulation in mice, *J. Pharmacol. Exp. Ther.*, 243, 281, 1987.
45. Pettit, H. O., Pan, H.-T., Parsons, L. H., and Justice, Jr., J. B., Extracellular concentrations of cocaine and dopamine are enhanced during chronic cocaine administration, *J. Neurochem.*, 55, 798, 1990.
46. Pan, H.-T., Menacherry, S., and Justice, Jr., J. B., Differences in the pharmacokinetics of cocaine in naive and cocaine-experienced rats, *J. Neurochem.*, 56, 1299, 1991.
47. Cass, W. A. and Zahniser, N. R., Cocaine levels in striatum and nucleus accumbens: augmentation following challenge injection in rat withdrawn from repeated cocaine administration, *Neurosci. Lett.*, 152, 177, 1993.
48. Orona, R. A., Mayfield, R. D., Cline, E. J., and Zahniser, N. R., Repeated intravenous cocaine administration to rats produces behavioral sensitization without changing brain cocaine levels, *Neurosci. Lett.*, 167, 121, 1994.
49. Satel, S. L. and Edell, W. S., Cocaine-induced paranoia and psychosis, *Am. J. Psychiatry,* 148, 1708, 1991.
50. Wise, R. A. and Leeb, K., Psychomotor-stimulant sensitization: a unitary phenomenon? *Behav. Pharmacol.*, 4, 339, 1993.
51. Horger, B. A., Shelton, K., and Schenk, S., Preexposure sensitizes rats to the rewarding effects of cocaine, *Pharmacol. Biochem. Behav.*, 37, 707, 1990.
52. Robinson, T. E., and Berridge, K. C., The neural basis of drug craving: an incentive-sensitization theory of addiction, *Brain Res. Rev.*, 18, 247, 1993.
53. Akimoto, K., Hamamura, T., and Otsuki, S., Subchronic cocaine treatment enhances cocaine-induced dopamine efflux, studied by in vivo intracerebral dialysis, *Brain Res.*, 490, 339, 1989.
54. Akimoto, K., Hamamura, T., Kazahaya, Y., Akiyama, K., and Otsuki, S., Enhanced extracellular dopamine level may be the fundamental neuropharmacological basis of cross-behavioral sensitization between methamphetamine and cocaine—an in vivo dialysis study in freely moving rats, *Brain Res.*, 507, 344, 1990.
55. Kalivas, P. W. and Duffy, P., Effect of acute and daily cocaine treatment on extracellular dopamine in the nucleus accumbens, *Synapse,* 5, 48, 1990.
56. Pettit, H. O., Pan, H.-T., Parsons, L. H., and Justice, J. B., Jr., Extracellular concentrations of cocaine and dopamine are enhanced during chronic cocaine administration, *J. Neurochem.*, 55, 798, 1990.
57. Weiss, F., Paulus, M. P., Lorang, M. T., and Koob, G. F., Increases in extracellular dopamine in the nucleus accumbens by cocaine are inversely related to basal levels: effects of acute and repeated administration, *J. Neurosci.*, 12, 4372, 1992.
58. Kalivas, P. W. and Duffy P., Time course of extracellular dopamine and behavioral sensitization to cocaine. I. Dopamine axon terminals, *J. Neurosci.*, 13, 266, 1993.
59. Hurd, Y. L., Weiss, F., Koob, G. F., And, N.-K., and Ungerstedt, U., Cocaine reinforcement and extracellular dopamine overflow in rat nucleus accumbens: an in vivo microdialysis study, *Brain Res.*, 498, 199, 1989.
60. Hurd, Y. L., Weiss, F., Koob, G., and Ungerstedt, U., The influence of cocaine self-administration on in vivo dopamine and acetylcholine neurotransmission in rat caudate-putamen, *Neurosci. Lett.*, 109, 227, 1990.

61. Segal, D. S. and Kuczenski, R., Repeated cocaine administration induces behavioral sensitization and corresponding decreased extracellular dopamine responses in caudate and accumbens, *Brain Res.*, 577, 351, 1992.
62. Parsons, L. H., Smith, A. D., and Justice, J. B., Jr., Basal extracellular dopamine is decreased in the rat nucleus accumbens during abstinence from chronic cocaine, *Synapse*, 9, 60, 1991.
63. Robertson, M. W., Leslie, C. A., and Bennett, J. P., Jr., Apparent synaptic dopamine deficiency induced by withdrawal from chronic cocaine treatment, *Brain Res.*, 538, 337, 1991.
64. Weiss, F., Markou, A., Lorang, M. T., and Koob, G. F., Basal extracellular dopamine levels in the nucleus accumbens are decreased during cocaine withdrawal after unlimited-access self-administration, *Brain Res.*, 593, 314, 1992.
65. Ackerman, J. M. and White, F. J., Decreased activity of rat A10 dopamine neurons following withdrawal from repeated cocaine, *Eur. J. Pharmacol.*, 218, 171, 1992.
66. McDougle, C. J., Price, L. H., Palumbo, J. M., Kosten, T. R., Heninger, G. R., and Kleber, H. D., Dopaminergic responsivity during cocaine abstinence, *Psychiat. Res.*, 43, 77, 1992.
67. Kalivas, P. W. and Duffy, P., Effects of daily cocaine and morphine treatment on somatodendritic and terminal field dopamine release, *J. Neurochem.*, 50, 1498, 1988.
68. Johnson, K. M. and Snell, L. D., Sensitization to the behavioral effects of cocaine is associated with altered dopamine metabolism and release in rat brain, *Soc. Neurosci. Abstr.*, 13, 1718, 1987.
69. Peris, J., Boyson, S. J., Cass, W. A., Curella, P., Dwoskin, L. P., Larson, G., Lin, L.-H., Yasuda, R. P., and Zahniser, N. R., Persistence of neurochemical changes in dopamine systems after repeated cocaine administration, *J. Phamacol. Exp. Ther.*, 253, 38, 1990.
70. Liang, N. Y. and Rutledge, C. O., Comparison of the release of [^{3}H]dopamine from isolated corpus striatum by amphetamine, fenfluamine and unlabeled dopamine, *Biochem. Pharmacol.*, 31, 983, 1982.
71. Parker, E. M. and Cubeddu, L. X., Effects of *d*-amphetamine and dopamine synthesis inhibitors on dopamine and acetylcholine neurotransmission in the striatum. I. Release in the absence of vesicular transmitter stores, *J. Pharmacol. Exp. Ther.*, 237, 179, 1986.
72. Carboni, E., Imperato, A., Perezzani, L., and Di Chiara, G., Amphetamine, cocaine, phencyclidine and nomifensine increase extracellular dopamine concentrations preferentially in the nucleus accubmens of freely moving rats, *Neuroscience*, 28, 653, 1989.
73. Cass, W. A., Gerhardt, G. A., Mayfield, R. D., Curella, P., and Zahniser, N. R., Differences in dopamine clearance and diffusion in rat striatum and nucleus accumbens following systemic cocaine administration, *J. Neurochem.*, 59, 259, 1992.
74. Moghaddam, B. and Bunney, B. S., Differential effect of cocaine on extracellular dopamine levels in rat medial prefrontal cortex and nucleus accumbens: comparison to amphetamine, *Synapse*, 4, 156, 1989.
75. Missale, C., Castelletti, L., Govoni, S., Spano, P. F., Trabucchi, M., and Hanbauer, I., Dopamine uptake is differentially regulated in rat striatum and nucleus accumbens, *J. Neurochem.*, 45, 51, 1985.
76. Allard, P., Eriksson, K., Ross, S. B., and Marcusson, J. O., Unaltered [^{3}H]GBR-12935 binding after chronic treatment with dopamine active drugs, *Psychopharmacology*, 102, 291, 1990.
77. Izenwasser, S. and Cox, B. M., Daily cocaine treatment produces a persistent reduction of [^{3}H]dopamine uptake in vitro in rat nucleus accumbens but not in striatum, *Brain Res.*, 531, 338, 1990.
78. Sharpe, L. G., Pilotte, N. S., Mitchell, W. M., and DeSouza, E. B., Withdrawal of repeated cocaine decreases autoradiographic [^{3}H]mazindol-labelling of dopamine transporter in rat nucleus accumbens, *Eur. J. Pharmacol.*, 203, 141, 1991.
79. Benmansour, S., Tejani-Butt, S. M., and Brunswick, D. J., Lack of effect of high-dose cocaine on monoamine uptake sites in rat brain measured by quantitative autoradiography, *Psychopharmacology*, 106, 459, 1992.
80. Reith, M. E. A. and Selmeci, G., Cocaine binding sites in mouse striatum, dopamine autoreceptors, and cocaine-induced locomotion, *Pharmacol. Biochem. Behav.*, 41, 227, 1992.
81. Cass, W. A., Gerhardt, G. A., Gillespie, K., Curella, P., Mayfield, R. D., and Zahniser, N. R., Reduced clearance of exogenous dopamine in rat nucleus accumbens, but not in dorsal striatum, following cocaine challenge in rats withdrawn from repeated cocaine administration, *J. Neurochem.*, 61, 273, 1993.
82. Hitri, A. and Wyatt, R. J., Regional differences in rat brain transporter binding: function of time after chronic cocaine, *Clin. Neuropharmacol.*, 16, 525, 1993.

83. Aloyo, V. J., Harvey, J. A., and Kirifides, A. L., Chronic cocaine increases WIN 35428 binding in rabbit caudate, *Soc. Neurosci. Abstr.*, 19, 1843, 1993.
84. Baker, I., Wyatt, R. J., and Masserano, J. M., Evaluation of dopamine uptake into the substantia nigra, ventral tegmental area and striatum of the rat: effects of cocaine treatment, *Soc. Neurosci. Abstr.*, 19, 936, 1993.
85. Letchworth, S. R., Vogt, L. J., Migliarese, K., and Porrino, L. J., Effects of chronic cocaine administration on dopamine transporter density in the rat striatum, *Soc. Neurosci. Abstr.*, 19, 1842, 1993.
86. Pilotte, N. S., Sharpe, L. G., and Kuhar, M. J., Delayed reduction of dopamine transporter binding in nucleus accumbens after termination of repeated passive infusion of cocaine, *Soc. Neurosci. Abstr.*, 19, 1858, 1993.
87. Yi, S.-J. and Johnson, K. M., Effects of acute and chronic administration of cocaine on striatal uptake, compartmentalization and release of [^{3}H]dopamine, *Neuropharmacology*, 29, 475, 1990.
88. Wamsley, J. R. and Alburges, M. E., Cocaine causes a time dependent and dose dependent increase in D_1 receptors and dopamine transporters, *Proc. West. Pharmacol. Soc.*, 36, 277, 1993.
89. Alburges, M. E., Narang, N., and Wamsley, J. K., Alterations in the dopaminergic receptor system after chronic administration of cocaine, *Synapse*, 14, 314, 1993.
90. Farfel, G. M., Kleven, M. S., Woolverton, W. L., Seiden, L. S., and Perry, B. D., Effects of repeated injections of cocaine on catecholamine receptor binding sites, dopamine transporter binding sites and behavior in rhesus monkey, *Brain Res.*, 578, 235, 1992.
91. Cline, E. J., Pilotte, N. S., Scheffel, U., Mitchell, W. M., and Kuhar, M. J., Daily i.v. cocaine infusion decreases in vivo [3H]WIN 35,428 autoradiographic binding in behaviorally relevant brain areas following withdrawal, *NIDA Research Monograph*, 141, 100, 1993.
92. Ng, J. P., Hubert, G. W., and Justice, J. B., Jr., Increased stimulated release and uptake of dopamine in nucleus accumbens after repeated cocaine administration as measured by in vivo voltammetry, *J. Neurochem.*, 56, 1485, 1991.
93. Xia, Y., Goebel, D. J., Kapatos, G., and Bannon, M. J., Quantitation of rat dopamine transporter mRNA: effects of cocaine treatment and withdrawal, *J. Neurochem.*, 59, 1179, 1992.
94. Cerruti, C., Pilotte, N. S., Uhl, G., and Kuhar, M. J., Reduced VTA dopamine transporter mRNA occurs after cessation of chronic cocaine administration, *Soc. Neurosci. Abstr.*, 19, 1498, 1993.
95. Cass, W. A., Zahniser, N. R., Flach, K. A., and Gerhardt, G. A., Clearance of exogenous dopamine in rat dorsal striatum and nucleus accumbens: role of metabolism and effects of locally applied uptake inhibitors, *J. Neurochem.*, 61, 2269, 1993.
96. Marshall, J. F., O'Dell, S. J., Navarrete, R., and Rosenstein, A. J., Dopamine high affinity transport site topography in rat brain: major differences between dorsal and ventral striatum, *Neuroscience*, 37, 11, 1990.
97. Richfield, E. K., Quantitative autoradiography of the dopamine uptake complex in rat brain using [^{3}H]GBR 12935: binding characteristics, *Brain Res.*, 540, 1, 1991.
98. Miller, L. G., Koff, J. M., and Shuster, L., Chronic cocaine administration is associated with behavioral sensitization and dopamine transporter upregulation, *Soc. Neurosci. Abstr.*, 19, 1498, 1993.
99. Zeigler, S., Lipton, J., Toga, A., and Ellison, G., Continuous cocaine administration produces persisting changes in brain neurochemistry and behavior, *Brain Res.*, 552, 27, 1991.
100. King, G. R., Joyner, C., Lee, T., Kuhn, C., and Ellinwood, E. H., Jr., Intermittent and continuous cocaine administration: residual behavioral states during withdrawal, *Pharmacol. Biochem. Behav.*, 43, 243, 1992.
101. Izenwasser, S. and Cox, B. M., Inhibition of dopamine uptake by cocaine and nicotine: tolerance to chronic treatments, *Brain Res.*, 573, 119, 1992.
102. King, G. R., Kuhn, C., and Ellinwood, E. H., Jr., Dopamine efflux during withdrawal from continuous or intermittent cocaine, *Psychopharmacology*, 111, 179, 1993.
103. Yu, Z. J., Jin, C., Rockhold, R. W., Hoskins, B., and Ho, I. K., Site and mechanism of behavioral tolerance to cocaine: a study of dopamine release in Wistar-Kyoto and spontaneously hypertensive rats, *Neurochem. Res.*, 18, 1203, 1993.
104. Little, K. Y., Kirkman, J. A., Carroll, F. I., Clark, T. B., and Duncan, G. E., Cocaine use increases [^{3}H]WIN 35428 binding sites in human striatum, *Brain Res.*, 628, 17, 1993.
105. Staley, J., Flynn, D. D., Stitt, F., Wetli, C. V., and Mash, D. C., [^{3}H]WIN 35,428 binding to the dopamine transporter in cocaine overdose deaths, *Soc. Neurosci. Abstr.*, 19, 1843, 1993.

106. Hurd, Y. L. and Herkenham, M., Molecular alterations in the neostriatum of human cocaine addicts, *Synapse,* 13, 357, 1993.
107. Parsons, L. H. and Justice, J. B., Jr., Serotonin and dopamine sensitization in the nucleus accumbens, ventral tegmental area, and dorsal raphe nucleus following repeated cocaine administration, *J. Neurochem.,* 61, 1611, 1993.
108. Cunningham, K. A., Paris, J. M., and Goeders, N. E., Chronic cocaine enhances serotonin autoregulation and serotonin uptake binding, *Synapse,* 11, 112, 1992.
109. Nestler, E. J., Hope, B. T., and Widnell, K. L., Drug addiction: a model for the molecular basis of neural plasticity, *Neuron,* 11, 995, 1993.
110. Povlock, S. L. and Schenk, J. O., Repeated cocaine treatments followed by withdrawal alters dopamine transport via a Na^+-dependent mechanism, *Soc. Neurosci. Abstr.,* 19, 745, 1993.
111. Madras, B. K., Spealman, R. D., Fahey, M. A., Neumeyer, J. L., Saha, J. K., and Milius, R. A., Cocaine receptors labeled by [^{3}H]2b-carbomethoxy-3b-(4-fluorophenyl) tropane, *Mol. Pharmacol.,* 36, 518, 1989.
112. Parsons, L. H., Schad, C. A., and Justice, J. B., Jr., Co-administration of the D_2 antagonist pimozide inhibits up-regulation of dopamine release and uptake induced by repeated cocaine, *J. Neurochem.,* 60, 376, 1993.
113. Meiergerd, S. M., Patterson, T. A., and Schenk, J. O., D_2 receptors may modulate the function of the striatal transporter for dopamine: kinetic evidence from studies in vitro and in vivo, *J. Neurochem.,* 61, 764, 1993.
114. Cass, W. A. and Gerhardt, G. A., Direct *in vivo* evidence that D2 dopamine receptors can modulate dopamine uptake, *Neurosci. Lett.,* 176, 259, 1994.
115. Sibley, D. R. and Monsma, F. J., Jr., Molecular biology of dopamine receptors, *Trends Pharmacol. Sci.,* 13, 61, 1992.
116. Gingrich, J. A. and Caron, M. G., Recent advances in the molecular biology of dopamine receptors, *Annu. Rev. Neurosci.,* 16, 299, 1993.
117. Kadowaki, K., Hirota, K., Koike, K., Ohmichi, M., Kiyama, H., Miyake, A., and Tanizawa, O., Adenosine 3',5'-cyclic monophosphate enhances dopamine accumulation in rat hypothalamic cell culture containing dopaminergic neurons, *Neuroendocrinology,* 52, 256, 1990.
118. Lonart, G. and Zigmond, M. J., High glutamate concentrations evoke Ca^{2+}-independent dopamine release from striatal slices: a possible role of reverse dopamine transport, *J. Pharmacol. Exp. Ther.,* 256, 1132, 1991.
119. Lonart, G. and Johnson, K. M., Nitric oxide generators can induce both short- and long-term changes in the activity of rat striatal dopamine transporter, *Soc. Neurosci. Abstr.,* 19, 1177, 1993.
120. Kuhar, M. J. and Pogun, S., Modulation of dopamine transport by nitric oxide (NO), *Soc. Neurosci. Abstr.,* 19, 937, 1993.
121. Morgan, J. I. and Curran, T., Stimulus-transcription coupling in the nervous system: involvement of the inducible proto-oncogenes fos and jun, *Annu. Rev. Neurosci.,* 14, 421, 1991.
122. Bhat, R. V., Cole, A. J. and Baraban, J. M., Chronic cocaine treatment suppresses basal expression of *zif*268 in rat forebrain: *in situ* hybridization studies, *J. Pharmacol. Exp. Ther.,* 263, 343, 1992.
123. Hope, B., Kosofsky, B., Hyman, S. E., and Nestler, E. J., Regulation of immediate early gene expression and AP-1 binding in the rat nucleus accumbens by chronic cocaine, *Proc. Natl. Acad. Sci. USA,* 89, 5764, 1992.
124. Daunais, J. B., Roberts, D. C. S., and McGinty, J. F., Cocaine self-administration increases preprodynorphin, but not c-fos, mRNA in rat striatum, *NeuroReport,* 4, 543, 1993.

Section III

MOLECULAR APPROACHES TO STUDYING COCAINE ACTIVATION

Chapter 12

Cocaine and Dopamine Transporters

Michael J. Kuhar, John W. Boja, Amrat Patel, Nancy Pilotte, Catherine Cerruti, and John Lever

CONTENTS

I. THE REINFORCING PROPERTIES OF COCAINE: THE DOPAMINE HYPOTHESIS REVISITED

A. IS COCAINE UNIQUE AMONG DOPAMINE UPTAKE INHIBITORS?

Over the past 10 to 15 years, many laboratories have utilized animal models to represent human drug-taking behavior.[1] The animal model of drug self-administration in monkeys and rodents, in particular, has been used in conjunction with many anatomical and biochemical studies to elucidate the neurochemical mechanisms of drug reinforcement. Experimental data from various laboratories obtained with a variety of techniques have implicated the mesolimbic dopamine system as important for psychostimulants and quite possibly for many other abused drugs as well.[2-5] In our laboratory, we have utilized receptor binding techniques to provide evidence that the initial site of action of cocaine and its analogs is the dopamine transporter rather than other cocaine binding sites (Figure 1).[6,7] However, it seems clear that there are limitations to the notion that the dopamine transporter is a cocaine "receptor" related to the reinforcing properties of psychostimulant drugs. A major difficulty is that some drugs which inhibit the transporter are not very reinforcing, or at least do not show a high degree of abuse liability in humans (e.g., mazindol).[8] Ritz et al.[6,7] attempted to limit their correlational analysis between drug self-administration behavior and drug binding at the dopamine transporter to cocaine, its analogs, and other amide-linked local anesthetic drugs known to be self-administered. This restriction was made in the attempt to utilize drugs that were comparable in a wide variety of properties such as pharmacokinetics and lipid solubility.

Additional relevant evidence has been obtained by Rothman and colleagues[9] who showed that GBR 12909 can oppose the dopamine uptake inhibiting effect of cocaine as measured by microdialysis even though it binds to dopamine transporters and inhibits dopamine uptake. These data suggest a complicated interaction of various types of drugs with the dopamine transporter, and are compatible with the view that there may be something unique about the action of cocaine at the dopamine transporter compared to other drugs such as GBR 12909.

1-8493-8311-0/95/$0.00+$.50

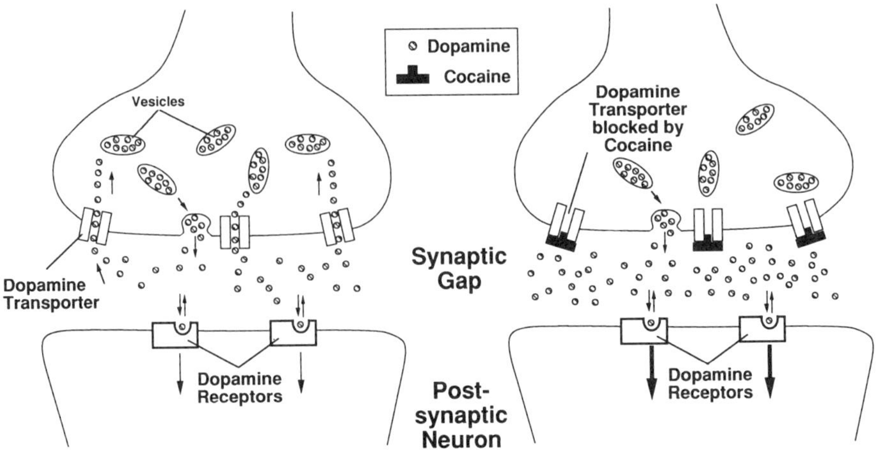

Figure 1. Dopamine hypothesis for reinforcing properties of cocaine. Cocaine binds to the dopamine transporter and blocks the reuptake of dopamine in the mesolimbocortical pathway. This potentiates dopaminergic neurotransmission and initiates the sequence of events that ultimately causes the rewarding effects of the drug.[4]

If there is some important feature of cocaine in addition to its ability to inhibit dopamine uptake, what could this unique feature be? One obvious possibility is that there is a totally different binding site for cocaine and its analogs (i.e., other than the dopamine transporter) that is critical to their reinforcing properties. However, while there are many binding sites for cocaine in the brain, none is as clearly associated with the reinforcing properties of cocaine as is the dopamine transporter.[6,7] Another possibility is that a site in addition to the dopamine transporter binds the drug concurrently. Cocaine could act on such a binding site to produce a combined inhibition of two functions and yield an especially potent reinforcing effect.

One of the unique features of cocaine is its rapid rate of entry into the brain. It is commonly said that drugs that enter the brain more rapidly have high abuse liability, which could contribute to the unusually high reinforcing potential of cocaine. By utilizing an *in vivo* receptor binding competition model, Pogun and colleagues have shown that cocaine enters the brain more rapidly than does mazindol or GBR 12909 (Figure 2).[10] This correlates with the evidence that cocaine has a higher abuse liability than mazindol.[11] However, the importance of the rate of entry into brain and the rate of occupancy of receptors needs to be explored further.

Other compounds which do not appear to be reinforcing in spite of the fact that they inhibit uptake might have other properties which preclude high abuse liability. Centrally acting drugs are well known to have multiple receptor binding properties and, in this sense, these compounds could be less "pure" than cocaine.

B. COCAINE AND ITS ANALOGS HAVE HIGH AFFINITY SITES

A relatively unique feature of cocaine and its analogs found in binding experiments is the presence of a high as well as a low affinity binding site (Figure 3). Radiolabeled cocaine,[12] WIN 35428,[13] WIN 35065-2,[14,15] RTI-55,[16] and RTI 121 (unpublished observations) all show both high and low affinity binding sites associated with the dopamine transporter, whereas nonphenyltropane analogs, such as methylphenidate,[17] GBR 12935,[18,19] mazindol,[20] and nomiphensine[21] do not show high and low affinity binding sites. Also, there is evidence that compounds which are structurally unlike cocaine bind to the low-affinity site associated with cocaine binding. Madras et al.[13] have noted that these noncocaine

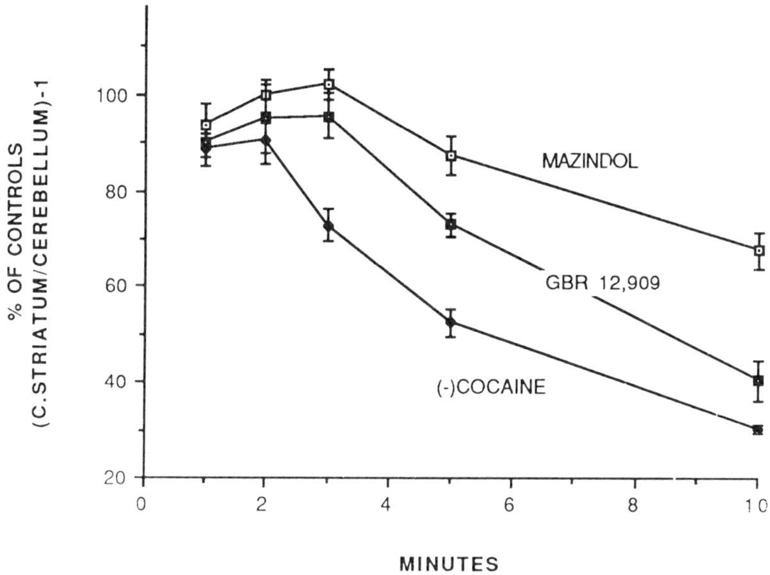

Figure 2. *In vivo* rate of displacement of [^{3}H]WIN 35428 binding in rat striatum. The rate of displacement reflects the rate of entry of compounds into the brain and occupancy of transporters. Data represent mean ± SEM percent of control (as vehicle injected animals). The data indicate that cocaine enters the brain and occupies transporters more rapidly than does GBR 12909 or mazindol.[10]

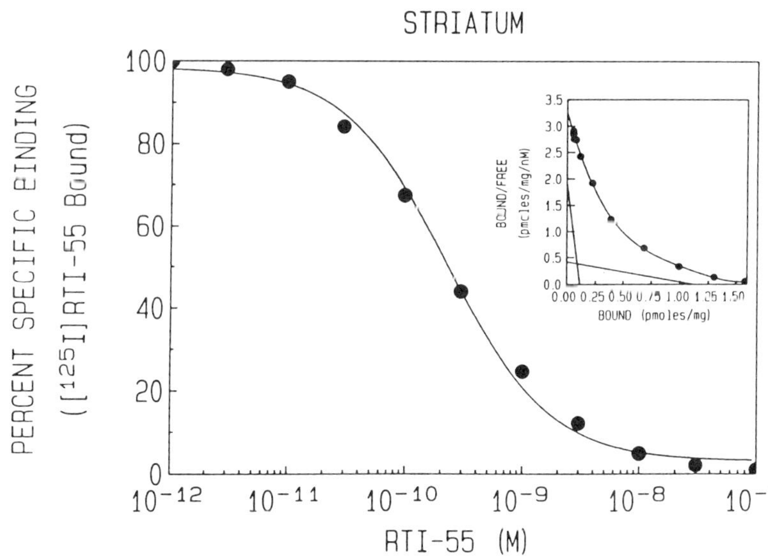

Figure 3. The binding of RTI-55, a cocaine analog, shows two sites in rat striatum. Striatal membranes were incubated with a fixed concentration of [^{125}I]RTI-55 (15 pM) and increasing concentrations of unlabeled RTI-55 (1 pM to 100 nM). Nonspecific binding was determined by the addition of 30 mM (–)cocaine. The results of a typical experiment are presented. The fit of this plot was determined using EBDA, and a two-site fit was significantly better than a one-site model. Insert: Scatchard transformation of the resulting data.[74]

Table 1. Potency of Cocaine in Competing for ^{3}H-Ligand Binding at the Dopamine Transporter

Ligand	K_i for Cocaine (nM)	Reference
[^{3}H]GBR 12935	782	Janowsky et al.[17]
	444	Anderson[18]
[^{3}H]Mazindol	540	Javitch et al.[20]
[^{3}H]Nomifensine	360	Dubocovich & Zahnizer[21]
[^{3}H]Methylphenidate	15.5	Schweri et al.[41]

analogs displace slightly fewer sites in cocaine binding experiments than do cocaine and related compounds, suggesting that these compounds are only associated with the low affinity site which accounts for most of their binding. Also, the K_i for cocaine binding in competition with other ^{3}H-ligands (Table 1) corresponds better to the low affinity binding site for cocaine than to the high affinity binding site (Table 2), at least in most cases. Thus, the presence of a high affinity binding site for cocaine and its analogs could be significant, and might reflect some unique neurochemical action that renders cocaine and related compounds with a higher abuse liability. This could occur simply because cocaine and these analogs are more potent at the high affinity site and therefore affect uptake at lower concentrations.

Some investigators observe only one site for cocaine analog binding, and suggest that additional sites represent an artifact of freezing and thawing,[22] although this has been refuted.[23] However, other investigators show that both sites are clearly associated with the transporter, as both are reduced by the injection of 6-hydroxydopamine, a neurotoxin which destroys dopaminergic nerve terminals.[15] Also, transfection of COS cells, which normally exhibit neither dopamine uptake nor cocaine analog binding, with a cDNA encoding the dopamine transporter from rat brain confers upon these cells both dopamine uptake and high and low affinity binding to a cocaine analog.[24,25] However, it is not known whether these high and low affinity sites represent an equilibrium of interconvertible states or whether in fact some other alteration of the protein such as glycosylation could produce these differences of affinity. It is known that the dopamine transporter is heterogeneous with regard to glycosylation.[26-28] For example, the molecular weight of the transporter in the nucleus accumbens is different from that of the transporter in the striatum, and this appears to be due to differences in glycosylation.[28] Thus, while some posttranslational processing difference might produce the high affinity form of the transporter, the significance is as yet unclear.

C. OTHER ISSUES

Yet another possibility is that the unique properties of cocaine in producing reinforcement or dependence are caused by a combination of several factors. It would indeed be difficult to show experimentally whether the simultaneous combination of binding site occupancy, rapid rate of entry into the brain, and interaction with a unique high affinity site is required to elicit the highly efficacious reinforcing properties of this drug. However, the recent cloning of cDNA encoding the dopamine transporter in the rat, cow, and human permit the exploration of the molecular properties of dopamine transporters to a degree and in

Table 2. Dissociation Constant of Cocaine at High (K_{dh}) and Low (K_{dl}) Affinity Binding Sites on Dopamine Transporters

K_{dh} (nM)	K_{dl} (nM)	Reference
16	660	Calligaro & Eldefrawi[57]
19	1120	Madras et al.[12]
52	890	Madras et al.[13]

a fashion that has not been previously possible. The ability to produce site-directed mutations in these cDNA molecules has revealed the likely presence of specific amino acid moieties that bind dopamine or cocaine and its analogs, which suggests that cocaine analog binding can be separated from dopamine transport (see below).[29]

It should be noted that the dopamine hypothesis was developed mainly based on data derived from the drug self-administration paradigm in animals. Thus, this hypothesis would apply to human addiction only in so far as the animal model applies to human addiction. Indeed, the main challenge of the dopamine hypothesis is to determine the extent of its role in human drug abuse.

D. HUMAN CLINICAL STUDIES

If the dopamine hypothesis is relevant to human cocaine abuse, then there are various pharmacological implications for treatment strategies. If potentiation of dopaminergic neurotransmission (Figure 1) is critical to cocaine abuse in humans, then blockade of dopamine receptors should attenuate the effects of cocaine in humans as it does in animals.[30] Early studies with dopamine receptor blockers showed promise,[31,32] but more recent studies, while showing a reduction in "drug liking", have not revealed prominent effects.[33]

If increased dopaminergic neurotransmission is a significant factor in cocaine addiction, and its reduction causes craving for cocaine, then dopamine receptor agonists should reduce craving. However, this has not been borne out.[34-36] In fact, craving might be highest immediately after receiving cocaine,[37] which further suggests the inadequacy of the dopamine model in dealing with human craving. Nevertheless, inhibition of the dopamine transporter might be important for initiation of the process of reinforcement, abuse, and dependence.

II. PHARMACOLOGY OF DOPAMINE TRANSPORT

The pharmacological properties of the dopamine transporter have been studied extensively.[38,39] More recently, a large number of cocaine analogs have been synthesized and tested at the dopamine, as well as at other transporters.[40] Irreversible acylating agents for the dopamine transporter have been synthesized.[41,42] The naturally occurring isomer of cocaine is the most potent compared to all seven other isomers.[43] Structure-activity studies indicate that the most important factor for activity at the dopamine transporter is configuration.[44] Probably the second most important factor is the substituent at C-2 which is normally a carbomethoxy group. Stereochemistry at C-2 is very important; substitution at the nitrogen atom can have large effects, particularly when election density is altered. Substitution at C-3 also can have major effects (e.g., RTI-31 is two orders of magnitude more potent than cocaine). The nitrogen atom can be moved to other positions in the ring system without great loss of potency.[40,43]

Some of these newer analogs are behaviorally active[45] and are more selective for the dopamine transporter than cocaine.[46] They are also useful as iodinated ligands for *in vitro* and *in vivo* binding studies, and for PET and SPECT imaging studies (Figure 4).[44,45,47–49] Numerous laboratories are currently synthesizing and testing yet additional cocaine analogs that are interesting and promising as experimental tools and potential medications.[50-53]

III. BIOCHEMICAL PROPERTIES OF THE DOPAMINE TRANSPORTER FROM RAT BRAIN

A. PHOTOAFFINITY PROBE STUDIES

[I^{125}]DEEP is an iodoazido-containing analog of GBR 12935 that has high affinity and reasonable specificity for the dopamine transporter. In binding experiments, [I^{125}]DEEP can be cross-linked to the transporter by photolysis. The properties of the [I^{125}]DEEP-dopamine transporter protein complex can be examined in SDS-page gels,[54] which reveal

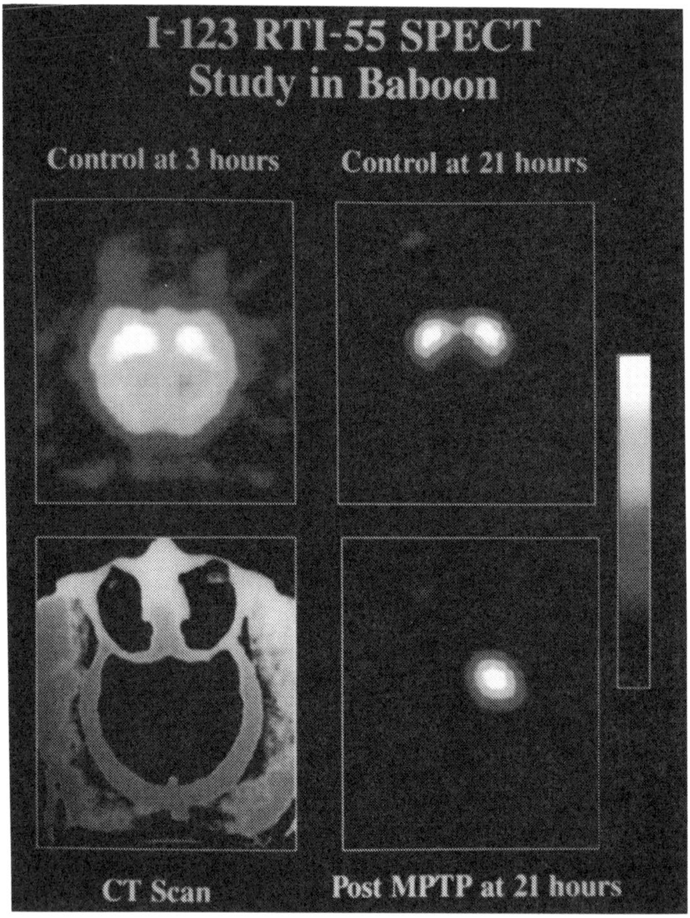

Figure 4. Top Left: [^{123}I]RTI-55 SPECT image at 3 hr postinjection, showing symmetrical tracer distribution in striata and cortical structures of control baboon. Top Right: [^{123}I]RTI-55 SPECT image at 21 hr postinjection, showing symmetrical binding in left and right striata of the control baboon. Bottom Left: CAT scan showing anatomical structures of the imaging plane that was used. Bottom Right: [^{123}I]RTI-55 SPECT image at 25 hr postinjection, showing lack of tracer accumulation in the right striatum (the side of MPTP pretreatment).[49]

both major and minor peaks of DEEP-labeled proteins (Figure 5). A major protein band (shown between the two arrows) of molecular weight 80 KDa, is present in the striatum and not in the cerebellum. Furthermore, mazindol which blocks DEEP binding to the dopamine transporter, prevents the formation of the DEEP protein complex while desipramine, a blocker of norepinephrine transporters, and citalopram, a blocker of serotonin transporters, were without effect. Thus, this 80 KDa protein has the properties of the dopamine transporter molecule.

Subsequent studies have shown that the transporter protein has an apparent molecular weight of 80 KDa, and is a glycoprotein containing sialic acid residues as well as N-linked glycosylation.[54-57] These findings are compatible with data from cloning experiments which reveal that the predicted amino acid sequence in the rat contains four possible glycosylation sites in an extracellular loop and has an amino acid composition consistent with this molecular weight.[24]

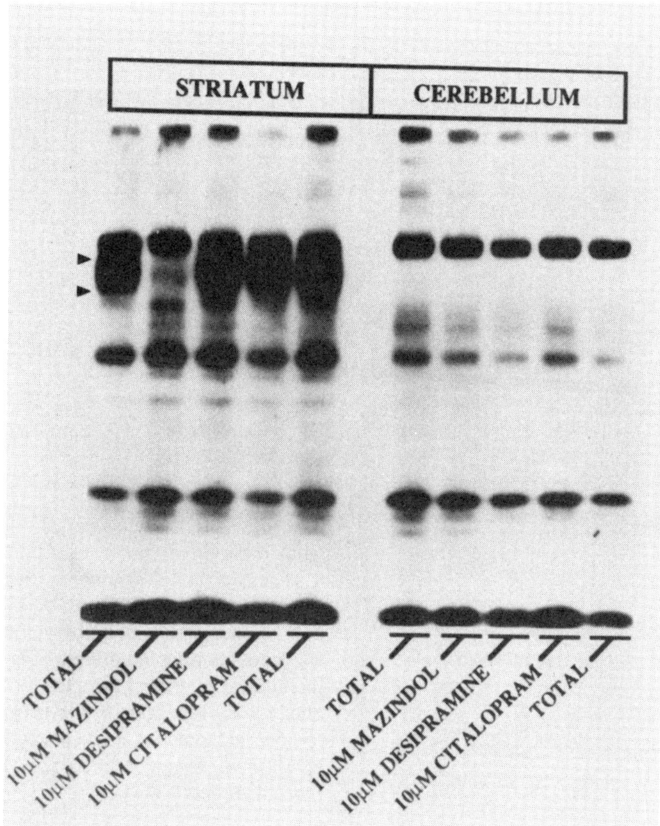

Figure 5. Covalent incorporation of [^{125}I]DEEP in rat striatal and cerebellar homogenates. Rat striatal and cerebellar homogenates were prepared in parallel, incubated with 1 to 2 nM [^{125}I]DEEP in the presence or absence of the compounds shown. Assays were conducted for 60 min at 0 to 4°C and the mixture was photolyzed. The autoradiogram shown here is overexposed to enhance the minor peaks of radioactivity in order to more thoroughly assess specificity; the bulk of radioactivity, however, localized within the 80 KDa band, shown by the arrows. Although identical photoincorporation of [^{125}I]DEEP to many nonspecific proteins was evident in both tissues, there was a conspicuous absence of any labeled proteins in the cerebellum with an apparent molecular weight of 80 KDa.[54]

B. TRANSPORTER HETEROGENEITY

In subsequent experiments, we examined the apparent molecular weights of the [^{125}I]DEEP-dopamine transporter complex from various brain regions.[28] We observed that the dopamine transporter from the nucleus accumbens appeared to have a slight but consistently higher molecular weight than the transporter in the striatum, and this was confirmed by the results of Western blot experiments using antibodies directed against the transporter.[58] This difference in the apparent molecular weight is not accompanied by large pharmacological or functional differences in the transporter from these two regions. While some laboratories have reported differences between dopamine transport in striatum vs. nucleus accumbens,[59,60] other groups[61] have produced data which is not completely in agreement with this. In fact, [^{3}H]GBR 12935 has a slightly different affinity for the dopamine transporter in the nucleus accumbens compared to the striatum in binding experiments (unpublished observations). While this difference is small, it was observed consistently in several experiments. Thus, the apparent difference in molecular weight of the dopamine transporter in different brain regions is not associated with a major difference in pharmacological properties.

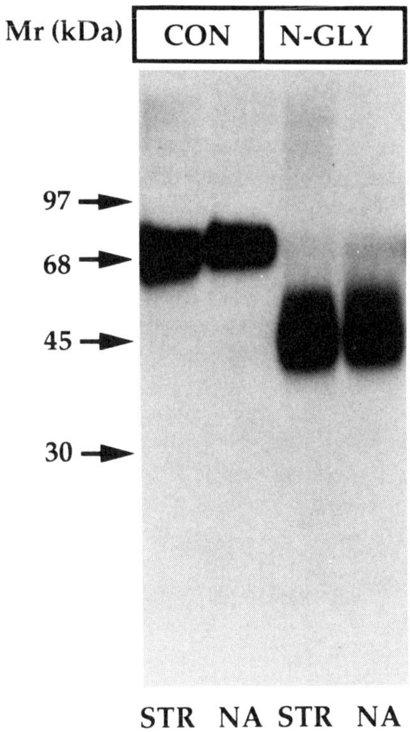

Figure 6. Effect of N-glycanase on [^{125}I]DEEP-labeled dopamine transporter from rat striatum (Str) and nucleus accumbens (Na). Treatment with N-glycanase, which removes N-linked carbohydrates, reduced the apparent molecular weight of the transporter from both brain regions (compared to control samples incubated in the absence of N-glycanase) and eliminated the difference in apparent molecular weight. Experiments were carried out four times with essentially the same results.[28]

The nucleus accumbens receives dopaminergic input mainly from limbic or midline ventral tegmental nucleus rather than the lateral neurons of the substantia nigra which supply the striatum.[63] Thus, dopamine transporters in the nucleus accumbens rather than those in the striatum are likely to be associated with the reinforcing effects of psychostimulant drugs since reinforcement has been associated with limbic brain regions.[5]

Are these regional differences of molecular weight due to different proteins expressed by different genes, or could differences in posttranslational processing account for these different molecular weights? Lew and co-workers[28] found that deglycosylation of transporters from the nucleus accumbens and striatum reduced the molecular weight of the DEEP-labeled transporter from both regions to a similar, apparently indistinguishable lower level (Figure 6). Thus, it appears that the heterogeneity is due at least partly to differences in glycosylation. Furthermore, the dopamine transporter changes its molecular weight during ontogeny (unpublished observations), and this change is reduced or eliminated when the transporters are deglycosylated with N-glycanase. However, these data do not rule out the presence of other modifications of the protein. While it is clear that these heterogeneities of molecular forms exist, neither their regional nor their ontogenetic functional significance is understood.

IV. EXPRESSION AND REGULATION OF THE DOPAMINE TRANSPORTER

The dopamine transporter has been thoroughly studied in meso-telencephalic dopamine systems. The distribution of transporters corresponds closely to the projections areas.[63,64] However, some dopamine-containing neurons, such as those in the arcuate nucleus, have much lower transport activity,[66] which is commensurate with their role in supplying

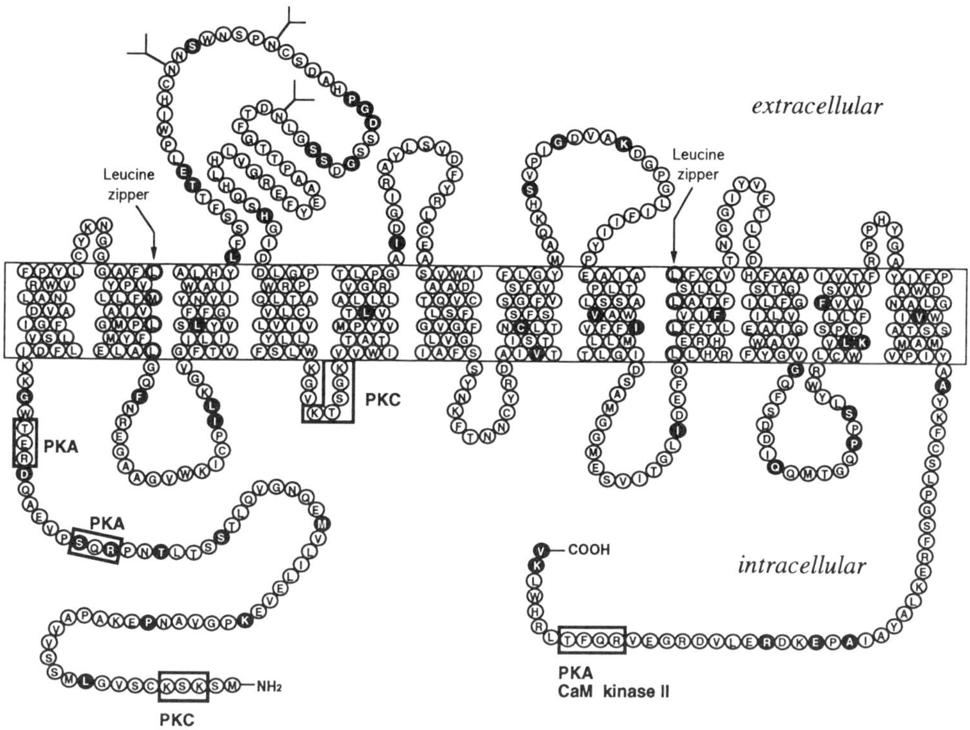

Figure 7. Proposed model of human dopamine transporter. Amino acid residues in black are those that differ between rat and human sequences. Y-shaped symbols represent potential sites for N-glycosylation. Boxed residues represent consensus phosphorylation sites for protein kinase A (PKA), protein kinase C (PKC), or Ca^{2+}-calmodulin-dependent kinase II (CaM kinase II). Disposition of the transporter protein with respect to the plasma membrane is putative. Two leucine zipper motifs are indicated.[71]

dopamine to the hypothalamohypophyseal portal system. With the success of cloning, suitable probes are now available for *in situ* hybridization histochemical experiments. Thorough studies of dopamine-containing cell groups reveal that dopaminergic cell bodies in the midbrain have significantly more mRNA for the transporter than other dopaminergic cells;[66,67] cells in the arcuate nucleus, retina, and olfactory bulb were found to have significantly lower quantities of mRNA.

There is evidence that neuronal transporters for choline and glutamate can be regulated by neuronal activity and nerve terminal depolarization.[68,69] These studies were carried using synaptosomes, and there is no detailed molecular knowledge of how these changes in transporter activity occur. However, recent cloning studies indicate that the predicted transporter proteins contain numerous phosphorylation sites (Figure 7).[70-72] While strong evidence is lacking that these sites are in fact phosphorylated and can cause regulation, relevant results are expected in the near future.

Another mechanism of transporter regulation is an alteration of gene expression. Many investigators have demonstrated changes in dopaminergic function during cocaine withdrawal. For example, Pilotte et al. have shown that withdrawal from repeated cocaine administration in Lewis rats downregulates dopamine transporter binding sites in the nucleus accumbens but not in the striatum after ten days of withdrawal (unpublished observations). This small change (25 to 45%) is highly consistent and its localization to the nucleus accumbens rather than the striatum suggests some role in the reinforcing or dependence-producing properties of cocaine. Cerruti et al.[73] have observed a reduction of

Table 3. Amount of Dopamine Transporter mRNA After Withdrawal of Cocaine

Nucleus	Projection Areas	% of Control
Substantia Nigra Compacta	Striatum	94
Paranigral N.	Septum	92
Rostral Linear N.	Septum	103
Interfascicular N.	Nuc Accumbens	78*
Caudal Linear N.	Nuc Accumbens	64*

* $p \leq 0.02$.

Data derived from Cerruti et al.[67] Labeling of dopamine transporter mRNA by *in situ* hybridization histochemistry was significantly reduced in the interfascular and caudal linear nuclei in animals withdrawn from cocaine, compared to saline-injected control animals. See text for additional details.

mRNA in some cell bodies as measured by *in situ* hybridization histochemistry in similarly treated animals (Table 3). Moreover, the reduction of mRNA occurs precisely in those nuclei that project to the nucleus accumbens and not in the substantia nigra zona compacta, which projects to the striatum. Thus, the biochemical data indicate a downregulation of dopamine transporter proteins during withdrawal and this regulatory change appears to be due to a change in gene expression.

Experimental data are beginning to accumulate suggesting that phosphorylation as well as changes in gene expression can regulate the production and activity of dopamine transporters. Further progress will be welcome in this rapidly developing area.

REFERENCES

1. Stolerman, I., Drugs of abuse, behavioral principals, methods and terms, *Trends Pharmacol. Sci.,* 13, 170, 1992.
2. Wise, R. A. and Bozarth, M. A., A psychomotor stimulant theory of addiction, *Psychol. Rev.,* 94, 469, 1987.
3. Koob, G. F. and Bloom, F. E., Cellular and molecular mechanisms of drug dependence, *Science,* 242, 715, 1988.
4. Kuhar, M. J., Ritz, M. C., and Boja, J. W., The dopamine hypothesis of the reinforcing properties of cocaine, *Trends Neurosci.,* 14, 299, 1991.
5. Robinson, T. and Berridge K. C., The neural basis of drug craving: an incentive-sensitization theory of addiction, *Brain Res. Rev.,* 18, 247, 1993.
6. Ritz, M. C., Lamb, R. J., Goldberg, S. R., and Kuhar, M. J., Cocaine receptors on dopamine transporters are related to self-administration of cocaine, *Science,* 237, 1219, 1987.
7. Ritz, M. C., Lamb, R. J., Goldberg, S. R., and Kuhar, M. J., Cocaine self-administration appears to be mediated by dopamine uptake inhibition, *Prog. Neuro-Psychopharmacol. Biol. Psychiatr.,* 12, 233, 1988.
8. Chait, L. D., Uhlenherth, E. H., and Johanson, C. E., The reinforcing and subjective effects of several anorectics in normal human volunteers, *J. Pharmacol. Exp. Ther.,* 242, 777, 1987.
9. Rothman, R. B., High affinity dopamine reuptake inhibitors as potential cocaine antagonists: a strategy for drug development, *Life Sci.,* 46, 17, 1990.
10. Pogun, S., Scheffel, U., and Kuhar, M. J., Cocaine displaced [^{3}H]WIN 35,428 binding to dopamine uptake sites *in vivo* more rapidly than mazindol or GBR 12909, *Eur. J. Pharmacol.,* 198, 203, 1991.
11. Johanson, C. E. and Fischman, M. W., The pharmacology of cocaine related to its abuse, *Pharmacol. Rev.,* 41, 3, 1989.
12. Madras, B. K., Spealman, R. D., Fahey, M. A., Neumeyer, J. L., Saha, J.K., and Milius, R.A., Cocaine receptors labeled by [^{3}H]2B-carbomethoxy-^{3}B-(4-fluorophenyl)tropane, *Mol. Pharmacol.,* 36, 518, 1989a.
13. Madras, B. K., Fahey, M. A., Bergman, J., Canfield, D. R., and Spealman, R. D., Effects of cocaine and related drugs in non-human primates. I. [^{3}H]cocaine binding sites in caudate-putamen, *J. Pharmacol. Exp. Ther.,* 251, 131, 1989.

14. Ritz, M. C., Boja, J. W., Zaczek, R., Carroll, F. I., and Kuhar, M. J., ^{3}H WIN 35,065-2: a ligand for cocaine receptors in rat striatum, *Soc. Neurosci. Abstr.,* 15, 1092, 1989.
15. Ritz, M. C., Boja, J. W., Zaczek, R., Carroll F. I., and Kuhar, M. J., ^{3}H WIN 35,065-2: a ligand for cocaine receptors in striatum, *J. Neurochem.,* 55, 1556, 1990.
16. Boja, J. W., Patel, A., Carroll, F. I., Rahman, M. A., Phillip, A., Lewin, A. H., Kopajtic, T. A., and Kuhar, M. J., [^{125}I]RTI-55: a potent ligand for dopamine transporters, *Eur. J. Pharmacol.,* 194, 133, 1991.
17. Janowsky, A., Schweri, M., Berger, P., Long, R., Skolnick, P., and Paul, S., The effects of chemical and surgical lesions on striatal [^{3}H]Threo-(±)-methylphenidate binding: correlation with [^{3}H]dopamine uptake, *Eur. J. Pharmacol.,* 108, 187, 1985.
18. Anderson, P. H., Biochemical and pharmacological characterization of [^{3}H]GBR 12935 binding *in vitro* to rat striatal membranes: labeling of the dopamine uptake complex, *J. Neurochem.,* 48, 1887, 1987.
19. Berger, P., Janowsky, A., Vocci, F., Skolnick, P., Schweri, M. M., and Paul, S. M., [^{3}H] GBR 12935: a specific high affinity ligand for labeling the dopamine transport complex, *Eur. J. Pharmacol.,* 107, 289, 1985.
20. Javitch, J. A., Blaustein, R. O., and Snyder, S. J., [^{3}H]Mazindol binding associated with neuronal dopamine uptake sites in corpus striatum membranes, *Eur. J. Pharmacol.,* 90, 461, 1983.
21. Dubocovich, M. L. and Zahniser, N. R., Binding characteristics of the dopamine uptake inhibitor [^{3}H]nomifensine to striatal membranes, *Biochem. Pharmacol.,* 34, 1137, 1985.
22. Kirifides, A. L., Harvey, J. A., and Aloyo, V. J., The low affinity binding site for the cocaine analog, WIN 35,428 is an artifact of freezing caudate tissue, *Life Sci.,* 50, 139, 1992.
23. Izenwasser, S., Rosenberger, J. G., and Cox B. M., The cocaine analog WIN 35,428 binds to two sites in fresh rat caudate-putamen: significance of assay procedures, *Life Sci.,* 52, 141, 1993.
24. Shimada, S., Kitayama, S., Lin, C. -L., Patel, A., Nathankumar, E., Gregor, P., Kuhar, M., and Uhl, G., Cloning and expression of cocaine-sensitive dopamine transporter complementary DNA, *Science,* 254, 576, 1991.
25. Boja, J. W., Markham, L., Patel, A., Uhl, G., and Kuhar, M. J., Expression of a single dopamine transporter cDNA can confer two cocaine binding sites, *NeuroReport,* 3, 247, 1992.
26. Lew, R., Grigoriadis, D. E., Wilson, A. A., and Kuhar, M. J., Dopamine transporter in rat caudate-putamen is glycosylated, *Soc. Neurosci. Abstr.,* 15, 107, 1989.
27. Lew, R., Vaughan, R., Simantov, R., Wilson, A., and Kuhar, M. J., Dopamine transporters in the nucleus accumbens and the striatum have different apparent molecular weights, *Synapse,* 8, 152, 1991.
28. Lew, R., Patel, A., Vaughan, R. A., Wilson, A., and Kuhar, M. J., Microheterogeneity of dopamine transporters in rat striatum and nucleus accumbens, *Brain Res.,* 584, 266, 1992.
29. Kitayama, S., Shimada, S., Xu, H., Markham, L., Donovan, D. M., and Uhl, G. R., Dopamine transporter site-directed mutations differentially alter substrate transport and cocaine binding, *Proc. Natl. Acad. Sci.USA,* 89, 7782, 1992.
30. Jonsson, L. E., Anggard, E., and Gunne, L. M., Blockade of intraveneous amphetamine euphoria in man, *Clin. Pharm. Ther.,* 12, 889, 1971.
31. Gunne, L. M., Anggard, E., and Jonsson L. E., Clinical trials with amphetamine-blocking drugs, *Psychiatr., Neuro., Neurochir.,* 75, 225, 1972.
32. Gawin, F. H., Allen, D., and Humblestone B., Outpatient treatment of "crack" cocaine smoking with flupenthixol decanoate. A primary report, *Arch. Gen. Psychiatr.,* 46, 322, 1989.
33. Sherer, M. A., Kumor, K. M., and Jaffe J. H., Effects of intravenous cocaine are partially attentuated by haloperidol, *Psychiatr. Res.,* 27, 117, 1989.
34. Weddington, W. W., Brown, B. S., Hoertyen, C. A., Hess, J. M., Mahaffey, J. M., Kolar, J. R., and Jeffe, A. F., Comparisons of antibodies and desipramine combined with psychotherapy for treatment of cocaine dependence, *Am. J. Drug Alcoh. Abuse,* 17, 137, 1991.
35. Gawin, F., Riordan, C., and Kleber H., Methylphenidate treatment of cocaine abusers without attention deficit disorder: a negative report, *Am. J. Drug Alcoh. Abuse,* 11, 193, 1985.
36. Preston, K., Sullivan, J. T., Berger, P., and Bigelow G. E., Effects of cocaine alone and in combination with mazindol in human cocaine abusers, *JPET,* 267(1), 296, 1993.
37. Jaffe, J. H., Cascella, N. G., Kumor, K. M., and Sherer M. A., Cocaine-induced cocaine craving, *Psychopharmacology,* 97, 59, 1989.
38. Horn, A. S., Characteristics of neuronal dopamine uptake, *Adv. Biochem. Psychopharmacol.,* 19, 25, 1978.

39. Horn, A. S., Dopamine uptake: a review of progress in the last decade, *Prog. Neurobiol.*, 34, 387, 1990.
40. Carroll, F. I., Lewin, A. H., Boja, J. W., and Kuhar, M. J., Cocaine receptor: biochemical characterization and structure-activity relationships of cocaine analogues at the dopamine transporter, *J. Med. Chem.*, 35, 969, 1992.
41. Schweri, M. M., Skolnick, P., Rafferty, M. F., Rice, K. C., Janowsky, A. J., and Paul, S. M., [^{3}H]Threo-(±)-methylphenidate binding to 3,4-dihydroxyphenylethylamine uptake sites in corpus striatum correlation with the stimulant properties of ritalinic acid esters, *J. Neurochem.*, 45, 1062, 1985.
42. Boja, J. W., Rahman, M. A., Philip, A., Lewin, A., Carroll, F. I., and Kuhar, M. J., Isothiocyanate derivatives of cocaine: irreversible inhibition of ligand binding at the dopamine transporter, *Mol. Pharmacol.*, 39, 339, 1991.
43. Carroll, F. I., Lewin, A. H., Abraham, P., Parham, K., Boja, J. W., and Kuhar, M. J., Synthesis and ligand binding of cocaine isomers at the cocaine receptor, *J. Med. Chem.*, 34, 883, 1991.
44. Carroll, F. I., Rahman, M. A., Abraham, P., Parham, K., Lewin, A. H., Dannals, R. F., Shaya, E., Scheffel U., Wong, D. F., Boja, J. W., and Kuhar, M. J., [^{123}I] 3-(4-Iodephenyl)tropan-2-carboxylic acid methyl ester (RTI-55), a unique cocaine receptor ligand for imaging the dopamine and serotonin transporters in vivo, *Med. Chem. Res.*, 1, 289, 1991.
45. Boja, J. W., Carroll, F. I., Rahman, M. A., Philip, A., Lewin, A. H., and Kuhar, M. J., New, potent cocaine analogs: ligand binding and transport studies in rat striatum, *Eur. J. Pharmacol.*, 184, 329, 1990.
46. Boja, J. W., McNeill, R. M., Lewin, A. H., Abraham, P., Carroll, F. I., and Kuhar, M. J., Selective dopamine transporter inhibition by cocaine analogs, *NeuroReport,* 3, 984, 1992.
47. Boja, J. W., Patel, A., Phillip, A., Lewin, A. H., Kopajtic, T. A., and Kuhar, M. J., [^{125}I] RTI-55: a potent ligand for dopamine transporters, *Eur. J. Pharmacol.*, 194, 133, 1991.
48. Innis, R., Baldwin, R., Sybirska E., Zea, Y., Laruell, M., Al-Tikriti, M., Charney, D., Zoghbi, S., Wisniewski, G., Hoffer, P., Wang, S., Millius, R., and Neumeyer, J., Single photon emission computed tomography imaging of monoamine reuptake sites in primate brain with [^{123}I] CIT. *Eur. J. Pharmacol.*, 200, 369, 1991.
49. Shaya, E. K., Scheffel, U., Dannals, R. F., Ricaurte, G. A., Carroll, F. I., Wagner, H. N., Jr., Kuhar, M. J., and Wong, D. F., *In vivo* imaging of dopamine reuptake sites in the primate brain using single photon emission computed tomography (SPECT) and iodine-123 labeled RTI-55, *Synapse,* 10, 169, 1992.
50. Davies, H. M. L., Saikali, E., Sexton, T., and Childers, S., Novel 2-substituted cocaine analogs: binding properties at dopamine transport sites in rat striatum, *Eur. J. Pharmacol. - Mol. Pharmacol. Sect.*, 244, 93, 1993.
51. Kozikowski, A.P., Roberti, M., Xiang, L., Bergmann, J. S., Callahan, P. M., Cunningham, K., and Johnson, K. M., Structure-activity relationship studies of cocaine: replacement of the C-2 ester group by vinyl argues against H-bonding and provides an esterase-resistant, high-affinity cocaine analogue, *J. Med. Chem.*, 35, 4764, 1992.
52. Kline, R. H., Jr., Wright, J., Fox, K. M., and Eldefrawi, M. E., Synthesis of 3-arylecgonine analogues as inhibitors of cocaine binding and dopamine uptake, *J. Med. Chem.*, 33, 2024, 1990.
53. Meltzer, P. C., Liang, A. Y., Brownell, A.-L., Elmaleh, D. R., and Madras, B. K., Substituted 3-Phenyltropane analogs of cocaine: synthesis, inhibition of binding at cocaine recognition sites, and positron emission tomography imaging, *J. Med. Chem.*, 36, 855, 1993.
54. Grigoriadis, D. E., Wilson, A. A., Lew, R., Sharkey, J. S., and Kuhar, M. J., Dopamine transport sites selectively labeled by a novel photoaffinity probe: ^{125}I-DEEP, *J. Neurosci.*, 9(8), 2664, 1989.
55. Sallee, F. R., Fogel, E. L., Schwartz, E., Choi, S. M., Curran, D. P., and Niznik, H. B., Photoaffinity labeling of the mammalian dopamine transporter, *FEBS Lett.*, 256, 219, 1989.
56. Berger, P., Martenson, R., Laing, P., Thurcaug, A., DeCosta, B., Rice, K. C., and Paul, S. M., Photoaffinity labeling of the dopamine reuptake carrier protein using a novel high affinity ayidoderivative of GBR-12935, *Soc. Neurosci. Abstr.*, 16, 13, 1990.
57. Calligaro, D. O. and Eldefrawi, M. E., High affinity stereospecific binding of [^{3}H]cocaine in striatum and its relationship to the dopamine transporter, *Membr. Biochem.*, 7, 87, 1988.
58. Vaughan, R. A., Uhl, G., and Kuhar, M. J., Recognition of dopamine transporters by antipeptide antibodies, *Mol. Cell. Neurosci.*, 4, 209, 1993.

59. Missale, C., Castelletti, L., Govoni, S., Spano, P. F., Trabucchi, M., and Hanbauer, I., Dopamine uptake is differentially regulated in rat striatum and nucleus accumbens, *J. Neurochem.*, 45, 51, 1985.
60. Izenwasser, S. and Cox, B. M., Daily cocaine treatment produces a persistent reduction of ^{3}H-dopamine uptake *in vitro* in rat nucleus accumbens but not in striatum, *Brain Res.*, 531, 338, 1990.
61. Boja, J. W. and Kuhar, M. J., ^{3}H-Cocaine binding and inhibition of ^{3}H-dopamine uptake is similar in both rat striatum and nucleus accumbens, *Eur. J. Pharmacol.*, 173, 215, 1989.
62. Bjorklund, A. and Lindvall, O., Dopamine-containing systems in the CNS, in *Handbook of Chemical Neuroanatomy, Vol. 2: Classical Transmitters in the CNS,* Bjorklund, A. and Hokfelt, T., Eds., Elsevier, New York, 1984, 55.
63. Kaufman, M. J. and Madras, B. K., Distribution of cocaine recognition sites in monkey brain: ex-vivo autoradiography with [^{3}H]CFT and [^{125}I]RTI-55, *Synapse,* 12, 99, 1992.
64. Canfield, D. R., Spealman, R. D., Kaufman, M. J., and Madras B. K., Autoradiographic localization of cocaine binding sites in [^{3}H]CFT in the monkey brain, *Synapse,* 6, 189, 1990.
65. Demarest, K. T. and Moore, K. E., Lack of a high affinity transport system for dopamine in the median eminence and posterior pituitary, *Brain Res.*, 171, 545, 1979.
66. Shimada, S., Kitayama, S., Walther, D., and Uhl, G., Dopamine transporter mRNA: dense expression in ventral midbrain neurons, *Mol. Brain Res.*, 13, 359, 1992.
67. Cerruti, C., Walther, D. M., Kuhar, M. J., and Uhl, G. R., Dopamine transporter mRNA expression is intense in rat midbrain neurons and modest outside midbrain, *Mol. Brain Res.*, 18, 181, 1993.
68. Murrin, L.C. and Kuhar, M. J., *In vitro* activation of high affinity choline uptake by depolarizing agents, *Mol. Pharmacol.*, 12, 1082, 1976.
69. Murrin, L. C., Lewis, M. S., and Kuhar, M. J. Amino acid transport: alterations due to synaptosomal depolarization, *Life Sci.*, 22, 2009, 1978.
70. Amara, S. and Kuhar M. J., Neurotransmitter transporters: recent progress, *Annu. Rev. Neurosci.*, 16, 73, 1993.
71. Uhl, G. R. and Hartig, P. R., Transporter explosion: update on uptake, *Trends Pharmacol. Sci.*, 13(12), 421, 1992.
72. Giros, B. and Caron, M. G., Molecular characterization of the dopamine transporter, *Trends Pharmacol. Sci.*, 14, 43, 1993.
73. Cerruti, C., Pilotte, N., Uhl, G., and Kuhar, M. J., Reduction in dopamine transporter mRNA after cessation of repeated cocaine administration, *Mol. Brain Res.*, 1994.
74. Boja, J. W., Mitchell, W. M., Patel, A., Kopajtic, T. A., Carroll, F. I., Lewin, A. H., Abraham, P., and Kuhar, M. J., High affinity binding of [^{125}I]RTI-55 to dopamine and serotonin transporters in the rat brain, *Synapse,* 12, 27, 1992.

Chapter 13

Effects of Cocaine on Immediate-Early Gene Response in Striatal Neurons

Ann M. Graybiel, Sabina Berretta, Rosario Moratalla, Fu-Chin Liu, and Bulent Elibol

CONTENTS

I. INTRODUCTION

Cocaine has powerful acute psychomotor activating effects and, when given repeatedly, can induce prolonged changes in behavior and behavioral sensitization.[1,2] How cocaine exposure can lead to long-term effects on behavior is a major unsolved issue. Here, we summarize the results of experiments aimed at identifying some of the molecular events associated with this form of neuroplasticity.

Cocaine binds with high affinity to the dopamine transporter, and, as a consequence, blocks the principal pathway for clearance of dopamine released by neurons.[3] This blockade produces large elevations in dopamine at dopaminergic nerve terminals. This strong dopaminergic stimulation is thought to account for many of the reinforcing effects of the drug and for its actions as a psychomotor activator.[4,5] Cocaine also blocks uptake of serotonin and norepinephrine. The role of these monoamines in cocaine-induced behaviors has not been established clearly, but selective uptake inhibitors of serotonin and of norepinephrine have not been found to have cocaine-like behavioral effects.[6] The potency of cocaine-like drugs as positive reinforcers correlates with their potency as inhibitors of dopamine uptake, but not of either serotonin or norepinephrine.[3]

This and other evidence indicates that the dopaminergic system is a principal neural substrate for the reinforcing aspects of addictive behavior. Similarly, the acute stimulant effects of cocaine on locomotion and stereotyped behaviors, and the sensitization to these effects by repeated treatment, are thought to be related to increased dopaminergic transmission in mesolimbic and nigrostriatal dopaminergic pathways (see References 1, 7-13).

II. OVERVIEW OF IMMEDIATE-EARLY GENE REGULATION BY ACUTE COCAINE

A new line of research on the neurobiological effects of cocaine has begun with the discovery that cocaine and other psychomotor stimulants regulate the expression of immediate-early genes in neurons. These genes are induced rapidly following stimulation, and are known to encode DNA binding proteins (transcription factors) that regulate

1-8493-8311-0/95/$0.00+$.50

the expression of target genes. The fact that cocaine can change the expression pattern of such genes suggests a mechanism whereby not only the short-term activity of these neurons, but also their long-term functioning, could be modified significantly by drug exposure.[13-29] The precise patterns of regulation of transcription factor expression determine which cascades of altered neuronal gene expression occur after exposure to the drug. These changes, in turn, mediate some of the long-term behavioral effects of cocaine.[14,16,30,31]

Using methods suitable for detecting individual transcription factors and their protein products, it has become possible to identify single neurons and populations of neurons activated to express these molecules in response to drug exposure. It is clear that the activation of particular immediate-early genes is dose-dependent and selective in terms of the responsive neurons and genes following cocaine administration in rats. Included in these activated neural systems are those known to participate in the control of movement and motivated behaviors. Various laboratories are engaged actively in studying the regulation of immediate-early gene expression in brain by psychomotor stimulants. We summarize here some of the work from our own laboratory aimed at analyzing patterns of gene activation in the striatum.

III. THE RESPONSE TO ACUTE EXPOSURE TO COCAINE

Following a single intraperitoneal injection of cocaine in adult rats or mice, a rapid and stereotyped pattern of immediate-early gene activation occurs in the brain. For example, as assayed by Northern blotting, mRNA for genes of the *fos-jun* family appear within an hour. This response is brief, with mRNA levels returning to baseline or near baseline by approximately 3 to 6 hr.[32] The striatum is affected prominently, as is the cerebral cortex. Low levels of activation occur in the hippocampus and in some other brain sites. Neurons in the substantia nigra and ventral tegmental area do not show detectable activation of these genes.

A. SELECTIVE INDUCTION OF IMMEDIATE-EARLY GENES

To screen for immediate-early genes activated by acute exposure to cocaine, we first assayed the striatum for expression of the protein products of such genes with polyclonal antisera raised against peptide sequences in Fos-, Fos/Fra-, JunB-, and FosB-like proteins. We also used oligonucleotide probes complementary to the sequences encoding c-*fos*, c-*jun*, *jun*B, *jun*D, and *NGFI*-A (*zif*268 or *egr*1) to study mRNA expression in the striatum.[14,18,25,32-34] Our results suggest that there is considerable selectivity of immediate-early gene induction in the striatum elicited by acute exposure to cocaine. Experiments using Northern blotting and *in situ* hybridization histochemistry indicate that the expression of c-*jun* mRNA does not increase after cocaine injection, whereas the expression of c-*fos*, *jun*B, and *NGFI*-A mRNA is rapidly enhanced in a dose-dependent manner (Figure 1). Northern analyses suggests that *jun*D is expressed constitutively at sufficiently high levels to obscure drug-induced changes. These findings are in accord with those of others (see References 16, 17, 26, and 35).

Although most emphasis in the literature on immediate-early gene regulation has been placed on Fos-like protein, the response to acute cocaine treatment detectable with probes for c-*fos* mRNA appears to be weaker than that observed using probes for *jun*B or *NGFI*-A mRNA (Figure 2).[25,33] Immunohistochemical staining of Fos-like protein is also generally weaker than that of JunB- or Fos/Fra-like proteins. This suggests further differential induction among the genes responsive to acute cocaine challenge.

B. NEUROTRANSMITTER RECEPTORS MEDIATING THE ACUTE EFFECT

Dopamine receptors of both the D_1-like and D_2-like families are stimulated by the release of dopamine in the striatum. Pharmacological evidence strongly suggests that the induction

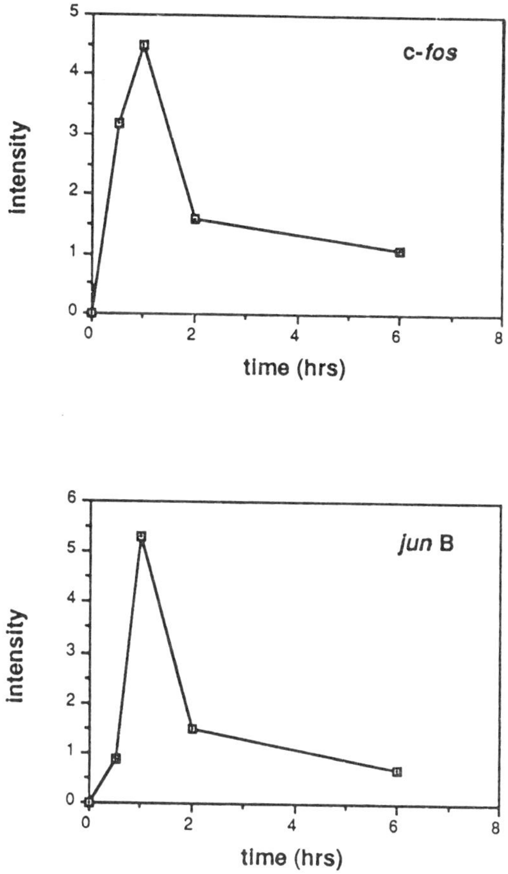

Figure 1. Time course of induction of c-*fos* and *jun*B mRNA in the striatum after a single systemic injection of cocaine (25 mg/kg, i.p.). Striatal RNA was isolated and analyzed by Northern blotting using a phosphorimager to measure the intensity of expression at different times following cocaine injection.

of those transcription factors analyzed thus far requires, at a minimum, stimulation of D_1-like receptors.[14-16,18-20,23,26] Responses to *in vivo* cocaine treatment at doses as high as 25 mg/kg are blocked effectively by pretreatment with the D_1-antagonist SCH23390 at doses thought to be selective for D_1-like receptors.

The dependence of these cocaine effects on D_1-like receptors suggests that adenylate cyclase activation might mediate part of the induction of *fos-jun* family genes and other immediate-early genes by cocaine. This view is consistent with the evidence that forskolin induces c-*fos* in striatal cell cultures[36] and Fos/Fra-like proteins in organotypic striatal slice cultures.[37,38] Considering the putative role of adenylate cyclases and cAMP as convergence points for intracellular cascades,[39] the selective changes in expression of cAMP-regulated transcription factors might indicate a specific cellular signal integration evoked by acute cocaine exposure.

In PC12 cells, membrane depolarization induces c-*fos* through calcium influx and a Ca^{++}-calmodulin-dependent protein kinase-mediated mechanism. Phosphorylation, and therefore activation, of the cAMP response element binding protein (CREB) at ser^{133} has been implicated in this effect.[40–42] CREB also has been shown to undergo phosphorylation in striatal slice cultures treated with agonists of D_1-like dopamine receptors, cAMP analogs, or forskolin treatment.[43] Preliminary evidence suggests that there is rapid phosphorylation of CREB *in vivo* following exposure to the psychomotor stimulant amphetamine,[44,45] which is known to release dopamine in the striatum. CREB also is phosphorylated following *in vivo* intrastriatal injections of forskolin.[46] Direct examination of CREB activation by exposure to cocaine has not, however, been reported.

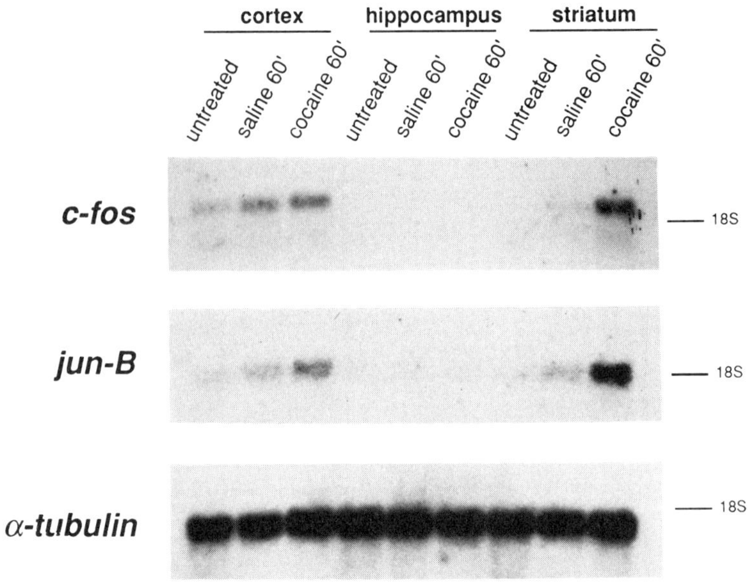

Figure 2. Northern hybridization with c-*fos* and *jun*B radiolabeled probes shows the induction of c-*fos* and *jun*B mRNA in rats treated with 25 mg/kg of cocaine, compared to saline-treated and untreated animals. RNA samples from cerebral cortex, hippocampus, or striatum were blotted and hybridized with the c-*fos* probe. The blot was then stripped, rehybridized with α-tubulin probe, and then again with *jun*B probe.

Receptors other than D_1-like dopamine receptors also appear to be involved in the induction of immediate-early genes in the striatum following cocaine treatment; D_2-like dopamine receptors also participate in the response. Although D_2-like receptor antagonists themselves induce Fos-like protein in the striatum, such compounds appear to *reduce* the amount of Fos-like protein present following cocaine exposure,[14] suggesting another example of dopamine D_1/D_2 synergism.[27,47] Blockade of serotonin receptors also reduces the induction of striatal c-*fos* and *NGFI*-A expression by cocaine,[19] indicating that one or more subtypes of the serotonin receptor also might contribute to the observed pattern of gene regulation. Glutamate receptors of both the NMDA and non-NMDA classes have been implicated by experiments using antagonists.[48–51]

1. Selective Activation of Specific Neural Populations in the Striatum

Immediate-early gene induction in the striatum following acute cocaine treatment is especially strong in the central and medial regions of the caudatoputamen and parts of the nucleus accumbens-ventral striatum complex. With increasing dose, the number of labeled nuclei increases, and the field of induction becomes larger. Within these striatal regions, many neurons are responsive, including neurons in both striosome and matrix compartments. Double immunolabeling demonstrates striking selectivity in the striatal neuronal subpopulations that show enhanced expression of Fos-like protein.[15,18,24,52] These include principally the dopamine and cyclic-AMP-regulated phosphoprotein (DARPP-32)-positive projection neurons and exclude, with rare exceptions, enkephalin-positive projection neurons (Figure 3). Many of the responsive DARPP-32 positive neurons can be identified as dynorphin-containing projection neurons (unpublished results).[24] It is striking that almost no striatal interneurons show expression of Fos-like protein after cocaine treatment,[18] although other *in vivo* treatments can activate Fos/Jun-like proteins in these neurons.[53,54]

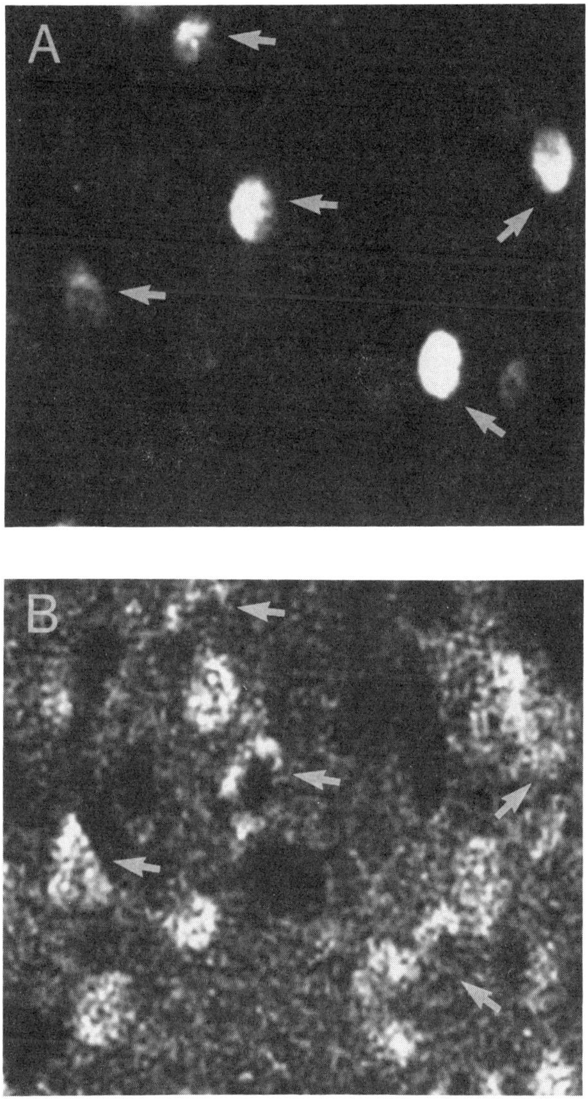

Figure 3. Confocal microscope images of a single region of the caudoputamen showing (A) Fos-like immunolabeling following a single injection of cocaine (25 mg/kg) and (B) DARPP-32-like immunoreactivity. Arrows indicate neurons positive for both Fos and DARPP-32.

These findings suggest that the main groups of striatal neurons responding to acute cocaine administration are the dynorphin-containing neurons of striosomes, which project robustly to the substantia nigra, and neurons giving rise to the "direct" pathways of the basal ganglia (which also contain dynorphin).[24,52,55,56] This result is of considerable interest given the psychomotor stimulant effect of these drugs. The "direct" dynorphin-containing pathway from the striatum to the pallidonigral complex is thought to facilitate movement, and striosomes are aligned more closely with the limbic system.[57,58] The striosomal system may predominantly influence dopaminergic neurons of the substantia nigra.

IV. IMMEDIATE-EARLY GENE ACTIVATION IN THE STRIATUM FOLLOWING CHRONIC EXPOSURE TO COCAINE

The picture emerging from the results discussed above is that the response of striatal neurons to acute cocaine treatment is rapid, brief, and selective for particular genes in discrete subpopulations of striatal neurons. There is growing evidence that chronic cocaine treatment produces marked changes in these response patterns.[21,28,29,33–35,59,60]

First, there is differential down-regulation of different *fos-jun* gene products. We, like others,[21,28,29,35,59] have observed that chronic cocaine treatment leads to a dramatic reduction of detectable Fos-like protein induced by a subsequent cocaine challenge. The same chronic treatment (25 mg/kg twice a day for 4 to 7 days in our experiments), however, evokes less complete down-regulation of Fra-like, JunB-like, and FosB-like proteins. Second, the activation of some *fos-jun* gene products is prolonged. We compared the expression of Fos-like and JunB-like proteins, and found that the Fra-like and FosB-like proteins persist for several days in striatal neurons without further challenge.[33] These results are in accord with those of Iadarola and colleagues[29] and Nestler and colleagues (B. Hope and E. Nestler, personal communication), who have suggested that these persistent proteins are newly synthesized as a result of the chronic treatments and that one or more might correspond to forms of truncated FosB.[61] Similar suggestions have been made for proteins induced in the dopamine-depleted striatum by chronic apomorphine,[35] and in the normal striatum by clozapine.[62] Finally, we have observed that chronic exposure to cocaine changes the anatomical distribution of striatal neurons expressing *fos-jun* gene products in response to a further cocaine challenge, and that these changes persist for extended periods of time (unpublished results).

These findings suggest that during a period of chronic cocaine exposure, new patterns of Fos/Jun protein expression occur, and that new combinations of these proteins may bind to regulatory elements in target genes in response to repeated doses of cocaine. This possibility is also suggested by the fact that elevated levels of AP-1 binding activity remain after chronic cocaine exposure, despite down-regulation of c-*fos*.[21] Such differential kinetics of immediate-early gene regulation in the striatum suggest that there might be selective but shifting modulation of other genes as well during the course of cocaine exposure.

One interesting candidate as a target gene of this regulatory mechanism is that encoding the neuropeptide, dynorphin. Dynorphin is expressed in those striatal neurons in which immediate-early genes are induced in response to cocaine, and the gene coding for dynorphin contains AP-1-like and cAMP response element (CRE) binding sites.[63,64] Moreover, chronic treatment with cocaine results in increased expression of dynorphin and preprodynorphin mRNA in striatal neurons.[59,65–67] As Fos levels decline, dynorphin levels rise.[28] Parallel findings have been obtained after chronic treatment with other dopamine receptor agonists.[35,68–71] Particularly marked increases in dynorphin expression in the striatum have been found following dopamine receptor agonist treatments in animals with 6-hydroxydopamine lesions of the nigrostriatal system, and these increases have been shown to be correlated closely with increases in the expression of Fra-like proteins having molecular weights in the 35 kDa range.[35] There is still no proof that dynorphin is regulated by *fos-jun* family transcription factors, but such changes in dynorphin expression exemplify the types of differential regulation that might underlie long-term behavioral changes induced by chronic cocaine.

REFERENCES

1. Post, R. M. and Rose, H., Increasing effects of repetitive cocaine administration in the rat, *Nature*, 260, 731, 1976.
2. Kalivas, P. W. and Stewart, J., Dopamine transmission in drug- and stress-induced behavioral sensitization, *Brain Res. Rev.*, 16, 223, 1991.

3. Ritz, M. C., Lamb, R. J., Goldberg, S. R., and Kuhar, M. J. Cocaine receptors on dopamine transporters are related to self-administration of cocaine, *Science,* 237, 1219, 1987.
4. Koob, G. F. and Bloom, F. E., Cellular and molecular mechanisms of drug dependence, *Science,* 242, 715, 1988.
5. Kuhar, M. J., Ritz, M. C., and Boja, J. M., The dopamine hypothesis of the reinforcing properties of cocaine, *Trends Neurosci.,* 14, 299, 1991.
6. Woolverton, W. L. and Johnson, K. M., Neurobiology of cocaine abuse, *Trends Pharmacol. Sci.,* 13, 193, 1992.
7. Petit, H. O., Pan, H.-T., Parsons, L. H., and Justice, J. B., Jr., Extracellular concentrations of cocaine and dopamine are enhanced during chronic cocaine administration, *J. Neurochem.,* 55, 798, 1990.
8. Akimoto, K., Hamamura, T., and Otsuki, S., Subchronic cocaine treatment enhances cocaine-induced dopamine efflux, studied by in vivo intracerebral dialysis, *Brain Res.,* 490, 339, 1989.
9. Peris, J., Boyson, S. J., Cass, W. A., Curella, P., Dwoskin, L. P., Larson, G., Lin, L. H., Yasuda, R. P., and Zahniser, N. R., Persistence of neurochemical changes in dopamine systems after repeated cocaine administration, *J. Pharmacol. Exp. Ther.,* 253, 38, 1990.
10. Kalivas, P. W. and Duffy, P., Time course of extracellular dopamine and behavioral sensitization to cocaine. I. Dopamine axon terminals, *J. Neurosci.,* 13, 266, 1993.
11. Henry, D. J., Greene, M. A., and White, F. J., Electrophysiological effects of cocaine in the mesoaccumbens dopamine system: repeated administration, *J. Pharmacol. Exp. Ther.,* 251, 833, 1989.
12. Henry, D. J. and White, F. J., Repeated cocaine administration causes persistent enhancement of D1 dopamine receptor sensitivity within the rat nucleus accumbens, *J. Pharmacol. Exp. Ther.,* 258, 882, 1991.
13. Grzanna, R. and Brown, R. M., Activation of immediate early genes by drugs of abuse, *National Institute on Drug Abuse Research Monograph Series,* Vol. 125, Rockville: U.S. Department of Health and Human Services, 1993.
14. Graybiel, A. M., Moratalla, R., and Robertson, H. A., Amphetamine and cocaine induce drug-specific activation of the c-fos gene in striosome-matrix and limbic subdivisions of the striatum, *Proc. Natl. Acad. Sci. USA,* 87, 6912, 1990.
15. Robertson, H. A., Graybiel, A. M., and Paul, M. L., Activation of rat striatal c-fos by direct infusion of dopaminergic agonists and forskolin, *Soc. Neurosci. Abstr.,* 16, 1232, 1990.
16. Young, S. T., Porrino, L. J., and Iadarola, M. J., Cocaine induces striatal c-Fos-immunoreactive proteins via dopaminergic D1 receptors, *Proc. Natl. Acad. Sci. USA,* 88, 1291, 1991.
17. Robertson, H. A., Paul, M. L., Moratalla, R., and Graybiel, A. M., Expression of the immediate early gene c-fos in basal ganglia: induction by dopaminergic drugs, *Can. J. Neurol. Sci.,* 18, 380, 1991.
18. Berretta, S., Robertson, H. A., and Graybiel, A. M., Dopamine and glutamate agonists stimulate neuron-specific transcriptional activation of c-fos in the striatum, *J. Neurophysiol.,* 68, 767, 1992.
19. Bhat, R. V., Cole, A. J., and Baraban, J. M., Role of monoamine systems in activation of zif268 by cocaine, *J. Psychiatr. Neurosci.,* 17, 94, 1992.
20. Cole, A. J., Bhat, R. V., Patt, C., Worley, P. F., and Baraban, J. M., D_1 dopamine receptor activation of multiple transcription factor genes in rat striatum, *J. Neurochem.,* 58, 1420, 1992.
21. Hope, B., Kosofsky, B., Hyman, S. E., and Nestler, E. J., Regulation of immediate early gene expression and AP-1 binding in the rat nucleus accumbens by chronic cocaine, *Proc. Natl. Acad. Sci. USA,* 88, 5764, 1992.
22. Nguyen, T. V., Kosofsky, B. E., Birnbaum, R., Cohen, B. M., and Hyman, S. E., Differential expression of c-Fos and Zif268 in rat striatum after haloperidol, clozapine, and amphetamine, *Proc. Natl. Acad. Sci. USA,* 89, 4270, 1992.
23. Pennypacker, K. R., Zhang, W. Q., Ye, H., and Hong, J. S., Apomorphine induction of AP-1 DNA binding in the rat striatum after dopamine depletion, *Mol. Brain Res.,* 15, 151, 1992.
24. Zhang, W. Q., Pennypacker, H., Ye, H. , Merchenthaler, I. J., Grimes, L., Iadarola, M. J., and Hong, J. S. A 35 kDa Fos-related antigen is co-localized with substance P and dynorphin in striatal neurons, *Brain Res.,* 577, 312, 1992.
25. Moratalla, R., Robertson, H. A., and Graybiel, A. M., Dynamic regulation of *NGFI*-A (*zif*268, *egr*1) gene expression in the striatum, *J. Neurosci.,* 12, 2609, 1992.
26. Dilts, R. P., Helton, T. E., and McGinty, J. F., Selective induction of Fos and Fra immunoreactivity within the mesolimbic and mesostriatal dopamine terminal fields, *Synapse,* 13, 251, 1993.
27. LaHoste, G. J., Jen, Y., and Marshall, J. F., Striatal Fos expression is indicative of dopamine D_1/D_2 synergism and receptor supersensitivity, *Proc. Natl. Acad. Sci. USA,* 90, 7451, 1993.

28. Steiner, H. and Gerfen, C. R., Cocaine-induced c-fos messenger RNA is inversely related to dynorphin expression in striatum, *J. Neurosci.*, 13, 5066, 1993.
29. Rosen, J. B., Chuang, E., and Iadarola, M. J., Differential induction of Fos protein and a Fos-related antigen following acute and repeated cocaine administration, *Mol. Brain Res.*, 25, 168, 1994.
30. Nestler, E. J., Molecular mechanisms of drug addiction in the mesolimbic dopamine pathway, *Sem. Neurosci.*, 5, 369, 1993.
31. Nestler, E. J., Hope, B. T., and Widnell, K. L., Drug addiction: a model for the molecular basis of neural plasticity, *Neuron*, 11, 995, 1993.
32. Moratalla, R., Vickers, E. A., Robertson, H. A., Cochran, B. H., and Graybiel, A. M., Coordinate expression of c-fos and jun B is induced in the rat striatum by cocaine, *J. Neurosci.*, 13, 423, 1993.
33. Moratalla, R., Elibol, B., and Graybiel, A. M., Acute and chronic psychomotor stimulants induce different sets of immediate-early genes in the striatum, *Soc. Neurosci. Abstr.*, 20, 1994.
34. Moratalla, R., Elibol, B., and Graybiel, A. M., in preparation.
35. Bronstein, D. M., Ye, H., Pennypacker, K. R., Hudson, P. M., and Hong, J.-S., Role of a 35 kDa fos-related antigen (FRA) in the long-term induction of striatal dynorphin expression in the 6-hydroxydopamine lesioned rat, *Mol. Brain Res.*, 23, 191, 1994.
36. He, X. P. and Hong, J. S., Dopamine receptor agonists induce expression of dynorphin and c-Fos in primary cell culture of rat, *Soc. Neurosci. Abstr.*, 17, 381, 1991.
37. Liu, F.-C., Takahashi, H., McKay, R. D. G., and Graybiel, A. M., Dopaminergic regulation of gene expression in striatal slice culture, *Soc. Neurosci. Abstr.*, 19, 132, 1993.
38. Liu, F.-C., Takahashi, H., McKay, R. D. G., and Graybiel, A. M., Dopaminergic regulation of transcription factor activation in organotypic culture of developing striatum, *J. Neurosci.*, in press.
39. Anholt, R. R. H., Signal integration in the nervous system: adenylate cyclases as molecular coincidence detectors, *Trends Pharmacol. Sci.*, 17, 37, 1994.
40. Sheng, M. and Greenberg, M. E., The regulation and function of c-fos and other immediate early genes in the nervous system, *Neuron*, 4, 477, 1990.
41. Sheng, M., McFadden, G., and Greenberg, M. E., Membrane depolarization and calcium induce c-fos transcription via phosphorylation of transcription factor CREB, *Neuron*, 4, 571, 1990.
42. Sheng, M., Thompson, M. A., and Greenberg, M. E., CREB: a Ca(2+)-regulated transcription factor phosphorylated by calmodulin-dependent kinases, *Science*, 252, 1427, 1991.
43. Liu, F.-C. and Graybiel, A. M., Dopamine receptors and calcium channel activation regulate phosphorylation of CREB in organotypic cultures of striatum, *Soc. Neurosci. Abstr.*, 20, 1994.
44. Berretta, S. and Graybiel, A. M., Rapid phosphorylation of CREB-like proteins in striatum occurs in response to amphetamine, *Soc. Neurosci. Abstr.*, 19, 129, 1993.
45. Konradi, C., Cole, R. L., Heckers, S., and Hyman, S. E., Amphetamine regulates gene expression in rat striatum via transcription factor CREB, *J. Neurosci.*, 14, 5623, 1994.
46. Simpson, J. N. and McGinty, J. F., Forskolin increases phosphorylated-CREB and Fos immunoreactivity in rat striatum, *NeuroReport*, 5, 1213, 1994.
47. Paul, M. L., Graybiel, A. M., David, J.-C., and Robertson, H. A., D1-like and D2-like dopamine receptors synergistically activate rotation and c-fos expression in the dopamine-depleted striatum in a rat model of Parkinson's disease, *J. Neurosci.*, 12, 3729, 1992.
48. Snyder-Keller, A. M., Striatal c-fos induction by drugs and stress in neonatally dopamine-depleted rats given nigral transplants: importance of NMDA activation and relevance to sensitization phenomena, *Exp. Neurol.*, 113, 155, 1991.
49. Torres, G. and Rivier, C., Cocaine-induced expression of striatal c-fos in the rat is inhibited by NMDA receptor antagonists, *Brain Res. Bull.*, 30, 173, 1993.
50. Dragunow, M., Logan, B., and Laverty, R., 3,4-Methylenedioxyamphetamine induces fos-like proteins in rat basal ganglia: reversal with MK-801, *Eur. J. Pharmacol.*, 206, 255, 1991.
51. Singh, N. A., Midgley, L. P., Bush, L. G., Gibb, J. W., and Hanson, G. R., N-methyl-D-aspartate receptors mediate dopamine-induced changes in extrapyramidal and limbic dynorphin systems, *Brain Res.*, 555, 233, 1991.
52. Cenci, M. A., Campbell, K., Wictorin, K., and Björklund, A., Striatal c-fos induction by cocaine or apomorphine occurs preferentially in output neurons projecting to the substantia nigra in the rat, *Eur. J. Neurosci.*, 4, 376, 1992.
53. Berretta, S., Parthasarathy, H. B., and Graybiel, A. M., Differential induction of immediate-early genes (IEGs) by cortical activation in striatal neuronal subpopulations, *Soc. Neurosci. Abstr.*, 20, 1994.

54. Parthasarathy, H. B., Berretta, S., and Graybiel, A. M., Cortical stimulation induces selective patterns of Jun B expression in the striatum, *Soc. Neurosci. Abstr.*, 20, 1994.
55. Robertson, G. S., Vincent, S. R., and Fibiger, H. C., Striatonigral projection neurons contain D_1 dopamine receptor-activated c-*fos*, *Brain Res.*, 523, 288, 1990.
56. Robertson, G. S., Vincent, S. R., and Fibiger, H. C., D_1 and D_2 dopamine receptors differentially regulate c-*fos* expression in striatonigral and striatopallidal neurons, *Neuroscience*, 49, 285, 1992.
57. DeLong, M. R., Primate models of movement disorders of basal ganglia origin, *Trends Neurosci.*, 13, 281, 1990.
58. Graybiel, A. M., Neurotransmitters and neuromodulators in the basal ganglia, *Trends Neurosci.*, 13, 244, 1990.
59. Daunais, J. B., Roberts, D. C. S., and McGinty, J. F., Cocaine self-administration increases preprodynorphin, but not c-fos mRNA in rat striatum, *NeuroReport*, 4, 543, 1993.
60. Bhat, R. V., Cole, A. J., and Baraban, J. M., Chronic cocaine treatment suppresses basal expression of zif268 in rat forebrain: in situ hybridization studies, *J. Pharmacol. Exp. Ther.*, 263, 343, 1992.
61. Nakabeppu, Y. and Nathans, D., A naturally occurring truncated form of FosB that inhibits Fos/Jun transcriptional activity, *Cell*, 64, 751, 1991.
62. Hiroi, N. and Graybiel, A. M., Effects of chronic clozapine treatment on induction of Fos-like proteins in the striatum, *Soc. Neurosci. Abstr.*, 18, 1043, 1992.
63. Naranjo, J. R., Mellstrom, B., Achaval, M., and Sassone-Corsi, P., Molecular pathways of pain: Fos/Jun-mediated activation of a noncanonical AP-1 site in the prodynorphin gene, *Neuron*, 6, 607, 1991.
64. Douglass, J., McKinzie, A. A., and Pollock, K. M., Identification of multiple DNA elements regulating basal and protein kinase A-induced transcriptional expression of the rat prodynorphin gene, *Mol. Endocrinol.*, 8, 333, 1994.
65. Sivam, S. P., Cocaine selectively increases striatonigral dynorphin levels by a dopaminergic mechanism, *J. Pharmacol. Exp. Ther.*, 250, 818, 1989.
66. Hurd, Y. L., Brown, E. E., Finlay, J. M., Fibiger, H. C., and Gerfen, C. R., Cocaine administration differentially alters mRNA expression of striatal peptides, *Mol. Brain Res.*, 13, 165, 1992.
67. Hurd, Y. L. and Herkenham, M., Molecular alterations in the neostriatum of human cocaine addicts, *Synapse*, 13, 357, 1993.
68. Li, S. J., Sivam, S. P., McGinty, J. F., Jiang, H.-K., Douglass, J., Calavetta, L., and Hong, J. S., Regulation of the metabolism of striatal dynorphin by the dopaminergic system, *J. Pharmacol. Exp. Ther.*, 246, 403, 1988.
69. Hanson, G. R., Merchant, K. M., Letter, A. A., Bush, L., and Gibb, J. W., Methamphetamine-induced changes in the striatal-nigral dynorphin system: role of D-1 and D-2 receptors, *Eur. J. Pharmacol.*, 144, 245, 1987.
70. Jian, M., Staines, W. A., Iadarola, M. J., and Robertson, G. S., Destruction of the nigrostriatal pathway increases Fos-like immunoreactivity predominantly in striatopallidal neurons, *Mol. Brain Res.*, 19, 156, 1993.
71. Gerfen, C. R., Engber, T. M., Mahan, L. C., Suscl, Z., Chase, T. N., Monsma, F. J., Jr., and Sibley, D. R., D_1 and D_2 dopamine receptor-regulated gene expression of striatonigral and striatopallidal neurons, *Science*, 250, 1429, 1990.

Chapter 14

Cocaine Regulation of Signal Transduction Pathways

Lawrence W. Fitzgerald and Eric J. Nestler

CONTENTS

I. INTRODUCTION

The mechanisms by which cocaine and other drugs of abuse induce addiction, and associated states of tolerance, sensitization, and dependence, have long been a subject of great interest. Since these phenomena, for the most part, develop gradually over time in response to prolonged drug exposure and last long after cessation of drug exposure, it is thought that their underlying mechanisms involve long-term and persistent adaptations in neural function. This chapter will review the increasing evidence that alterations in postreceptor signal transduction, including gene expression, contribute significantly to these adaptive changes. Because monoaminergic nerve terminals, the initial targets of cocaine, are distributed widely throughout the mammalian central nervous system, cocaine has the potential to influence processes of signal transduction in a vast number of neurons. However, we will focus here on the molecular mechanisms underlying acute and chronic cocaine action on the mesolimbic dopamine (DA) system based upon the knowledge that this pathway is a major locus of the reinforcing and addictive properties of cocaine. Because the behavioral, neurochemical, and electrophysiological characteristics of the mesolimbic DA system have been analyzed extensively over the last 20 years, it is possible to understand the molecular adaptations that result from chronic cocaine use within a functional context. Before considering the specific signal transduction pathways

1-8493-8311-0/95/$0.00+$.50

by which acute and chronic cocaine treatment influences brain function, we will provide a brief overview of the major signal transduction pathways in the central nervous system.

II. OVERVIEW OF BRAIN SIGNAL TRANSDUCTION PATHWAYS

When a neurotransmitter binds to a receptor on the surface of a cell, it sets into motion signaling cascades which can have a myriad of intracellular consequences (Figure 1). With the exception of ligand-gated ion channels (e.g., ionotropic glutamate or GABA receptors) or receptors with intrinsic enzymatic activity, signals from extracellular stimuli are transduced across the plasma membrane through membrane-bound regulatory GTP-binding proteins (G proteins) to influence intracellular processes.[1] G proteins are heterotrimers, consisting of single α, β, and γ subunits, and are typically classified according to their α subunits. Activation of G protein-linked receptors by ligand binding leads to a conformational change in the G protein. This in turn leads to the binding of GTP to the α subunit and the displacement of GDP. The binding of GTP facilitates the dissociation of the α subunit from its $\beta\gamma$ constituents. Free α subunits bound to GTP, as well as the free $\beta\gamma$ subunits, regulate various effector proteins depending on the type of subunit and cell type involved. In some instances, the activated α (or $\beta\gamma$) subunit binds and gates certain ion channels (e.g., Ca^{2+} and K^+ channels), thereby mediating some of the electrophysiological effects of the initial neurotransmitter stimulus. For example inhibitory, pertussis toxin-sensitive G proteins (G_i and/or G_0) couple opioid, α_2-adrenergic, muscarinic cholinergic, D_2 dopaminergic, 5-HT_{1A} serotonergic, somatostatinergic, and $GABA_B$ receptors to the activation of inward rectifying K^+ channels or to the inhibition of voltage-dependent Ca^{2+} channels.[2]

In many other instances, the G protein mediates the generation of second messengers. One of the best-studied second messengers, adenosine 3′5′-cyclic monophosphate (cyclic AMP or cAMP), is generated by adenylyl cyclase.[2] The precise mechanism by which G proteins stimulate adenylyl cyclase is now firmly established. Activation of receptors which couple to G_s (e.g., D_1 DA receptors) leads to the formation of free $G_{s\alpha}$ subunits which bind to and activate adenylyl cyclase. The exact manner by which G proteins inhibit adenylyl cyclase is more equivocal. $G_{i\alpha}$ may directly inhibit adenylyl cyclase, although it has been difficult to demonstrate this in cell-free reconstitution experiments. In addition, it is likely that $\beta\gamma$ subunits play a role in adenylyl cyclase regulation, as different forms of the enzyme can be stimulated or inhibited by these subunits *in vitro*.[2] G proteins also play a central role in generating a variety of other second messengers in response to a neurotransmitter stimulus. Other second messengers in brain include guanosine 3'5'-cyclic monophosphate [cyclic GMP or cGMP], Ca^{2+}, nitric oxide [NO], and many byproducts of arachidonic acid or inositol phosphate metabolism.[2]

Each second messenger mediates numerous physiological responses of the original neurotransmitter stimulus, responses that are mediated largely through alterations in protein phosphorylation[3,4] (Figure 1). This often is accomplished via the direct activation of specific protein kinases by second messengers, as seen for cAMP-dependent, cGMP-dependent, and calcium-calmodulin-dependent protein kinases, and protein kinase C. In addition, extracellular stimuli and second messengers can alter the phosphorylation of proteins by regulating specific protein phosphatases. Not all protein kinases, however, are under the direct control of a second messenger, as neurons express numerous second messenger-independent protein kinases that also play important roles in cell function, as exemplified by the MAP-kinases, also referred to as ERKs (*E*xtracellular signal-*R*egulated *K*inases). These enzymes were identified in part because of their association with and phosphorylation of microtubule-associated proteins (MAPs), and have since been shown to phosphorylate a number of other proteins in brain including tyrosine hydroxylase and several DNA-binding proteins.[5,6] Although not influenced directly by second messengers, the catalytic activity of MAP-kinase can be regulated, through several

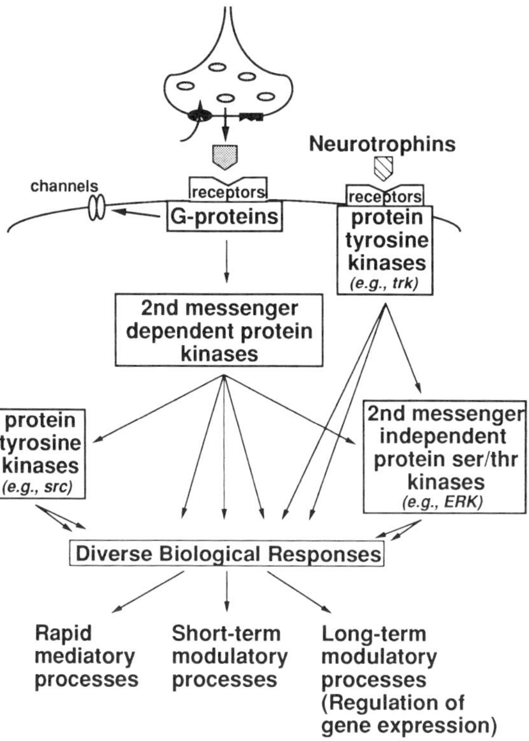

Figure 1. A working model of synaptic transmission. Molecular neuroscience has constructed a complex view of synaptic transmission which includes intracellular messenger systems involving coupling factors (G proteins), second messengers (e.g., cAMP, cGMP, Ca^{2+}, NO, and metabolites of phosphotidylinositol and arachidonic acid), and the phosphorylation of phosphoproteins by protein kinases and their dephosphorylation by protein phosphatases. These signaling cascades mediate the multiple actions of the initial neurotransmitter stimulus. Second messenger-dependent protein kinases (those activated by cAMP or Ca^{2+}) are classified as protein serine/threonine kinases. The brain also contains numerous protein serine/threonine kinases that are not regulated directly by second messengers (e.g., ERKs or MAP-kinases). Moreover, there are numerous types of protein tyrosine kinases, some of which reside in the receptors for neurotrophins and most other growth factors (e.g., the trk proteins), and others that are not associated with growth factor receptors (e.g., *src* kinase). The receptor-associated protein tyrosine kinases are activated upon the binding of the growth factor to its receptor. The second messenger-independent protein serine/threonine kinases and protein tyrosine kinases that are not receptor associated seem to be regulated indirectly via the second messenger-dependent and growth factor-dependent pathways depicted herein. The brain also contains numerous types of protein serine/threonine and protein tyrosine phosphatases (not shown) which are also subject to regulation by extracellular and intracellular stimuli. Second messenger-dependent and -independent protein kinases phosphorylate a specific array of substrate proteins (which can be considered as third messengers) that elicit multiple biological responses. These biological responses can be divided into three main categories. First, intracellular messengers can mediate the actions of neurotransmitters rapidly in gating particular ion channels. Second, intracellular messengers can mediate the short-lived modulation of the general metabolic state of the neuron, its ability to synthesize or release neurotransmitter, and the functional sensitivity of various receptors and ion channels to synaptic inputs. Third, other responses are relatively long-lived and are achieved through the regulation of gene expression in target neurons. Thus, neurotransmitters, through the regulation of intracellular messenger pathways and alterations in gene transcription and protein synthesis, alter the numbers and types of receptors and ion channels in target neurons, the functional activity of the intracellular messenger systems in these neurons, and even the shape and numbers of synapses formed by these neurons Modified from Hyman and Nestler, 1993.[10]

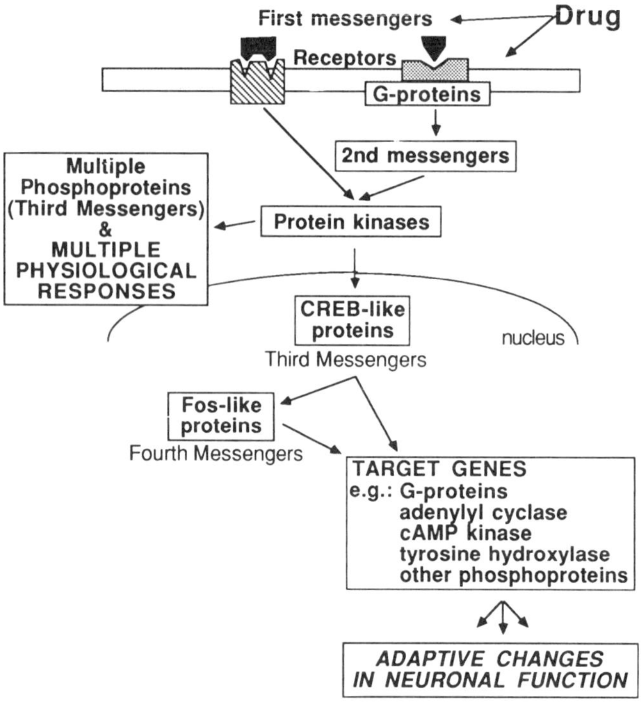

Figure 2. Schematic illustration of the intracellular messenger pathways through which cocaine and other extracellular signals regulate gene expression in target neurons. See text for discussion. From Nestler, 1992, with permission.[78]

intervening steps, by cAMP and Ca^{2+}-dependent protein phosphorylation. Thus, any neurotransmitter or drug that influences cellular levels of cAMP or Ca^{2+} would be expected to lead to the regulation of the MAP-kinase systems. Another example of a second messenger-independent signal transduction mechanism is provided by the protein tyrosine kinases[7] (Figure 1). The best characterized protein tyrosine kinases are those associated with membrane receptors for growth factors, such as insulin, epidermal growth factor, and nerve growth factor and related neurotrophins. For example, the neurotrophins activate a class of receptor-kinases known as trk proteins. The binding of a growth factor to its receptor leads to the stimulation of protein tyrosine kinase activity intrinsic to the receptor. This activity results in the phosphorylation of the receptor itself as well as other proteins with Src homology (SH) adaptor domains. The phosphorylated receptor then interacts with other proteins which ultimately leads to the activation of several intracellular signaling pathways, including the MAP-kinase pathway. It should be apparent from the above discussion that very different intracellular messenger pathways can interact functionally through complex signaling cascades in order to regulate cell function in a highly coordinated fashion.

Protein phosphorylation, then, represents a fundamental regulatory process by which extracellular stimuli influence virtually all neuronal processes, including neurotransmitter synthesis, neurotransmitter release, generation of synaptic potentials, electrical excitability of neurons, axoplasmic transport, elaboration of dendritic and axonal processes, and the development and maintenance of the differentiated characteristics of neurons.[3,4] Phosphorylation promotes these effects by altering the functional characteristics of proteins already present (e.g., changing conductance properties of channels or catalytic activity of enzymes), or by inducing nascent protein synthesis by neuron-specific gene expression.

Neurotransmitter and drug regulation of gene expression is known to occur via the phosphorylation and/or induction of transcription factors (Figure 2). Transcription factors are proteins that bind to 5′ response elements in the promoter regions of genes and regulate the rate at which these genes are transcribed.[8-10] There are at least two general mechanisms by which this might occur. Second messenger- or growth factor-elicited phosphorylation can regulate factors already present in the cell. This is exemplified by the cAMP response element binding (CREB) proteins, which are phosphorylated and activated by cAMP- or Ca^{2+}-dependent protein kinases. These proteins presumably are regulated by neurotransmitters that alter the intracellular concentration of cAMP or Ca^{2+}. CREB and CREB-like proteins bind to a cAMP response element or CRE present in certain genes and thereby influence the transcription of these genes. Alternatively, a second messenger or growth factor can regulate gene expression via the induction of transcription factors, such as Fos and Jun, coded by immediate early genes (IEGs), such as c-*fos* and c-*jun*. Unlike CREB, these proteins are expressed at very low levels under basal conditions, and are induced in response to an acute extracellular stimulus via increased transcription and mRNA translation.[11] Such induction can be mediated by cAMP- or Ca^{2+}-dependent phosphorylation and activation of CREB or CREB-like proteins, or by growth factor-stimulated phosphorylation of other transcription factors. The newly synthesized IEG products then enter the nucleus of the cell and regulate the expression of other target genes by binding to DNA response elements called AP-1 sites. Zif268 is an additional IEG transcription factor that is regulated in a similar fashion to *c-fos* and *c-jun,* but binds to a distinct DNA response element. IEG mechanisms are slower than the more direct action of phospho-CREB, because they require protein synthesis; a newly synthesized IEG product can appear within 30 to 60 min after the initial stimulus. Fos, Jun, Zif268, and CREB-like proteins represent only a small portion of the many transcription factor families that likely mediate the effects of extracellular stimuli on gene expression in the brain.

III. OVERVIEW OF THE DOPAMINE SYNAPSE AND DOPAMINE-MEDIATED SIGNAL TRANSDUCTION

Significant progress has been made in the last decade toward identifying the anatomical and neurochemical bases for the acute and chronic actions of cocaine in the brain. It is now widely accepted that many of the locomotor and reinforcing properties of cocaine result from its acute influence on the mesolimbic DA system. This neural pathway consists of DA neurons located in the ventral tegmental area (VTA) of the midbrain that project largely to the nucleus accumbens (NAc), olfactory tubercle, and medial prefrontal and cingulate cortices. Cocaine also acts on other major DA systems (e.g., the nigrostriatal) that generally are not implicated in brain reward processes.

Cocaine is an indirect-acting DA, norepinephrine (NE), and serotonin (5-HT) agonist: it potentiates the synaptic actions of these monoamines by blocking their reuptake through transporters located on the plasma membrane of monoaminergic nerve terminals (Figure 3). It is generally accepted, however, that it is the specific action of cocaine at the DA transporter which accounts most for its reinforcing and drug addicting properties. This conclusion is based on several lines of evidence. First, it was found that the abuse liability or behavioral potency in the self-administration paradigm of numerous drugs with varied potency in blocking DA, NE, or 5-HT uptake is correlated most closely with their affinity for the DA transporter.[12] This relationship was not observed with respect to the propensity of these drugs to inhibit 5-HT or NE uptake. Second, while DA agonists promote and antagonists suppress cocaine reward, selective noradrenergic and serotonergic compounds do not possess these properties.[13-15] Finally, lesions of the mesolimbic DA system block cocaine self-administration whereas specific ablation of central serotonergic and noradrenergic systems do not.[16,17] While these data clearly support the importance of

DA in reward processes, it should be pointed out that not all agents which increase dopaminergic transmission can produce motivational dependence. This conceivably could depend on the kinetics of the agents' interaction with the DA system and subsequent activation of receptor-mediated signal transduction pathways. It also, however, leaves open a possible role for 5-HT or NE in mediating some of the short- and long-term effects of cocaine exposure.

A structurally homologous family of cocaine-sensitive DA, NE, and 5-HT transporters have been cloned and expressed *in vitro* (see Chapter 12).[18-20] Using energy of the sodium gradient generated by Na^+/K^+ ATPase, the DA transporter regulates the synaptic concentration of DA and, consequently, the activation of DA receptors. Within the nigrostriatal or mesolimbic nerve terminal, cytoplasmic DA (derived from reuptake or *de novo* synthesis by tyrosine hydroxylase, the rate-limiting enzyme of catecholamine biosynthesis) is pumped into synaptic vesicles by a vesicle-associated transporter (VAT).[21] This transport depends on the maintenance of a transmembrane gradient (inside positive) via a proton pump (vacuolar H^+-ATPase).[22] It has been hypothesized that an additional action of cocaine and other psychostimulants is their ability to reduce this gradient and cause a diminution of DA uptake by the VAT.[23] Biochemical and molecular studies of cloned transporters should facilitate a better understanding of these basic transport mechanisms as well as determine exactly how drugs like cocaine influence these transport proteins.

Because cocaine is an indirect agonist, cocaine-induced increases of synaptic DA levels presumably lead to an increased activation of all subtypes of DA receptor (Figure 3). It was initially believed that DA acted through only two receptor subtypes, termed D_1 and D_2.[24] Cloning of the D_1 and D_2 receptor genes[25] confirmed that these receptors are G protein-coupled. Unlike the D_1 receptor gene, the D_2 gene encodes two variants of the receptor (D_{2A} and D_{2B}) derived by alternative splicing. D_1 receptors are positively coupled to adenylyl cyclase. They increase cAMP levels via the G_s type of G protein, and are localized to the cell bodies, dendrites, and terminals of dopaminoceptive neurons. Generally, it is believed that D_1 receptors are not expressed by dopaminergic neurons themselves, although this has not been established definitively. D_2 receptors are negatively coupled to adenylyl cyclase via subtypes of G_i and G_o, and activate inward rectifying K^+ channels; it is possible that they also inhibit voltage-dependent Ca^{2+} channels. D_2 receptors are located on the cell bodies, dendrites, and terminals of dopaminoceptive cells as well as on most DA neurons. D_1 or D_2 receptor-mediated alterations of cAMP, in turn, lead to alterations of cAMP-dependent protein kinase activity, which consequently lead to changes in the phosphorylation state of numerous phosphoprotein substrates, including receptors, ion channels, enzymes, and cytoskeletal proteins[4] (Figure 3). Paradoxically, however, the opposing actions of D_1 and D_2 receptors on adenylyl cyclase contrasts with the general observation that both D_1 and D_2 receptor occupancy are required to elicit the full electrophysiological and motoric effects of DA agonists.[26] Whether D_1/D_2 cooperativity results from crosstalk between co-expressed D_1 and D_2 receptor subtypes within single neurons and/or from converging motor output pathways independently regulated by the two receptors remains an intriguing area of debate.[25,27,112] The D_2 receptors located on DA neurons may serve as inhibitory autoreceptors in that they, reduce neuronal firing, DA synthesis, and release.

DA receptors have been reported to influence other effector pathways in addition to adenylyl cyclase depending on the cellular environment; these include cellular Ca^{2+} levels, inositol phosphate hydrolysis, arachidonic acid release, and Na^+-H^+ exchange.[28-31] Moreover, DA regulation of cAMP and other second messengers would be expected to influence MAP kinases and protein tyrosine phosphorylation, and these effects would undoubtedly also mediate some of the many biological effects of DA.

The molecular details of effector coupling to DA receptors are derived mainly from *in vitro* studies of transfected receptors. Such *in vitro* expression systems of cloned

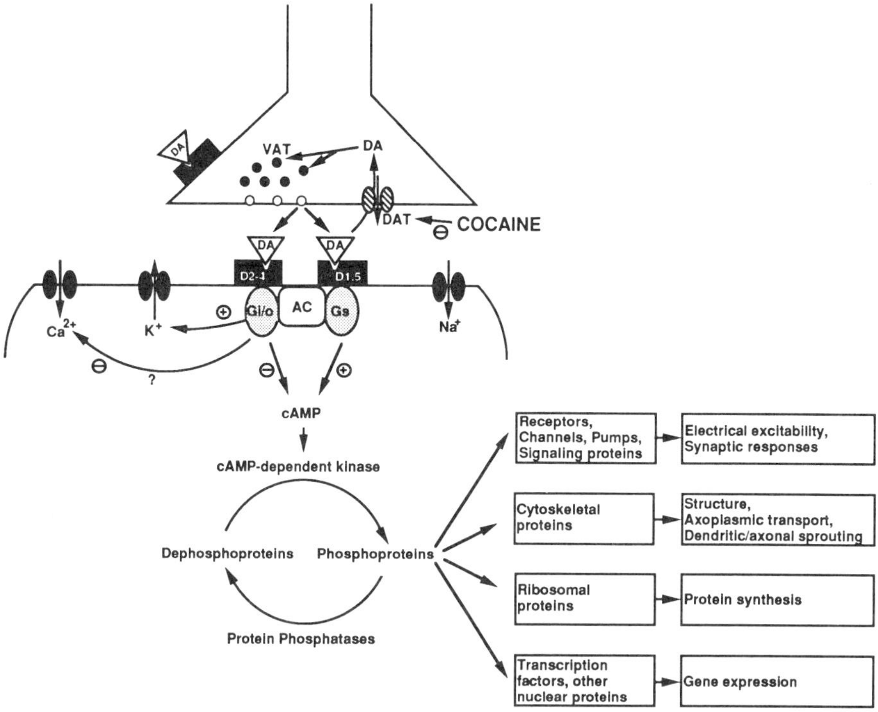

Figure 3. Schematic illustration of dopamine-regulated signal transduction pathways. Cocaine acutely increases the synaptic levels of DA by blocking reuptake of DA through the DA transporter (DAT). The DA taken up is either metabolized or subsequently re-released either as free DA (via the DAT) or in vesicles (via Ca^{2+}-dependent exocytosis). DA is pumped into storage vesicles by the vesicular amine transporter (VAT) using a H^+-ATPase. Cocaine-induced increase of the synaptic DA level leads to the stimulation of postsynaptic G protein-coupled DA receptors (D_1-D_5), as well as presynaptic autoreceptors that exhibit a D_2-like (D_2-D_4) pharmacologic profile. D_1 (and D_5 or D_{1B}) and D_2 (and D_4) receptors can positively (via G_s) and negatively (via G_{i10}) couple to adenylyl cyclase, respectively; however, this has been shown for the D_4 and D_5 (D_{1B}) receptors only *in vitro*. The coupling of the D_3 receptor to adenylyl cyclase is equivocal at this time but is likely to be inhibitory. D_2-like receptors can also stimulate K^+ channels, and possibly inhibit Ca^{2+} channels. DA receptor-regulated alteration of adenylyl cyclase leads to changes in cAMP levels. This in turn leads to the modulation of cAMP-dependent protein kinase, which together with protein-phosphatases, regulates the phosphorylation state of a myriad of protein substrates that influence diverse aspects of neuronal function. DA receptors have been reported to influence second messenger pathways (not shown) in addition to cAMP depending on the cellular environment; these include Ca^{2+} levels, inositol phosphate metabolism, arachidonic acid release, and Na^+-H^+ exchange. Such actions would alter other phosphorylation pathways, which could result in many additional biological responses to DA.

receptors have greatly enhanced our understanding of the intracellular signal transduction pathways through which DA and other neurotransmitters exert their physiological effects. They also have been important in identifying common motifs in signal transduction; one being that a single receptor subtype can couple, via similar or different G proteins, to distinct effector pathways. This arrangement affords the neuron a greater specialization of function, and perhaps, added plasticity to respond in a more sophisticated fashion to the demands placed by the environment. For example, a recent study reported that stimulation of protein kinase C redirects the effector output of transfected D_2 receptors from a preponderance of G_i-mediated inhibition of adenylyl cyclase to the G_i-mediated release of arachidonic acid by phospholipase A_2.[32]

The recent cloning of multiple DA receptor genes has necessitated an expansion of the DA receptor classification to include the D_3, D_4, and D_5 (or D_{1B}) receptor subtypes.[25,33] The D_1 through D_5 receptor subtypes have been classified into two families of receptors, D_1-like and D_2-like, based on similarities in signal transduction mechanisms and pharmacology as determined *in vitro,* as well as sequence homology. However, because of a paucity of selective drugs, little is known about the transduction or regulatory characteristics of these novel subtypes *in vivo*. For example, while clozapine showed early promise as a D_4-selective compound,[34] there is some evidence that it has high affinity for other dopaminergic subtypes.[35] *In vitro,* D_4 and D_5 receptors negatively and positively couple to adenylyl cyclase, respectively. The signal transduction characteristics of the D_3 receptor are established less firmly. Reliable D_3-mediated inhibition of adenylyl cyclase has not yet been shown. Moreover, guanyl nucleotide inhibition of agonist binding to D_3 receptors *in vitro* is variable, and insensitivity to these agents *ex vivo* has been demonstrated by receptor autoradiography.[36-39] Despite these uncertainties, however, the D_3 receptor is clearly a member of the G protein-coupled receptor superfamily and likely acts via similar mechanisms as do the D_2 and D_4 receptors. The D_3 receptor also remains an attractive candidate for mediating at least some of the locomotor and addictive qualities of cocaine, given its unique anatomical localization. Unlike the D_1-like and the D_2 and D_4 receptors, the D_3 receptor is located preferentially in terminal regions of the mesolimbic DA system (e.g., the nucleus accumbens, olfactory tubercle, islands of Calleja). Its potential importance for addiction is emphasized further by the observation that compounds that have higher affinity for D_3 than D_2 receptors (e.g., quinpirole and 7-hydroxy-N,N-di-n-propyl-2-aminotetralin or 7-OH-DPAT) potently affect cocaine self-administration in rats at doses that are not by themselves reinforcing.[40]

IV. ACUTE EFFECTS OF COCAINE *IN VIVO*

A. NEUROCHEMICAL AND BEHAVIORAL ASPECTS OF ACUTE COCAINE ACTION

The acute inhibition of monoaminergic uptake by cocaine results in a number of physiological effects in the peripheral and central nervous systems. In the periphery, cocaine acts as a sympathomimetic agent and a powerful vasoconstrictor because of its potent inhibition of NE uptake. Centrally, the acute behavioral and neurochemical actions of cocaine and other psychostimulants on mesolimbic and nigrostriatal DA function have been well documented.[41] In brief, a single injection of cocaine increases synaptic levels of DA in the striatum, NAc, and medial prefrontal cortex as measured by *in vivo* microdialysis.[42-44] The acute effects of cocaine are transient due to its short half-life; following an intraperitoneal injection, synaptic levels of DA rise markedly within minutes, peak within a half-hour, and return to normal within an hour.[45] These pharmacokinetic properties (i.e., rapid on- and off-rates) of cocaine probably contribute greatly to its profound abuse potential.

Behaviorally, acute cocaine administration reliably increases locomotor activity, an effect typical of psychostimulants (e.g., *d*-amphetamine). At low-to-moderate doses, generalized increases in activity consist mainly of increased forward ambulation, rearing, grooming, and exploratory behavior, while at higher doses it takes the form of increased stereotypical behavior (e.g., gnawing, sniffing).[41] The locomotor responses are mediated largely by the mesolimbic DA system since they are blocked by pretreatment with DA receptor antagonists or lesioning of the NAc by 6-hydroxydopamine (6-OHDA).[16,46] Moreover, the locomotor activating effect can be evoked by intra-accumbens administration of cocaine.[47] In contrast, cocaine-induced stereotypy is thought to be mediated largely by the nigrostriatal DA system, although both DA systems probably contribute to both behavioral responses.

B. ELECTROPHYSIOLOGICAL EFFECTS OF ACUTE COCAINE

The electrophysiological effects of cocaine on neurons postsynaptic to DA nerve terminals and on dopaminergic cell bodies have been studied extensively. DA neurons in the substantia nigra and the VTA exhibit low discharge rates and wide action potentials, characteristics which distinguish them from neighboring nondopaminergic cells.[48] DA and DA agonists inhibit these neurons at low concentrations due to the stimulation of D_2-like somatodendritic autoreceptors. Cocaine exerts an indirect inhibitory effect on these neurons due to the blockade of dendritic DA transporters and the subsequent stimulation of D_2 autoreceptors, as mentioned above. Systemic cocaine also might inhibit VTA neurons by recruitment of an inhibitory feedback pathway from the NAc to the VTA.[49] The DA cells in the VTA, however, are not homogeneous; those projecting to the prefrontal and cingulate cortices appear to differ from those projecting to the NAc in that they are not regulated as potently by autoreceptors.[50]

In the terminal regions of the nigrostriatal and mesocorticolimbic DA systems, the electrophysiological effects of cocaine can be considerably more complicated due to the multiple DA receptor subtypes located pre- and postsynaptically, as well as the heterogeneity of cell populations intrinsic to these regions. In the NAc, cocaine generally inhibits neurons via both D_1-like and D_2-like receptor subtypes in anesthetized rats.[51] In contrast in an *in vitro* slice preparation from the rat NAc, DA activation of D_1 receptors inhibits NAc neurons, whereas D_2 receptor activation excites them.[52] Similar findings have been reported in guinea pig slices.[53] The discrepancies observed for the effects of D_2-like receptor stimulation of NAc neurons may be due, in part, to differences in the electrophysiological preparations (*in vitro* vs. *in vivo*) or to the difficulties associated with sampling from a heterogenous nucleus with multiple cell types and receptor subtypes. In the striatum, the issue of whether DA is inhibitory or excitatory has been debated for many years. The general impression remains that DA is generally inhibitory to striatal neurons, although some cells do indeed show excitation.

V. CHRONIC EFFECTS OF COCAINE *IN VIVO*

A. BEHAVIORAL SENSITIZATION

In addition to the acute actions of cocaine on neural functions, various distinct alterations are found only after chronic exposure to the drug. In animal studies, one of the best-documented phenomena associated with repeated cocaine (or *d*-amphetamine) use is behavioral sensitization. A more comprehensive discussion of this phenomenon is provided by several recent reviews[54,55] (see Chapter 6). In brief, behavioral sensitization or reverse tolerance refers to a progressive and enduring enhancement in the behavioral responses (increased locomotion and stereotypy) to an acute challenge with cocaine (or other psychostimulants) after repeated preexposure to the drug.[56,57] All psychostimulants cross-sensitize to each other and to the effects of stress, suggesting that similar neurochemical and molecular mechanisms might underlie the different forms of behavioral sensitization.[54] A widely observed neurochemical alteration associated with chronic cocaine (and other psychostimulant) exposure is the enhanced ability of cocaine to increase DA release in the striatum and NAc; this has been observed by *in vivo* microdialysis[45] (see Chapter 6) and in slice preparations in vitro (see Chapter 11).[58] There is also some evidence suggesting that rats which self-administer psychostimulants sensitize to the reinforcing properties of the drug.[59] These data corroborate the clinical impression that craving for these drugs by humans (as well as certain other effects, such as psychosis) generally increases with the chronicity of the exposure. It is not clear, however, whether the distinct behavioral phenomena of addiction and locomotor sensitization share common neurochemical or molecular substrates. Behavioral sensitization is an enduring and persistent phenomenon, since it can last for weeks or months after withdrawal from repeated cocaine or amphetamine administration.[54,55,58] Since many acute actions of cocaine and other psychostimulants occur at the level of the mesolimbic DA system, there has been an intensive research effort to identify cellular and molecular adaptations in this system which might underlie the effects of chronic psychostimulant exposure.

B. ELECTROPHYSIOLOGICAL EFFECTS OF CHRONIC COCAINE

NAc neurons in anesthetized rats are supersensitive to the inhibitory influences of DA 12 to 24 hr after chronic cocaine administration (see Chapter 7). This phenomenon is persistent, lasting up to 1 month, and is receptor subtype-specific, since NAc neurons were supersensitive to the effects of the D_1-like agonist SKF 38393, but not the D_2-like agonist quinpirole.[60] In addition, VTA neurons are subsensitive to autoreceptor-stimulating effects of intravenous apomorphine 12 to 24 hr after their last cocaine injection.[61-63] Cocaine-treated rats had a greater number of spontaneously active VTA neurons which also showed higher firing rates. However, unlike the D_1 supersensitivity in the NAc, autoreceptor subsensitivity in the VTA is more transient lasting only 4 days. Similarly, autoreceptor subsensitivity resulting from chronic *d*-amphetamine administration was observed in VTA neurons after 3 to 4 days of withdrawal, but not after 10 to 14 days.[64,65] Thus, autoreceptor subsensitivity of VTA neurons may be important for the initial development of behavioral sensitization, whereas D_1 supersensitivity in NAc neurons may be necessary for its enduring expression. The importance of the VTA in the development and early manifestation of sensitization is underscored by the finding that while *d*-amphetamine injected acutely into the VTA has no immediate behavioral effects, repeated injection of amphetamine into the VTA, but not the NAc, sensitizes animals to subsequent amphetamine challenge.[66,67]

VI. MOLECULAR AND BIOCHEMICAL ADAPTATIONS TO CHRONIC COCAINE

A. REGULATION OF DOPAMINE LEVELS, RECEPTORS, AND UPTAKE

A number of studies have examined the regulation of DA D_1 and D_2 receptors and transporters after chronic cocaine administration.[58,68-72] Unfortunately, the results taken as a whole are variable with reports of increases, decreases, or no changes in the levels of these proteins. The inconsistency of these findings is probably due, in part, to differences in the treatment durations, drug doses, postwithdrawal intervals, and brain regions evaluated. A systematic consideration of these parameters should contribute to our understanding of the important role these proteins might play in cocaine addiction and/or abstinence syndromes. Part of the difficulty also might be the reliance on radioligand binding techniques, given certain ambiguities associated with measuring multiple DA receptors and possibly transporters. These include the lack of selective radioligands that can distinguish the various DA receptor subtypes, and the uncertainty regarding the exact physical nature of cocaine binding sites on the DA transporter protein. Recent molecular cloning studies have enabled a more precise investigation of these proteins. Indeed, chronic cocaine administration reportedly produced a persistent decrease of DA transporter mRNA levels in the VTA (and transporter binding in the NAc).[70]

Catecholamine and metabolite levels also have been examined extensively after chronic cocaine or amphetamine treatment. Generally, it has been observed that chronic cocaine (as well as amphetamine, morphine, and ethanol) treatment decreases regional extracellular DA levels under conditions of abstinence that can persist for days.[71-75] The possibility that chronic cocaine might induce a relative DA deficiency is attractive, as it might be expected to elicit a compensatory supersensitivity to the electrophysiological actions of DA.[60] However, the absence of any consistent changes in D_1 receptors per se following chronic cocaine exposure suggests that the observed electrophysiological supersensitivity to D_1 receptor stimulation stems from adaptations in post-receptor signal transduction mechanisms.

B. REGULATION OF GLUTAMATE RECEPTORS IN THE VENTRAL TEGMENTUM

While the DA system likely plays an important role in the functional adaptations to chronic cocaine, there is growing evidence that other neurotransmitter systems (e.g., *glutamate,* 5-HT, and neuropeptides) are important as well. One neurotransmitter of particular importance is the excitatory amino acid, glutamate, which acts upon the *N-methyl-D-aspartate* (NMDA) and non-NMDA receptors. The antagonism of glutamate's actions in the mesolimbic DA system has been shown to modulate behavioral sensitization and cocaine self-administration.[54,76] In this regard, it is interesting that repeated treatment with cocaine (2 × 15 mg/kg for 2 wks; or 1 × 20 mg/kg for 1 wk) as well as other agents that exhibit sensitizing qualities (e.g., ethanol, stress), upregulates the expression of specific NMDA and non-NMDA receptor subunits in the VTA but not other components of the mesolimbic and nigrostriatal DA systems (Fitzgerald and Nestler, unpublished observations). These changes are consistent with an increased state of activation of VTA neurons as inferred from the observed increase in tyrosine hydroxylase following cocaine, stress, and other drugs of abuse (see next section). These data may also explain observations recently made by White and colleagues[77] which demonstrate an increased electrical sensitivity of VTA neurons to glutamate as well as an increased susceptability of the neurons for depolarization blockade. Such alterations in the electrical excitability of VTA dopamine neurons in the chronic drug-treated state could influence locomotor activity and sensitization as well as modulate reinforcement (i.e., craving) mechanisms underlying drug addiction.

C. REGULATION OF G PROTEINS AND THE cAMP PATHWAY

Work from this laboratory and others has shown that compensatory changes of post-receptor signal transduction elements are important responses to chronic treatment with cocaine and other drugs of abuse.[78,79] Chronic cocaine administration (15 mg/kg twice daily for 14 days) upregulates the cAMP pathway in the NAc at several levels of signal transduction, producing increased adenylyl cyclase and cAMP-dependent kinase activity, and decreased levels of the inhibitory G proteins, $G_{i\alpha}$ and $G_{o\alpha}$, as measured by immunoblotting and pertussis toxin (PTX)-mediated ADP-ribosylation.[80,81] $G_{i\alpha}$ is also reduced in the rat NAc 14 days, but not 1 day, after cessation of chronic cocaine treatment (15 to 30 mg/kg for 5 days).[82] This latter finding contrasts slightly with the reported reduction of G_i (and G_o) 1 hr after the last cocaine treatment; this difference could be explained by differences in the two treatment regimens. Since D_1 receptors stimulate the cAMP pathway, the observed increase in adenylyl cyclase and cAMP-dependent protein kinase activity, together with the decrease of inhibitory G proteins (with no change in G_s), could provide a molecular substrate for the persistent D_1 receptor supersensitivity observed electrophysiologically. In this regard, it is interesting that rodents whose cAMP system in the NAc is activated for sustained periods (e.g., by cholera toxin which ADP-ribosylates and activates G_s) display sensitized locomotor responses to acute and chronic cocaine.[83,113] Adaptations of G proteins might also underlie the D_1 receptor supersensitivity (i.e., increased DA-mediated adenylyl cyclase activity) observed in rats whose nigrostriatal DA pathway is lesioned with 6-OHDA. Such animals show an upregulation in the stimulatory G proteins, $G_{s\alpha}$ and $G_{olf\alpha}$, without consistent changes in striatal D_1 receptor density.[84]

Chronic cocaine also decreases the level of $G_{i\alpha}$ and $G_{o\alpha}$ in the VTA.[80] Similarly, chronic cocaine treatment transiently (24 hr after last injection) reduced the ADP-ribosylation of G proteins in this brain region *in vitro,* and the degree of this reduction was correlated negatively with the locomotor response to a cocaine challenge.[85] The same study also observed that intra-VTA injection of PTX (which ADP-ribosylates and inactivates $G_{i\alpha}$ and $G_{o\alpha}$) mimics the ability of chronic cocaine to facilitate the locomotor-activating effect of an acute cocaine challenge. Since D_2 receptors transduce their signals via PTX-sensitive G proteins, a decreased level of these G proteins in the VTA could account for the D_2 receptor subsensitivity observed in electrophysiological studies.[61] It is interesting to note that the reported short-lived decline of inhibitory G proteins coincides with the transient nature of the autoreceptor subsensitivity. These effects contrast with the more resilient changes reported postsynaptically in the NAc, and provide further evidence that the VTA may be an important initiation site for behavioral sensitization, as well as for the development of cocaine dependence.

In addition to changes in adenylyl cyclase, cAMP-dependent kinase, and G proteins, the effects of chronic cocaine treatment on the next step in this signal transduction pathway, namely, on cAMP-dependent protein phosphorylation, also have been investigated. One of the first phosphoproteins studied was tyrosine hydroxylase.[86] Chronic, but not acute, cocaine treatment increased the level of tyrosine hydroxylase phosphorylation and immunoreactivity in the VTA, an effect that was recently replicated.[87,88] Increases of tyrosine hydroxylase mRNA also have been reported in the VTA,[87] although this effect was not seen with a different cocaine treatment regimen.[88] Since tyrosine hydroxylase activity correlates with the activity of DA-containing neurons, increased tyrosine hydroxylase expression after chronic cocaine could reflect the increased spontaneous activity and discharge rates of VTA neurons observed electrophysiologically.[61] In contrast to the effect in the VTA, tyrosine hydroxylase phosphorylation was reduced in the NAc, although no change was observed in enzyme immunoreactivity.[86] Since phosphorylation of tyrosine hydroxylase (via Ca^{2+} and cAMP-dependent protein kinases) is associated with increased catalytic activity, reduced phosphorylation of the enzyme would suggest a relative deficiency of DA synthetic capacity in dopaminergic nerve terminals in the NAc, and could account for the decrease of basal DA levels seen by microdialysis.[71]

Our laboratory also has found that neurofilament (NF) proteins (e.g., NF-200, NF-160, NF-68) are specifically reduced in the VTA after chronic, but not acute, cocaine administration.[89] NF proteins are major cytoskeletal constituents that play an important role in axonal caliber and transport, neuronal shape, and dendritic sprouting.[90] Cocaine-induced alterations of NF proteins could therefore impede neurotransmission in the mesoaccumbens pathway and local DA action in the VTA, and perhaps reflect negatively on the structural integrity of the dopaminergic neurons themselves. For example, decreased NF levels following chronic cocaine exposure would be expected to reduce axonal transport in the VTA-NAc pathway. It is tempting to speculate that this effect might contribute to the relative presynaptic DA deficiency in the NAc suggested by reduced tyrosine hydroxylase phosphorylation and by microdialysis studies. The finding that cocaine regulates such basic elements underscores the notion that psychotropic agents, through divergent and cascading signal transduction pathways, can chronically influence neuronal function in ways not easily predicted based upon their acute actions.

D. COMMON BIOCHEMICAL EFFECTS OF CHRONIC COCAINE AND OTHER DRUGS OF ABUSE

Most of the aforementioned biochemical alterations seen with chronic cocaine treatment have also been observed following chronic morphine treatment[81,86] (Figure 4). This may seem surprising since the acute actions of cocaine and morphine are so divergent. Pharmacologically, cocaine is a monoamine reuptake inhibitor while morphine is an opioid receptor agonist. Whereas cocaine inhibits VTA DA neurons, morphine generally excites them (via indirect inhibition of GABAergic interneurons). Behaviorally, cocaine causes behavioral activation, while morphine causes behavioral depression (although at low doses, morphine can have activating effects). These disparate effects contrast with the observations that both drugs are acutely reinforcing and can produce addiction (i.e., drug craving). Moreover, cross-sensitization between cocaine (or amphetamine) and morphine has been observed.[54,67] These common actions have been attributed to the ability of both drugs to increase synaptic levels of DA in the NAc acutely and to produce facilitated increases chronically. This idea, however, is not supported fully by the findings that 6-OHDA lesions of the VTA-NAc pathway, or the focal application of DA antagonists into the NAc, completely block cocaine but only partially attenuate heroin self-administration.[91,92] This suggests that nondopaminergic mechanisms, such as the direct effect of opiates on opioid receptors located on NAc neurons, contribute to opiate reinforcement.

Moreover, the finding that D_2-like and opioid receptors exert similar effects via G_i/G_o proteins raises the possibility of some shared postreceptor mechanisms. Indeed, it recently has been shown that G_i/G_o proteins in the NAc represent one common biochemical substrate for the acute reinforcing properties of both drugs since ADP-ribosylation and inactivation of G_i/G_o in the NAc by PTX injections is paralleled by commensurate increases in cocaine and heroin self-administration.[93] Also, cAMP analogs injected directly into the NAc regulate cocaine and heroin self-administration in a similar manner.[94]

Postreceptor signal transduction proteins also appear to represent common biochemical substrates for *chronic* cocaine and morphine exposure, since these drugs can elicit a number of common biochemical and molecular alterations of specific signal transduction elements in the mesolimbic DA system. Similar to cocaine, chronic morphine administration has been shown to increase adenylyl cyclase and cAMP-dependent protein kinase and to decrease G_i and the phosphorylation state of tyrosine hydroxylase in the NAc. Moreover, like chronic cocaine, chronic morphine increases tyrosine hydroxylase level and decreases NF level in the VTA. This latter effect has been shown to be physiologically relevant, since chronic morphine impairs axonal transport in the VTA-NAc pathway, but not in other pathways of rat brain.[95] We also have observed that chronic treatment with another drug of abuse, ethanol, increases the level of cAMP-dependent protein kinase in the NAc, and increases the level of tyrosine hydroxylase and decreases the level of NF proteins in the VTA.[79] In contrast, chronic administration of several classes of drugs without reinforcing properties do not produce these changes. It is conceivable that these common molecular alterations might account for some of the common behavioral features associated with chronic drug use, such as drug craving and behavioral sensitization.

E. MOLECULAR MECHANISMS UNDERLYING BIOCHEMICAL ADAPTATIONS TO COCAINE

Most of the adaptations in the signal transduction elements described above involve an alteration in the amount of protein present. There are three general mechanisms by which this might occur: regulation of RNA translation and turnover, regulation of gene transcription, and regulation of protein turnover. While most attention has been focused on changes of gene transcription after chronic drug treatments, the potential importance of posttranscriptional processes should not be overlooked. For example, while some drugs regulate G protein mRNA and protein levels in a coordinate manner,[96] it is becoming equally clear that changes in G proteins (e.g., via translational regulation or proteolysis) can occur without mRNA changes.[97]

Studies of the regulation of gene expression have focused on two families of transcription factors, the CREB protein and related proteins and the IEG products which probably

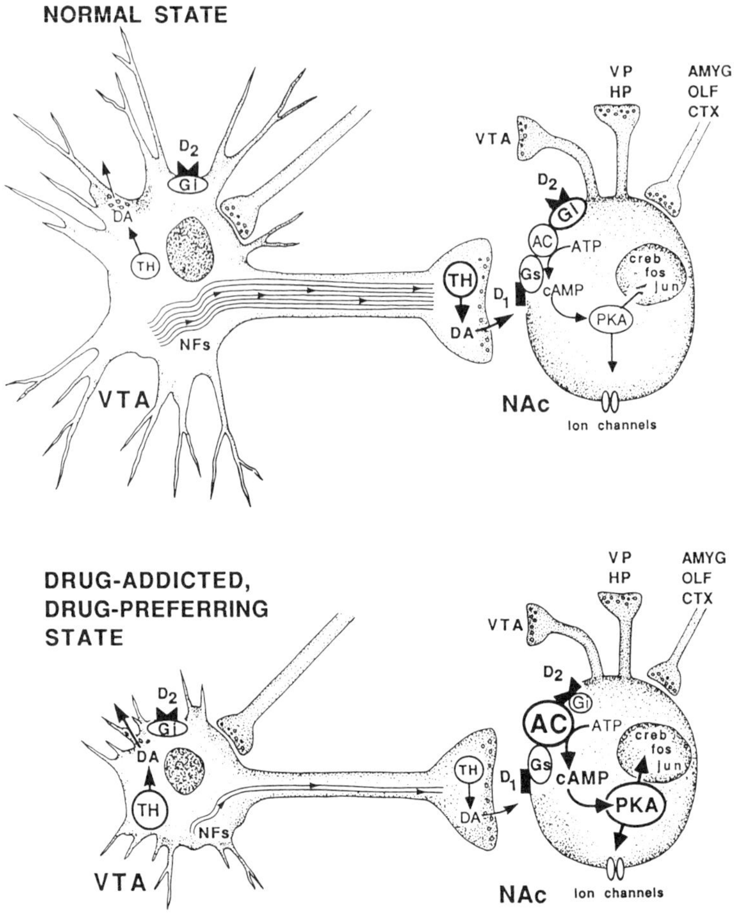

Figure 4. Schematic summary of similar biochemical manifestations of the drug-addicted and drug-preferring state. The top figure depicts a normal VTA neuron projecting to the NAc. Shown in the VTA cell are tyrosine hydroxylase (TH), DA, presynaptic DA receptors (D_2 or D_2-like) coupled to G proteins (G_i), and neurofilaments (NFs). Shown in the NAc are, in addition to TH and DA, DA receptors (D_1- and D_2-like), G proteins (G_i and G_s), components of the intracellular cAMP system (adenylyl cyclase activity, AC; cAMP-dependent protein kinase activity, PKA; and possible substrates for the kinase—ion channels and nuclear transcription factors, CREB, Fos and Jun), as well as major inputs and outputs of this region (ventral pallidum, VP; hippocampus, HP; amygdala, AMYG; olfactory cortex, OLF; other cortical regions, CTX). The bottom figure depicts a VTA neuron projecting to the NAc after chronic morphine or cocaine treatment, or in a Lewis (drug-preferring) rat as compared to a Fischer 344 rat. In the drug-addicted or drug-preferring animal, TH levels are increased in the VTA and decreased in the NAc (owing to either decreased phosphorylation due to chronic morphine or cocaine, or decreased enzyme levels in Lewis vs. F344 rats). NF levels are decreased in the VTA of the drug-addicted or drug-preferring rat. This decrease of NF might be associated with an alteration of neuronal structure, a decrease of axonal caliber, and/or a decrease of axonal transport in these cells. Decreased axonal transport has been observed in the VTA-NAc pathway of rats treated chronically with morphine. This decrease in axonal transport could conceivably account for the absence of increased TH level in dopaminergic terminals in the NAc. Decreased TH in the NAc implies decreased DA synthesis which could reduce dopaminergic neurotransmission in this brain region. In the NAc of the drug-addicted or drug-preferring rat, G_i is decreased, and adenylyl cyclase and cAMP-dependent protein kinase activities are increased—changes that could account for the D_1-like receptor supersensitivity observed electrophysiologically. It should be noted that alterations of

Figure 5. Scheme illustrating the regulation of AP-1 binding proteins by acute and chronic cocaine. Acute cocaine administration transiently induces several AP-1 binding proteins in the NAc and striatum, including Fos, Jun, Fos B, and Jun B. Induction of these transcription factors could mediate some of the rapid stimulatory (+) and inhibitory (–) effects of cocaine on the expression of neural genes (A-E) that contain AP-1 binding sites. In contrast, chronic cocaine administration produces a persistent induction of AP-1 binding activity, which is accounted for by different Fos-like (and possibly Jun-like) proteins, including recently identified proteins designated chronic FRAs (Fos-related antigens). Such chronic, persistent AP-1 complexes of altered composition might be expected to exert different transcriptional effects compared to the acute AP-1 complexes and could mediate some of the long-term effects of cocaine on gene expression.

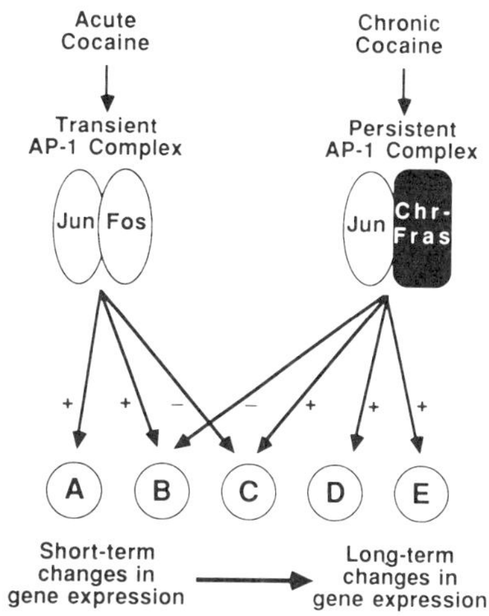

represent only a small fraction of the transcription factors that play an important role in mediating effects of extracellular stimuli on gene transcription. The ultimate goal of molecular studies of chronic cocaine action is to relate changes of specific transcription factors to the altered expression of target genes in brain regions relevant to the chronic behavioral phenomena of interest. While this task appears daunting, it is best achieved through investigations of these regulatory mechanisms within a functional context, that is, where specific putative target proteins (e.g., G proteins, adenylyl cyclase, and cAMP-dependent protein kinase) have been identified in behaviorally relevant brain regions (e.g., NAc).

The acute administration of cocaine (or amphetamine) has been shown to increase the expression of c-*fos*, *fos B*, c-*jun*, and *jun B*, and levels of AP-1 binding, in the NAc and striatum[98-102] (Figure 5). Induction of *zif268* also has been observed. In contrast to the effects of acute treatment, chronic cocaine treatment abolishes or desensitizes the ability of a subsequent cocaine challenge to induce c-*fos* and other IEGs in the NAc and striatum.[101] However, chronic cocaine treatment leads to the formation of a persistent AP-1 complex, which is present for at least 7 days after the last cocaine injection. This suggests that chronic cocaine leads to a concomitant induction of novel Fos- or Jun-like proteins. Indeed, we have determined that the resulting persistent AP-1 binding is associated with the induction of novel Fos-like proteins, or chronic Fos-related antigens (FRAs).[103] These chronic FRAs are induced gradually in response to repeated cocaine administration and exhibit a long half-life (approximately 7 days) following discontinuation of the drug (Figure 5). The induction of such long-lasting chronic FRAs, which appear to form qualitatively different AP-1 complexes, might be important in mediating some of the G protein and cAMP system changes discussed above.

Since cocaine influences the cAMP system, it would expected that the drug also would influence the CREB family of transcription factors both acutely and chronically. Indeed, acute amphetamine reportedly stimulates CREB phosphorylation in the striatum.[104]

Figure 4 (continued). dopaminergic transmission probably influence many cell types within the NAc, as well as other nerve terminals (e.g., 5-HT) in the NAc. Similarly, altered local dopaminergic transmission in the VTA could also influence other nerve terminals in this brain region. Thus, biochemical alterations in the mesolimbic DA system could potentially lead to altered neuronal function in many other brain regions as well. (From Beitner-Johnson, Guitart, and Nestler, "The Neurobiology of Drug and Alcohol Addiction," *Annu. N.Y. Acad Sci.*, 654. Copyright 1992 by the Annals of the New York Academy of Sciences. Reprinted with permission.)

Moreover, preliminary results in our laboratory indicate that chronic exposure to cocaine alters the phosphorylation state of CREB, overall levels of CREB immunoreactivity, and the level of CRE-binding in the NAc. Although it is generally believed that CREB is a constituitively expressed protein regulated solely by phosphorylation, these data provide the first direct evidence that total CREB protein can be regulated in the central nervous system.

VII. GENETIC DETERMINANTS OF DRUG ADDICTION

Genetic factors clearly play a role in a number of mental disorders. The potential involvement of predisposing genetic influences on addiction is less clear, although it is generally presumed that these factors contribute to individual differences in susceptibility to drug addiction in animals and humans. One model system used to study such genetic factors is that of Fischer 344 vs. Lewis rats, two inbred strains which differ in several behavioral responses to drugs of abuse. Lewis rats self-administer cocaine, opiates, and ethanol at much higher rates, and show greater conditioned place preferences to these drugs than do Fischer rats.[105-108] Lewis and Fischer rats also differ in their locomotor responses to acute and repeated cocaine exposure.[108,109] In an effort to understand the neurobiological basis for these behavioral differences, we have examined G proteins, the cAMP system, and several phosphoproteins which have been shown to be highly regulated in outbred rats chronically treated with drugs of abuse. Inherent differences in these proteins have been observed in the VTA-NAc pathway of Fischer and Lewis rats.[109–111] These studies have revealed that lower levels of G_i and tyrosine hydroxylase, and higher levels of adenylyl cyclase and cAMP-dependent kinase activity, are present in the NAc of Lewis than Fischer rats under drug-naive conditions. In the VTA, Lewis rats have higher levels of tyrosine hydroxylase and lower levels of NF proteins than do Fischer rats. Thus, the Lewis rat, which exhibits a relative drug preference, resembles an outbred Sprague-Dawley rat that has been treated chronically with morphine or cocaine (Figure 4). These findings raise the possibility that different levels of intracellular signaling proteins contribute not only to the chronic actions of drugs of abuse, but also to genetic differences in the inherent responsiveness to these drugs.

VIII. CONCLUSIONS

The studies described herein illustrate that systematic investigations of intracellular signal transduction pathways have contributed greatly to our understanding of the molecular mechanisms by which cocaine and other drugs of abuse induce changes in neuronal function. Three major challenges remain: (1) to identify additional molecular adaptations that arise in response to drugs of abuse; (2) to establish causal links between individual molecular and cellular adaptations and specific behavioral measurements of drug reward and craving; and (3) to delineate at the molecular level the precise mechanisms by which repeated drug exposure elicits these various adaptations.

Clearly, new approaches are required to facilitate the rather difficult task of relating the biochemical and molecular changes associated with chronic drug exposure with the complex phenomenon of addiction. For example, it would be useful to develop novel behavioral approaches that provide a clearer measure of drug craving, a core feature of addiction clinically. Moreover, newer methods in molecular biology, such as transgenic mouse technology or more targeted approaches involving gene transfer or antisense oligodeoxynucleotides, might afford the unique opportunity to selectively evaluate the role of individual proteins (e.g., G proteins, cAMP-dependent protein kinase, NF proteins) in the expression of, or genetic predisposition to, drug-seeking behavior. This approach would be interesting since some of the proteins encoded by genetic factors might be the same as those regulated pharmacologically by chronic exposure to drugs of abuse. The results of recent studies involving the intra-NAc injection of PTX or cAMP analogues indicate that these approaches could be employed to elucidate the intracellular mechanisms underlying complex behavioral phenomena.[93,94] It is also important to determine the exact molecular pathways through which chronic cocaine treatment alters intracellular messenger proteins in target neurons. Insofar as these changes are mediated at the level of gene expression, future efforts should examine the regulation of transcription factors in cocaine-responsive brain regions. *In vitro,* and ultimately *in vivo,* studies of transfected proteins could provide the causal, and still elusive, link between the activation of a transcription factor and the altered expression of a protein putatively associated with addiction.

Studies directed at unraveling the biochemical and molecular basis of drug addiction have a number of potential implications and benefits. An enhanced understanding of the biological bases of drug addiction will lead to the development of novel pharmacological interventions which might prevent or mitigate the actions of chronic drug exposure. Such drugs could not only treat physical abstinence syndromes associated with withdrawal from drugs of abuse, but also the craving for these substances. Moreover, studies directed at identifying some of the predisposing genetic factors will advance our overall understanding of addiction and, perhaps, help to identify those individuals at particular risk.

ACKNOWLEDGMENTS

This work was supported by USPHS grants DA08227, DA07359, DA00203, P32-NS07136, and by the Abraham Ribicoff Research Facilities of the Connecticut Mental Health Center, State of Connecticut Department of Mental Health.

REFERENCES

1. Gilman, A. G., G proteins: transducers of receptor-generated signals. *Annu. Rev. Biochem.*, 56, 615, 1987.
2. Nestler, E. J. and Duman, R. S., G proteins and cyclic nucleotides in the nervous system, in *Basic Neurochemistry: Molecular, Cellular, and Medical Aspects, 5th Ed.,* Seigel, G. J., Agranoff, B. W., Albers, R. W., and Molinoff, P. P., Eds., Raven Press, New York, 1994, 545.
3. Nestler, E. J. and Greengard, P., *Protein Phosphorylation in the Nervous System,* Wiley, New York, 1984.
4. Nestler, E. J., and Greengard, P., Protein phosphorylation and the regulation of neuronal function, in *Basic Neurochemistry: Molecular, Cellular, and Medical Aspects, 5th Ed.,* Seigel, G. J., Agranoff, B., Albers, R. W., and Molinoff, R. W., Eds., Raven Press, New York, 1994, 545.
5. Blenis, J., Signal transduction via the MAP kinases: proceed at your own RSK, *Proc. Natl. Acad. Sci. USA,* 90, 5889, 1993.
6. Davis, R. J., The mitogen-activated protein kinase signal transduction pathway, *J. Biol. Chem.*, 268, 14553, 1993.
7. Hunter, T. and Sefton, B. M., Eds. Protein phosphorylation. Parts A and B, *Methods in Enzymology,* Vols. 200 and 201, Academic Press, New York, 1991.
8. Mitchell, P. J. and Tjian, R., Transcriptional regulation in mammalian cells by sequence-specific DNA binding proteins, *Science,* 245, 371, 1989.
9. Montminy, M. R., Gonzales, G. A., and Yamamoto, K. K., Regulation of cAMP-inducible genes by CREB, *Trends Neurosci.,* 13, 184, 1990.
10. Hyman, S. E. and Nestler, E. J., *The Molecular Foundations of Psychiatry,* American Psychiatric Press, Washington, D.C., 1993.
11. Morgan, J. I. and Curran, T., Stimulus-transcription coupling in the nervous system, *Annu. Rev. Neurosci.,* 14, 421, 1991.
12. Ritz, M. C., Lamb, R. J., Goldberg, S. R., and Kuhar, M. J., Cocaine receptors on dopamine transporters are related to self-administration of cocaine, *Science,* 237, 1219, 1987.
13. Woolverton, W. L., Evaluation of the role of norepinephrine in the reinforcing effects of psychomotor stimulants in rhesus monkeys, *Pharmacol. Biochem. Behav.,* 26, 834, 1987.

14. Porrino, L. J., Ritz, R. C., Goodman, N. L., Sharpe, L. G., Kuhar, M. J., and Goldberg, S. R., Differential effects of the pharmacological manipulation of serotonin systems on cocaine and self-administration in rats, *Life Sci.,* 45, 1529, 1989.
15. Wise, R. A. and Bozarth, M. A., A psychomotor stimulant theory of addiction, *Psychol. Rev.,* 94, 469, 1987.
16. Roberts, D. C. S., Corcoran, M. E., and Fibiger, H. C., On the role of ascending catecholaminergic systems in the intravenous self-administration of cocaine, *Pharmacol. Biochem. Behav.,* 6, 615, 1977.
17. Leccese, A. P. and Lyness, W. H., The effects of putative 5-hydroxytryptamine receptor active agents on d-amphetamine self-administration in controls and rats with 5,7-dihydroxytryptamine median forebrain bundle lesion, *Brain Res.,* 303, 153, 1984.
18. Hoffman, B. J., Mezey, E., and Brownstein, M. J., Cloning of a serotonin transporter affected by antidepressants, *Science,* 254, 579, 1991.
19. Pacholczyk, T., Blakely, R. D., and Amara, S. G., Expression cloning of a cocaine- and antidepressant-sensitive human noradrenaline transporter, *Nature,* 350, 350, 1991.
20. Shimada S., Kitayama, S., Lin, C. L., Patel, A., Nanthakumar , E., Gregor, P., Kuhar, M., and Uhl, G., Cloning and expression of a cocaine-sensitive dopamine transporter complementary DNA, *Science,* 254, 576, 1991.
21. Johnson, R. G., Jr., Accumulation of biological amines into chromaffin granules: A model for hormone and neurotransmitter transport, *Physiol. Rev.,* 68, 233, 1988.
22. Nelson, N., Presynaptic events involved in neurotransmission, *J. Physiol.,* 87, 171, 1993.
23. Sulzer, D. and Rayport, S., Amphetamine and other psychostimulants reduce pH gradients in midbrain dopaminergic neurons and chromaffin cells: a mechanism of action, *Neuron,* 5, 797, 1990.
24. Kebabian, J. W. and Calne, D. B., Multiple receptors for dopamine, *Nature,* 277, 93, 1979.
25. Civelli, O., Bunzow, J. R., and Grandy, D. K., Molecular diversity of the dopamine receptors, *Annu. Rev. Pharmacol. Toxicol.,* 32, 281, 1993.
26. Wachtel, S. R., Hu, X.-T., Galloway, M. P., and White, F. J., D1 dopamine receptor stimulation enables the postsynaptic, but not the autoreceptor, effects of D2 dopamine agonists in nigrostriatal and mesoaccumbens dopamine systems, *Synapse,* 4, 327, 1989.
27. Gerfen, C. R., Engber, T. M., Mahan, L. C., Susel, Z., Chase, T. N., Monsma, F. J., and Sibley, D. R., D1 and D2 dopamine receptor-regulated gene expression of striatalnigral and striatopallidal neurons, *Science,* 250, 1429, 1990.
28. Mahan, L. C., Burch, R. M., Monsma, F. J., Jr., and Sibley, D. R., Expression of striatal D1 dopamine receptors coupled to inositol phosphate production and Ca^{2+} mobilization in Xenopus oocytes, *Proc. Natl. Acad. Sci. USA,* 87, 2196, 1990.
29. Ganz, M. B., Patcher, J. A., and Barber, D. L., Multiple receptors coupled to adenylate cyclase regulate Na-H exchange independent of cAMP, *J. Biol. Chem.,* 265, 8989, 1990.
30. Kanterman, R. Y., Mahan, L. C., Briley, E. M., Monsma, F. J., and Sibley, D. R., Axelrod, J., and Felder, C. C., Transfected D2 receptors mediated the potentiation of arachidonic acid release in Chinese hamster ovary cells, *Mol. Pharmacol.,* 39, 364, 1991.
31. Lledo, P. M., Homburger, V., Bockaert, J., and Vincent, J. D., Differential G-protein-mediated coupling of the D2 dopamine receptors to K^+ and Ca^{2+} currents in rat anterior pituitary cells, *Neuron,* 8, 455, 1992.
32. Di Marzo, V., Vial, D., Sokoloff, P., Schwartz, J.-C., and Piomelli, D., Selection of alternative G_i_mediated signaling pathways at the dopamine D2 receptor by protein kinase C, *J. Neurosci.,* 13, 4846, 1993.
33. Sibley, D. R. and Monsma, F. J., Molecular biology of dopamine receptors, *Trends Pharmacol. Sci.,* 13, 61, 1992.
34. Van Tol, H. H. M., Bunzow, J. R., Guan, H.-C., Sunahara, R. K., Seeman, P., Niznik, H. B., and Civelli, O., Cloning of the gene for a human dopamine D4 receptor with high affinity for the antipsychotic clozapine, *Nature,* 350, 610, 1991.
35. Malmberg, A., Jackson, D. M., Eriksson, A., and Mohell, N., Unique binding characteristics of antipsychotic agents interacting with human dopamine D_{2A}, D_{2B}, and D_3 receptors, *Mol. Pharmacol.,* 43, 749, 1993.
36. Sokoloff, P., Giros, B., Martres, M.-P., Bouthenet, M.-L., and Schwartz, J.-C., Molecular cloning and characterization of a novel dopamine receptor (D3) as a target for neuroleptics, *Nature,* 347, 146, 1990.

37. Sokoloff, P., Martres, M.-P., Giros, B., Bouthenet, M.-L., and Schwartz, J.-C., The third dopamine receptor (D3) as a novel target for antipsychotics, *Biochem. Pharmacol.,* 43, 659, 1992.
38. Castro S. W. and Strange, P. G., Coupling of D2 and D3 dopamine receptors to G-proteins, *FEBS,* 315, 223, 1993.
39. Gehlert, D. R., Quantitative autoradiography of Gpp(NH)p sensitive and insensitive [^{3}H]quinpirole binding sites in the rat brain, *Synapse,* 14, 113, 1993.
40. Caine, S. B. and Koob, G. F., Modulation of cocaine self-administration in the rat through D-3 dopamine receptors, *Science,* 260, 1814, 1993.
41. Johanson, C.-E. and Fischman, M. W., The pharmacology of cocaine related to its abuse, *Pharmacol. Rev.,* 41, 3, 1989.
42. Church, W. H. and Justice, J. B., Jr., Rapid sampling and determination of extracellular dopamine in vivo, *Anal. Chem.,* 59, 712, 1987.
43. Hurd, Y. L. and Ungerstedt, U., Cocaine: an in vivo microdialysis evaluation of its acute action on dopamine transmission in the rat striatum, *Synapse,* 3, 48, 1989.
44. Maisonneuve, I. M., Keller, R. W., and Glick, S. D., Similar effects of D-amphetamine and cocaine on extracellular dopamine levels in the medial prefrontal cortex of rats, *Brain Res.,* 535, 221, 1990.
45. Pettit, H. O., Pan, H., Parsons, L. H., and Justice, J. B., Jr., Extracellular concentrations of cocaine and dopamine are enhanced during chronic cocaine administration, *J. Neurochem.,* 55, 798, 1990.
46. Kelly, P. H., Seviour, P. W., and Iverson, S. D., Amphetamine and apomorphine responses in the rat following 6-OHDA lesions of the nucleus accumbens and striatum, *Brain Res.,* 94, 507, 1975.
47. Delfs, J. M., Schreibner, L., and Kelley, A. E., Microinjection of cocaine into the nucleus accumbens elicits locomotor activation in the rat, *J. Neurosci.,* 10, 303, 1990.
48. Bunney, B. S., Walters, J. R., Roth, R. H., and Aghajanian, G. K., Dopaminergic neurons: effects of antipsychotic drugs and amphetamine on single cell activity, *J. Pharmacol. Exp. Ther.,* 185, 560, 1973.
49. Einhorn, L. C., Johansen, P. A., and White, F. J., Electrophysiological effects of cocaine in the mesoaccumbens dopamine system: studies in the ventral tegmental area, *J. Neurosci.,* 8, 100, 1988.
50. Wolf, M. E. and Roth, R. H., Dopamine autoreceptors, in *Dopamine Receptors,* Creese, I., Ed., Alan Liss, New York, 1987, 45.
51. White, F. J. and Wang, R. Y., Electrophysiological evidence for both D1 and D2 dopamine receptors in the rat nucleus accumbens, *J. Neurosci.,* 6, 274, 1986.
52. Uchimura, N. and North, R. A., Actions of cocaine on rat nucleus accumbens neurones in vitro, *Br. J. Pharmacol.,* 99, 736, 1990.
53. Uchimura, N., Higasi, H., and Nishi, S., Hyperpolarizing and depolarizing actions of dopamine via D1 and D2 receptors on nucleus accumbens neurons, *Brain Res.,* 375, 368, 1986.
54. Robertson, T. E. and Becker, J. B., Enduring changes in brain and behavior produced by chronic amphetamine administration: a review and evaluation of animal models of amphetamine psychosis, *Brain Res. Rev.,* 11, 157, 1986.
55. Kalivas, P. W. and Stewart, J., Dopamine transmission in the initiation and expression of drug- and stress-induced sensitization of motor activity, *Brain Res. Rev.,* 16, 1991.
56. Post, R. M. and Rose, H., Increasing effects of repetitive cocaine administration in rats, *Nature,* 260, 731, 1976.
57. Kilbey, M. M. and Ellinwood, E. H., Administration of stimulant drugs: response modification, in *Cocaine and Other Stimulants,* Ellinwood, E. H. and Kilbey, M. M., Eds., Plenum Press, New York, 1977, 404.
58. Peris, J., Boyson, S. J., Cass, W. A., Curella, P., Dwoskin, L. P., Larson, G., Lin, L.-H., Yasuda, R. P., and Zahniser, N. R., Persistence of neurochemical changes in dopamine systems after repeated cocaine administration, *J. Pharmacol. Exp. Ther.,* 253, 38-44, 1990.
59. Lett, B. T., Repeated exposures intensifies rather than diminish the rewarding effects of amphetamine, morphine, and cocaine, *Psychopharmacology,* 98, 357, 1989.
60. Henry, D. J. and White, F. J., Repeated cocaine administration cause persistent enhancement of D1 dopamine receptor sensitivity within the rat nucleus accumbens, *J. Pharm. Exp. Ther.,* 258, 822, 1991.
61. Henry, D. J., Green, M. A., and White, F. J., Electrophysiological effects of cocaine in the mesoaccumbens dopamine system: repeated administration, *J. Pharmacol. Exp. Ther.,* 251, 833, 1989.

62. Ackerman, J. M. and White, F. J., A10 somatodendritic dopamine autoreceptor sensitivity following withdrawal from repeated cocaine treatment, *Neurosci. Lett.*, 117, 181, 1990.
63. Wolf, M. E., White, F., Nassar, R., Brooderson, R. J., and Khansa, M. R., Differential development of autoreceptor subsensitivity and enhanced dopamine release during amphetamine sensitization, *J. Pharmacol. Exp. Ther.*, 264, 249, 1993.
64. Kalivas, P. W. and Weber, B., Amphetamine injection into the ventral mesencephalon sensitizes rats to peripheral amphetamine and cocaine, *J. Pharmacol. Exp. Ther.*, 245, 1095, 1988.
65. Vezina, P. and Stewart, J., Amphetamine administered to the ventral tegmental area but not to the nucleus accumbens sensitizes rats to systemic morphine: lack of conditioned effects, *Brain Res.*, 516, 99, 1990.
66. Zahniser, N. R., Peris, J., Dwoskin, L. P., Curella, P., Yasude, R. P., O'Keefe, L., and Boyson, S. J., Sensitization to cocaine in the nigrostriatal dopamine system, *NIDA Res. Monogr.*, 88, 55, 1988.
67. Benmansour, S., Tejani-Butt, S. M., Hauptmann, M., and Brunswick, D. J., Lack of effect of high-dose cocaine on monoamine uptake sites in rat brain measured by quantitative autoradiography, *Psychopharmacology*, 106, 459, 1992.
68. Alburges, M. E. , Narang, N., and Wamsley, J. K., Alterations in the dopaminergic receptor system after chronic administration of cocaine, *Synapse*, 14, 314, 1993.
69. Little, K. Y., Kirkman, J. A., Carroll, F. I., Clark, T. B., and Duncan, G. E., Cocaine use increases [^{3}H]WIN 35428 binding sites in human striatum, *Brain Res.*, 628, 17, 1993.
70. Cerruti, C., Pilotte, N. S., Uhl, G., and Kuhar, M. J., Reduction in dopamine transporter mRNA after cessation of repeated cocaine administration, *Mol. Brain Res.*, 22, 132, 1994.
71. Parsons, L. H., Smith, A. D., and Justice, J. B., Jr., Basal extracellular dopamine is decreased in the rat nucleus accumbens during abstinence from chronic cocaine, *Synapse*, 9, 60, 1991.
72. Robertson, M. W., Leslie, C. A., and Bennett, J. P., Jr., Apparent synaptic dopamine deficiency induced by withdrawal from chronic cocaine treatment, *Brain Res.*, 538, 1991.
73. Imperato, A., Mele, A., Scrocco, M. G., and Puglisi-Allegra, S., Chronic cocaine alters limbic extracellular dopamine. Neurochemical basis for addiction, *Eur. J. Pharmacol.*, 212, 299, 1992
74. Rossetti, Z. L., Hmaidan, Y., and Gessa, G. L., Marked inhibition of mesolimbic dopamine release: a common feature of ethanol, morphine, cocaine, and amphetamine abstinence in rats, *Eur. J. Pharmacol.*, 221, 227, 1992.
75. Segal, D. S. and Kuczenski, R., Repeated cocaine administration induces behavioral sensitization and corresponding decreased extracellular dopamine responses in caudate and accumbens, *Brain Res.*, 577, 351, 1992.
76. Pulvirenti, L., Maldonado-Lopez, R., and Koob, G. F., NMDA receptors in the nucleus accumbens modulate intravenous cocaine but not heroin self-administration in the rat, *Brain Res.*, 594, 327, 1992.
77. White, F. J., Hu, X.-T., Zhang, X.-F., and Wolf, M. E., Repeated administration of cocaine or amphetamine alters neuronal responses to glutamate in the mesoaccumbens dopamine system, *J. Pharmacol. Exp. Ther.*, in press.
78. Nestler, E. J., Molecular mechanisms of drug addiction, *J. Neurosci.*, 12, 2439, 1992.
79. Nestler, E. J., Hope, B. T., and Widnell, K. L., Drug addiction: a model for the molecular basis of neural plasticity, *Neuron*, 11, 1, 1993.
80. Terwilliger, R., Beitner-Johnson, D., Sevarino, K. A., Crain, S. M., and Nestler, E. J., A general role for adaptations in G-proteins and the cyclic AMP system in mediating the chronic actions of morphine and cocaine on neuronal function, *Brain Res.*, 548, 100, 1991.
81. Nestler, E. J., Terwilliger, R. Z., Walker, J. R., Sevarino, K. A., and Duman, R. S., Chronic cocaine treatment decreases levels of the G protein subunits $G_{i\alpha}$ and $G_{o\alpha}$ in discrete regions of the rat brain, *J. Neurochem.*, 55, 1079, 1990.
82. Striplin, C. D. and Kalivas, P. W., Robustness of G protein changes in cocaine sensitization shown with immunoblotting, *Synapse*, 14, 10, 1993.
83. Cunningham, S. T. and Kelley, A. E., Hyperactivity and sensitization to psychostimulants following cholera toxin infusion in the nucleus accumbens, *J. Neurosci.*, 13, 2342, 1993.
84. Hervé, D., Lévi-Strauss, M., Marey-Semper, I., Verney, C., Tassin, J.-P., Glowinski, J., and Girault, J.-A., G_{olf} and G_s in the basal ganglia: possible involvement of G_{olf} in the coupling of dopamine D_1 receptor with adenylyl cyclase, *J . Neurosci.*, 13, 2237, 1993.
85. Steketee, J. D., Striplin, C. D., Murray, T. F., and Kalivas, P. W., Possible role for G-proteins in behavioral sensitization to cocaine, *Brain Res.*, 545, 287, 1991.

86. Beitner-Johnson, D. and Nestler, E. J., Morphine and cocaine exert common chronic action on tyrosine hydroxylase in dopaminergic brain regions, *J. Neurochem.*, 57, 344, 1991.
87. Vrana, S. L., Vrana, K. E., Koves, T. R., Smith, J. E., and Dworkin, S. I., Chronic cocaine administration increases CNS tyrosine hydroxylase enzyme activity and mRNA levels and tryptophan hydroxylase enzyme activity levels, *J. Neurochem.*, 61, 2262, 1993.
88. Sorg, B. A., Chen, S.-Y., and Kalivas, P. W., Time course of tyrosine hydoxylase expression after behavioral sensitization to cocaine, *J. Pharmacol. Exp. Ther.*, 266, 424, 1993.
89. Beitner-Johnson, D., Guitart, X., and Nestler, E. J., Neurofilament proteins and mesolimbic dopamine system: common regulation by chronic morphine and cocaine in the rat ventral tegmental area, *J. Neurosci.*, 12, 2165, 1992.
90. Hammerschlag, R., Cyr, J. L., and Brady, S. T., Axonal transport and the neuronal cytoskeleton, in *Basic Neurochemistry: Molecular, Cellular, and Medical Aspects, 5th Ed.*, Seigel, G., Agranoff, B., Albers, R .W., and Molinoff, P. B., Eds., Raven Press, New York, 1994, 545.
91. Ettenberg, A., Pettit, H. O., Bloom, F. E., and Koob, G. F., Heroin and cocaine intravenous administration in rats: mediation by separate neural systems, *Psychopharmacology*, 78, 204, 1982.
92. Pettit, H. O., Ettenberg, A., Bloom, F. E., and Koob, G. F., Destruction of dopamine in the nucleus accumbens selectively attentuates cocaine but not heroin self-administration in rats, *Psychopharmacology*, 84, 167, 1984.
93. Self, D. W., Terwilliger, R. Z., Nestler, E. J., and Stein, L., Inactivation of G_i and G_0 proteins in nucleus accumbens reduces both cocaine and heroin reinforcement, *J. Neurosci.*, 14, 6239, 1994.
94. Self, D. W. and Nestler, E. J., Molecular mechanisms of drug reinforcement and addiction, *Annu. Rev. Neurosci.*, 18, 463, 1995.
95. Beitner-Johnson, D. and Nestler, E. J., Chronic morphine impairs axoplasmic transport in the rat mesolimbic dopamine system, *NeuroReport*, 5, 57, 1993.
96. Saito, N., Guitart, X., Hayward, M., Tallman, J. F., Duman, R. S., and Nestler, E. J., Corticosterone differentially regulates the expression of $G_{s\alpha}$ and $G_{i\alpha}$ messenger RNA and protein in rat cerebral cortex, *Proc. Natl. Acad. Sci. USA*, 86, 3906, 1989.
97. Milligan, G., Agonist regulation of cellular G protein levels and distribution: mechanisms and functional implications, *Trends Pharm. Sci.*, 14, 413, 1993.
98. Graybiel, A. M., Moratella, R., and Robertson, H .A., Amphetamine and cocaine induce drug-specific activation of c-fos gene in striosome-matrix compartments and limbic subdivision of the striatum, *Proc. Natl. Acad. Sci. USA*, 87, 6912, 1990.
99. Young, S. T., Porrino, L. J., and Iadorola, M. J., Cocaine induces striatal c-fos immunoreactivity proteins via dopaminergic D1 receptors, *Proc. Natl. Acad. Sci. USA*, 88, 1291, 1991.
100. Cole, A. J., Bhat, R. V., Patt, C., Worley, P. F., and Baraban, J. M., D1 dopamine receptor activation of multiple transcription factor genes in rat striatum, *J. Neurochem.*, 58, 1420, 1992.
101. Hope, B., Kosofsky, B., Hyman, S. E., and Nestler, E. J., Regulation of immediate early gene expression and AP-1 binding in the rat nucleus accumbens by chronic cocaine, *Proc. Natl. Acad. Sci. USA*, 89, 5764, 1992.
102. Nguyen, T. V., Kosofsky, B. E., Birnbaum, R., Cohen, B. M., and Hyman, S. E., Differential expression of c-Fos and Zif268 in rat striatum after haloperidol, clozapine, and amphetamine, *Proc. Natl. Acad. Sci. USA*, 89, 4270, 1992.
103. Hope, B. T., Nye, H. E., Kelz, M. B., Self, D. W., Iadorola, M. J., Nakabeppu, Y., Duman, R. S., and Nestler, E. J., Induction of a long-lasting AP-1 complex composed of altered Fos-like proteins in brain by chronic cocaine and other chronic treatments, *Neuron*, 13, 1235, 1994.
104. Konradi, C., Cole, R. L., Heckers, S., and Hyman, S. E., Amphetamine regulates gene expression in rat striatum via transcription factor CREB, *J. Neurosci.*, 14, 5623, 1994.
105. George, F. R. and Goldberg, S. R., Genetic approaches to the analysis of addiction processes, *Trends Pharmacol. Sci.*, 10, 78, 1989.
106. Suzuki, T., George, F. R., and Meish, R. A., Differential establishment and maintenance of oral ethanol reinforced behavior in Lewis and Fischer 344 inbred rat strains, *J. Pharmacol. Exp. Ther.*, 245, 164, 1989.
107. Guitart, X., Beitner-Johnson, D., Marby, D. W., Kosten, T. A., and Nestler, E. J., Fischer and Lewis rat strains differ in basal levels of neurofilament proteins and their regulation by chronic morphine in the mesolimbic dopamine system, *Synapse*, 12, 242, 1992.

108. Kosten, T. A., Miserendino, M. J. D., Chi, S., and Nestler, E. J., Fischer and Lewis rat strains show differential cocaine effects in conditioned place preference and behavioral sensitization, but not in locomotor activity or conditioned taste aversion, *J. Pharmacol. Exp. Ther.*, 269, 137, 1994.
109. George, F. R., Porrino, L. J., Ritz, M. C., and Goldberg, S. R., Inbred rat strain comparisons indicate different sites of action for cocaine and amphetamine locomotr stimulant, *Psychopharmacology*, 104, 457, 1991.
110. Beitner-Johnson, D., Guitart, X., and Nestler, E. J., Dopaminergic brain reward regions of Lewis and Fischer rats display different levels of tyrosine hydroxylase and other morphine- and cocaine-regulated phosphoproteins, *Brain Res.*, 561, 147, 1991.
111. Guitart, X., Kogan, J. H., Berhow, M., Terwilliger, R. Z., Aghajanian, G. K., and Nestler, E. J., Lewis and Fischer rat strains display differences in biochemical, electrophysiological, and behavioral parameters: studies in the nucleus accumbens and locus coeruleus of drug naive and morphine-treated animals, *Brain Res.*, 611, 7, 1993.
112. Nestler, E. J., Hard target: Understanding dopaminergic neurotransmission, *Cell*, 79, 923, 1994.
113. Miserendino, M. J. D. and Nestler, E. J., Behavioral sensitization to cocaine: modulation by the cyclic AMP system in the nucleus accumbens, *Brain Res.*, in press.

INDEX

C

D

E

F

K

L

M

N

O

P

Q

R

S

T